Broadband Wireless Communications

3G, 4G and Wireless LAN

THE KLUWER INTERNATIONAL SERIES
IN ENGINEERING AND COMPUTER SCIENCE

BROADBAND WIRELESS COMMUNICATIONS
3G, 4G and Wireless LAN

by

Jiangzhou WANG
University of Hong Kong

SPRINGER SCIENCE+BUSINESS MEDIA, LLC

Library of Congress Cataloging-in-Publication Data

Wang, Jiangzhou.
Broadband wireless communications: 3G, 4G, and Wireless LAN / by Jiangzhou Wang.
p. cm.
Includes bibliographical references and index.
ISBN 978-1-4757-7534-1 ISBN 978-0-306-46999-2 (eBook)
DOI 10.1007/978-0-306-46999-2

1. Broadband communication systems. 2. Wireless communication systems. I. Title.

TK5103.4 .W36 2001
384.5'3—dc21

2001035426

Originally published by Kluwer Academic Publishers in 2001
Softcover reprint of the hardcover 1st edition 2001

Printed on acid-free paper.

To

Leiping, April, Angela and Larry

TABLE OF CONTENTS

Preface

The broadband wireless communications field is growing at an explosive rate, stimulated by a host of important emerging applications ranging from 3G, 4G and wireless LAN. Wideband CDMA and CDMA2000 will be used for 3G. OFDM+CDMA might be a good choice for 4G, CDMA overlay will possibly be used for new-generation broadband wireless LAN. For system planners and designers, the projections of rapidly escalating demand for such wireless services present major challenges and meeting these challenges will require sustained technical innovation on many fronts.

The text of this book has been developed through years of research by the author and his graduate students at the University of Hong Kong. The aim of this book is to provide a R&D perspective on the field of broadband wireless communications by describing the recent research developments in this area and also by identifying key directions in which further research is needed. As a background, I presume that the reader has a thorough understanding of digital communications and spread spectrum/CDMA.

The book is arranged into 13 chapters. In **chapter 1**, some key specifications of 3G WCDMA are described and discussed. These techniques include channel coding, rate matching, modulation and spreading, power control, cell search, transmit diversity, soft-handoff, and so son.

In **Chapter 2**, the coherent RAKE reception of Wideband CDMA signals with complex spreading is considered. A dedicated pilot channel, which is separate from data channels, is used for the purpose of channel estimation. Based on a digital implementation, the coherent demodulation scheme is presented. Pilot channel estimation error due to multiple access and multipath interference is studied. System performance is evaluated by means of bit error rate. The analysis shows that the error of channel estimation significantly degrades system performance and can be effectively suppressed by low pass filters (LPFs). A discussion on the envelope variation of complex spread signals is also included, which illustrates that the complex spread signal has more stable envelope than the dual-channel spread signal.

Multicode CDMA is a new transmission scheme for flexible and high-speed data communications. The basic idea of multicode CDMA is to assign multiple channelization codes to any given user. If these channel codes are orthogonal, the self-interference among them can be eliminated in an AWGN channel. However, in a multipath environment, these intra-user signals from different delay paths no longer maintain orthogonality and thus cause interference, i.e. multicode interference (MCI), to each other. In high-speed data networks, where the number of users is much less than in the voice networks, the MCI may represent a large portion of the total interference and has great impact on the system performance. In **Chapter 3**, a complex spreading multicode wideband CDMA receiver with RAKE structure and MCI cancellation is studied. By pilot aided channel estimation, the MCI associated with the reference user is regenerated and subtracted from the received signal. A complete and consolidated theoretical analysis is presented to show that the system performance is significantly improved by the MCI cancellation.

In **Chapter 4**, half-sine and triangular chip waveforms are proposed for the locally generated early and late despreading codes in a coherent DLL for a single user in a DS/SS systems. Both ideal and band-limited received signals are investigated. These non-rectangular waveforms reduce the effective loop noise power with only minor effect on the DLL sensitivity.

Chapter 5 is concerned with a direct sequence code division multiple access (DS-CDMA) system sharing common spectrum with a narrowband waveform for wireless LAN applications. A suppression filter at a receiver is employed to reduce the narrowband interference. The average up-link bit error rate (BER) performance is investigated how to be influenced by various parameters. **Chapter 6** presents the behavior of reflection coefficients of a stochastic gradient lattice (SGL) filter applied to a CDMA overlay system. Analytic expressions for coefficients for a two stage filter are derived with the presence of narrowband interference and additive white gaussian noise (AWGN). It is shown that the coefficients of the lattice filter exhibit separate tracking and convergent properties and that compared to a LMS filter, the lattice filter provides fast rate of convergence, while having good capability of narrowband interference suppression.

Adaptive power control has widely been used in DS/CDMA systems to overcome the so-called "near-far" problem. **Chapter 7** studies the adaptive open-loop power control of a cellular CDMA system, which is overlaid in downlink by a narrowband signal. The effects of downlink power allocation schemes to power control error in the presence of narrowband interference are analyzed and numerically computed. In order to get a minimum power control error in the CDMA overlay situations, an optimum downlink power allocation scheme is used, which works well for wide range of signal to narrow-band interference ratio. In **Chapter 8**, we investigate the effect of power control error of a closed-loop power controlled CDMA system with imperfect power estimation. By use of a simplified log-linear power control model, the power control error of a strength-based power control algorithm is studied in the presence of narrowband interference.

In **Chapter 9**, uplink multicarrier direct-sequence/slow-frequency-hopping (MC DS/SFH) CDMA systems are proposed, in which multiple carriers are modulated by the same DS waveform and hopped in frequency according to a random hopping pattern. The receiver dehops the received signal with the same pattern, provides RAKE receivers for each carrier and combines the outputs with a maximal ratio combiner (MRC). The performance of the proposed system is investigated over a frequency selective Rayleigh fading channel and compared to that of the MC DS-CDMA systems. This scheme is suitable for uplink transmission where BS can have more antennas. In **Chapter 10**, another type of the slow frequency hopping (SFH) technique to the multicarrier (MC) CDMA systems for downlink overlay situations. Using lower chip rate, which results in a narrower spectrum for each carrier and hopping the signal in frequency, the MC DS/SFH system achieves better performance than the MC DS CDMA system in most cases, especially when the bandwidth of the narrowband interference is narrower than one subchannel. This scheme is good at downlink where MS does not have multiple antennas.

Frequency hopping is an alternative technique for wireless mobile communications. **Chapter 11** is concerned with the bit error rate (BER) performance of a cellular one-hop/symbol frequency hopping multiple access (FHMA) system operating through a multipath Rayleigh fading channel The multiple access and adjacent cell interference of the cellular FHMA system has been studied. In **Chapter 12**, a novel FH/MFSK receiver with additional partial correlators is proposed by means of side information for asynchronous frequency hopping multiple-acess (FHMA) systems in the presence of Rayleigh fading. Performance analysis is carried out and compared for both the novel and conventional receivers.

Finally, with both economic growth and technology development of the world, multimedia services with even higher data rate (i.e. 20Mbps) will be provided by 4G. In **Chapter 13**, the forecast of mobile communications and key techniques of 4G will be presented.

Acknowledgements

Throughout my professional carrier, I have had the opportunity to work with and learn from the following people whom I would like to greatly acknowledge:

Professor Laurence B. Milstein	University of California at San Diego, USA
Professor Tung-Sang Ng	University of Hong Kong
Professor Fumiyuki Adachi	Tohoku University, Japan
Professor Marc Moeneclaey	University of Ghent, Belgium
Dr. Mamoru Sawahashi	NTT DoCoMo, Japan
Dr. Ker Zhang	Synchronization Inc., USA
Dr. Zhengmao Li	China Unicom

Professor Milstein introduced me to CDMA overlay during my postdoctoral research. Since then, I have published extensively in this area. Professor Ng has been giving me much encouragement and support in my professorial carrier. Professor Adachi selflessly introduced me to several key issues in wideband CDMA. Professor Moeneclaey effectively supervised my Ph.D. dissertation. Dr. Sawahashi helped me much with patience in the understanding of 3G standard during my visit to NTT DoCoMo. Dr. Zhang was a Group Leader when I worked in a US company. Dr. Li was my classmate during my early Ph.D. study.

During the past a few years, I have had pleasure to supervise a number of hardworking graduate students, without whom this book would have never appeared. I would like to express my gratitude to:

Chen Jiang	Jun Chen
Ai Yu	Xiangyang Wang
Vicknarajah Prahatheesan	Haidong Wu
Thayaparan Subramaniam	Yiqing Zhou
Hu Huang	Bin Xia
Xi Zhang	

This book is the product of funded research supported by NTT DoCoMo, Japan, the RGC of Hong Kong SAR Government and the University of Hong Kong. Their financial support is appreciated.

Finally, I wish to thank my wife, Leiping Xu, and three children, April Pingyang, Angela Pingyu, Larry Pinglei for their encouragement and love during the writing phases of this book.

Chapter 1

3G MOBILE COMMUNICATIONS

Abstract The standardization of third generation (3G) mobile systems is accelerating at a fast pace due to the proliferation of high-speed wireless multimedia communications and wireless internet services. There are three main standards for 3G: Wideband CDMA, CDMA2000 and EDGE. The first one is single carrier CDMA with bandwidth of 5MHz, whereas the second one is multicarrier CDMA based on US narrowband CDMA (IS-95). EDGE is TDMA based. Many enabling techniques including software radio, smart antennas and digital processing devices are greatly improving the spectral efficiency of 3G systems. In this chapter, some key specifications of only WCDMA are described and discussed.

Keywords: Wideband CDMA, Channel Coding, Rate Matching, Modulation and Spreading, Power Control, Cell Search, Transmit Diversity, Soft-handoff.

1.1 INTRODUCTION

Mobile communications services are penetrating into our society at an explosive growth rate. The major services the second generation mobile systems provide are limited to basic services (e.g. voice and low-bit rate data). We have entered the 21st century, when demands for a variety of wideband services (e.g. high speed Internet access and the video/high quality images transmission) will continue to increase. The third generation (3G) mobile systems (called IMT-2000, see Fig. 1.1) are being designed to support wideband services, at data rates as high as 2Mbps, with the same quality as fixed networks. It is everyone's wish that wireless could act like a wired connection. To realize true IMT-2000 systems, a new wideband wireless access technology incorporating as many recent technology developments as possible is necessary.

Wideband direct sequence code division multiple access (Wideband CDMA) is emerging as the predominant wireless access technology for the 3G systems and is being developed in the world. Wideband CDMA is designed to flexibly offer wideband services, such as wireless Internet services (i.e., peak rate of 384 kbps to download information from the Web) and video transmissions (data rate can be up to 2 Mbps) (see Figs. 1.1 and 1.2). Wideband is essentially about data rate. The physical limitations and impairments on radio channels (bandwidth constraints, multipath fading, noise and interference) present a fundamental technical challenge to reliable high data rate communications.

Wideband CDMA supports inter-cell asynchronous/synchronous operation, adaptive (variable rate) transmission and applications of advanced techniques such as adaptive antenna arrays and Turbo codes.

- Cell acquisition for inter-cell asynchronous operation, which does not require inter-base-station synchronization; therefore, there is no requirement on any external system such as GPS to allow for easy deployment in all types of environments.
- Adaptive (variable rate) transmission provides multimedia services. To achieve higher bit rates, multiple downlink physical channels (or multi-code) can be transmitted in parallel on one direction.
- The adaptive antenna array directs beam nulls to interference sources to maximize the signal to interference ratio of each user. In the case of voice-only services, the application of the antenna arrays is rather impractical because a large number of antenna elements is required. However, antenna arrays are particularly useful for multimedia communications in which different users are transmitting with different data rates. High rate users (Internet and video) although the number of them is not large, give significant interference to low rate users (in particular, voice users); thus, without adaptive antenna array, the link capacity would be significantly reduced. In Wideband CDMA, dedicated pilot symbols in both up- and downlink facilitate user-unique antenna patterns.

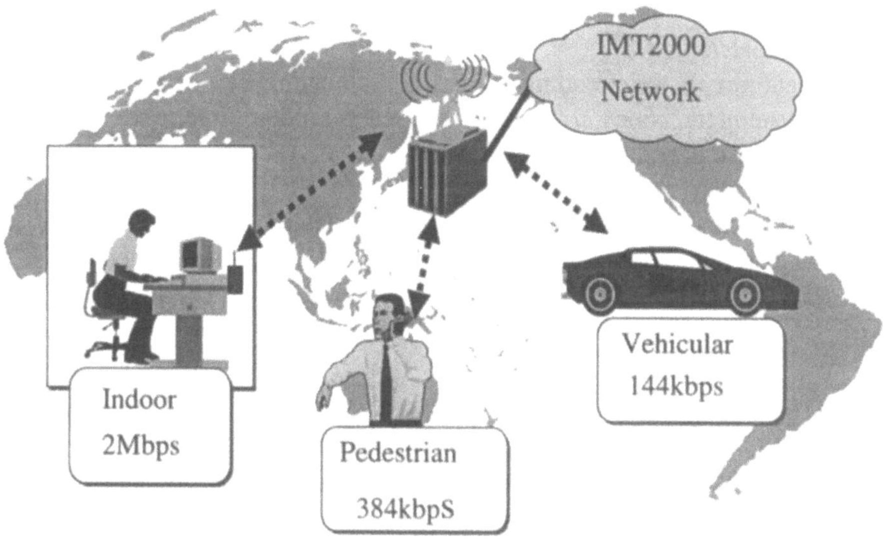

Fig. 1.1 3G mobile communications environment [12]

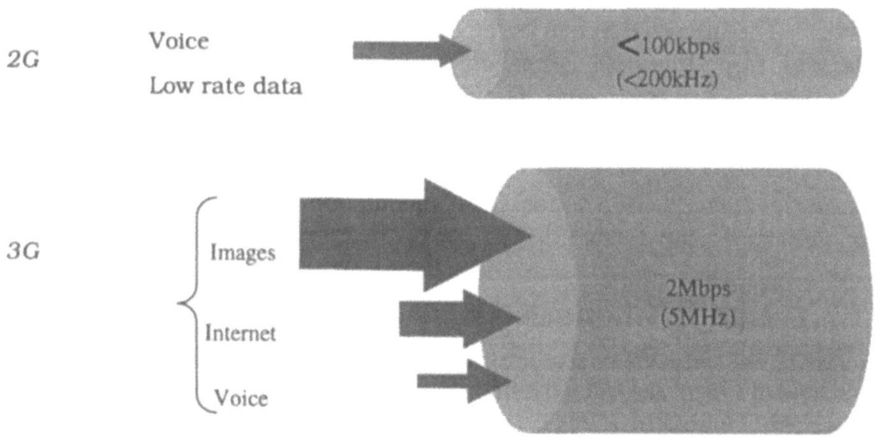

Fig. 1.2 Difference in data rate (bit pipe) between 2G and 3G systems.

- Turbo coding has been shown extremely good for high-rate high quality services because of its relatively large coding gain with reasonable computation complexity. Turbo codes are constructed by concatenating two or more recursive systematic codes. The main features of turbo coding are (a) concatenating two or more recursive systematic codes, (b) using a pseudo-random interleaver, and (c) iterative decoding algorithm.

This chapter describes the technical details of the WCDMA air interface. The description is primarily focused on the physical layer.

1.2 MULTIPLE ACCESS

The access scheme is Wideband Code Division Multiple Access (WCDMA) with information spread over approximately 5 MHz bandwidth and with FDD (Frequency Division Duplex) mode of paired bands. FDD is defined as follows:

> A duplex method whereby uplink and downlink transmissions use two separated radio frequencies. In the FDD, each uplink and downlink uses the different frequency band. A pair of frequency bands which have specified separation shall be assigned for the system. For WCDMA, the paired bands 1920--1980 and 2110—2170 MHz are used in FDD operation.

A 10 ms radio frame is divided into 15 slots (2560 chip/slot at the chip rate 3.84 Mcps). A physical channel is therefore defined as a code (or number of codes). The information rate of the channel varies with the symbol rate being derived from the 3.84 Mcps chip rate and the spreading factor (SF). Spreading factors are from 256 to 4 with FDD uplink and from 512 to 4 with FDD downlink. Thus the modulation symbol rate varies from 960 k symbols/s to 15 k symbols/s (7.5 k symbols/s) for FDD uplink (downlink), corresponding to BPSK for data modulation in uplink and QPSK in downlink, respectively.

1.2.1 Uplink Channels

There are two types of uplink dedicated physical channels: the uplink Dedicated Physical Data Channel (uplink DPDCH) and the uplink Dedicated Physical Control (or Pilot) Channel (uplink DPCCH). The DPDCH and the DPCCH are I/Q code multiplexed within each radio frame (see Chapter 2). The uplink DPDCH is used to carry the DCH transport channel. There may be zero, one, or several uplink DPDCHs on each radio link. When several uplink DPDCHs on each link are used, several codes need to be used. This is called multi-code CDMA (see Chapter 3). The uplink DPCCH is used to carry control information generated at the physical layer. The physical layer control information consists of known pilot bits to support channel estimation for coherent detection, transmit power-control (TPC) commands, feedback information (FBI), and an optional transport-format combination indicator (TFCI). The transport-format combination indicator informs the receiver about the instantaneous transport format combination of the transport channels mapped to the simultaneously transmitted uplink DPDCH radio frame. There is one and only one uplink DPCCH on each radio link, no matter how many codes are used for data channels. Figure 1.3 shows the frame structure of the uplink dedicated physical channels. Each radio frame of length 10 ms is split into 15 slots, each of length T_{slot} = 2560 chips, corresponding to one power-control period.

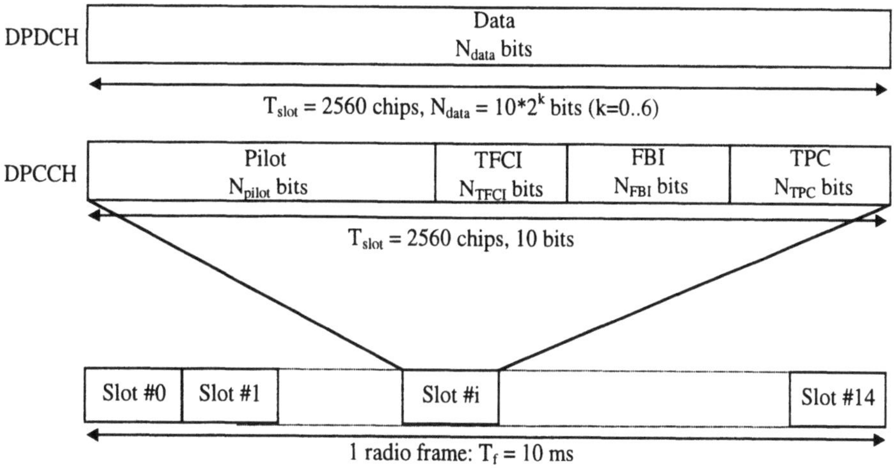

Figure 1.3 Frame structure for uplink DPDCH/DPCCH [4]

The parameter k in the figure determines the number of bits per uplink DPDCH slot. It is related to the spreading factor SF of the DPDCH as SF = $256/2^k$. The DPDCH spreading factor may range from 256 down to 4. The spreading factor of the uplink DPCCH is always equal to 256, i.e. there are 10 bits per uplink DPCCH slot. However, the exact number of bits of the uplink DPDCH and the different uplink DPCCH fields (N_{pilot}, N_{TFCI}, N_{FBI}, and N_{TPC}) is dependent on the channel data rate. For details, please see [4].

1.2.2 Downlink Channels

There is only one type of downlink dedicated physical channel, the Downlink Dedicated Physical Channel (downlink DPCH). Within one downlink DPCH, dedicated data, i.e. the dedicated transport channel (DCH), is transmitted in time-multiplex with control information (known pilot bits, TPC commands, and an optional TFCI). The downlink DPCH can thus be seen as a time multiplex of a downlink DPDCH and a downlink DPCCH, compared to uplink. Figure 1.4 shows the frame structure of the downlink DPCH. Each frame of length 10 ms is split into 15 slots, each of length T_{slot} = 2560 chips, corresponding to one power-control period. The spreading factor in the downlink may range from 512 down to 4.

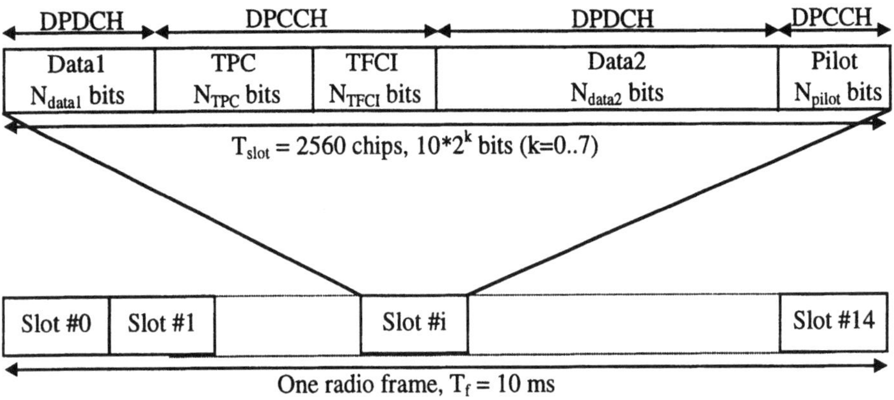

Figure 1.4 Frame structure for downlink DPCH

Common downlink pilot channels (CPICH)

In downlink, there are CPICHs. The CPICH is a fixed rate (30 kbps, SF=256) downlink physical channel that carries a pre-defined bit/symbol sequence. Figure 1.5 shows the frame structure of the CPICH.

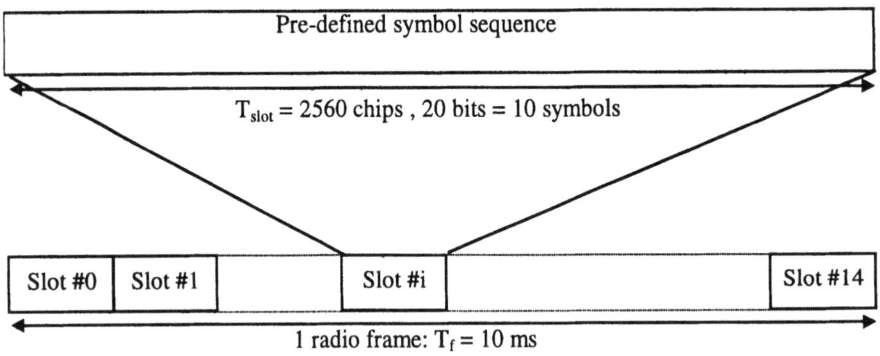

Figure 1.5 Frame structure for Common Pilot Channel

There are two types of Common pilot channels: the Primary and Secondary CPICH. They differ in their use and the limitations placed on their physical features. The Primary Common Pilot Channel (P-CPICH) has the following characteristics:

- The same channelization code is always used for the P-CPICH;
- The P-CPICH is scrambled by the primary scrambling code;
- There is one and only one P-CPICH per cell;
- The P-CPICH is broadcast over the entire cell.

The Primary CPICH is a phase reference for the downlink channels. A Secondary Common Pilot Channel (S-CPICH) has the following characteristics:

- An arbitrary channelization code of SF=256 is used for the S-CPICH;
- A S-CPICH is scrambled by either the primary or a secondary scrambling code;
- There may be zero, one, or several S-CPICH per cell;
- A S-CPICH may be transmitted over the entire cell or only over a part of the cell;

A Secondary CPICH may be a phase reference for a Secondary CCPCH carrying downlink access channels only and/or a downlink DPCH.

1.3 CHANNEL CODING

Different coding and interleaving schemes can be applied to a transport channel depending on the specific requirements in terms of error rates, delay, and so forth. Basically, the channel coding scheme is a combination of error detection, error correcting, rate matching, interleaving and transport channels mapping onto/splitting from physical channels. Here, we only describe error correcting. The following channel coding schemes can be applied to TrCHs [4]:

- Rate 1/3 (or ½) convolutional coding is typically applied for low delay services with moderate error rate requirements (BER=10^{-3});
- turbo coding can be applied for high-quality services (BER=10^{-6});
- no coding.

Usage of coding scheme and coding rate for the different types of TrCH is shown in Table 1.1. Defining K_i is the number of bits in each code block and

Y_i is the number of encoded bits, the values of Y_i in connection with each coding scheme are:

- convolutional coding with rate 1/2: $Y_i = 2*K_i + 16$; rate 1/3:

  $Y_i = 3*K_i + 24$, where "16" or "24" stands for the value of "tail".

  The known tail bits are used as reference at the end of decoding;
- turbo coding with rate 1/3: $Y_i = 3*K_i + 12$;
- no coding: $Y_i = K_i$.

Table 1.1: Usage of channel coding scheme and coding rate

Type of TrCH	Coding scheme	Coding rate
BCH	Convolutional coding	1/2
PCH		
RACH		
CPCH, DCH, DSCH, FACH		1/3, 1/2
	Turbo coding	1/3
	No coding	

1.3.1 Convolutional Coding

Convolutional coding has been widely used in 2G mobile systems. Convolutional codes are different from block codes by the existence of memory in the encoding scheme. In convolutional codes each block of k bits is mapped into a block of n bits to be transmitted over the channel, but these n bits are not only determined by the present k information bits but also by the previous information bits. This dependence on the previous information bits causes the encoder to be a finite state machine.

Convolutional codes with constraint length 9 and coding rates 1/3 and 1/2 are adopted. The configuration of the convolutional coder is presented in Figure 1.6. Output from the rate 1/3 convolutional coder shall be done in the order output0, output1, output2, output0, output1, output 2, output 0,...,output2. Output from the rate 1/2 convolutional coder shall be done in the order output 0, output 1, output 0, output 1, output 0, ..., output 1. 8 tail bits with binary value 0 shall be added to the end of the code block before encoding. The initial value of the shift register of the coder shall be "all 0" when starting to encode the input bits.

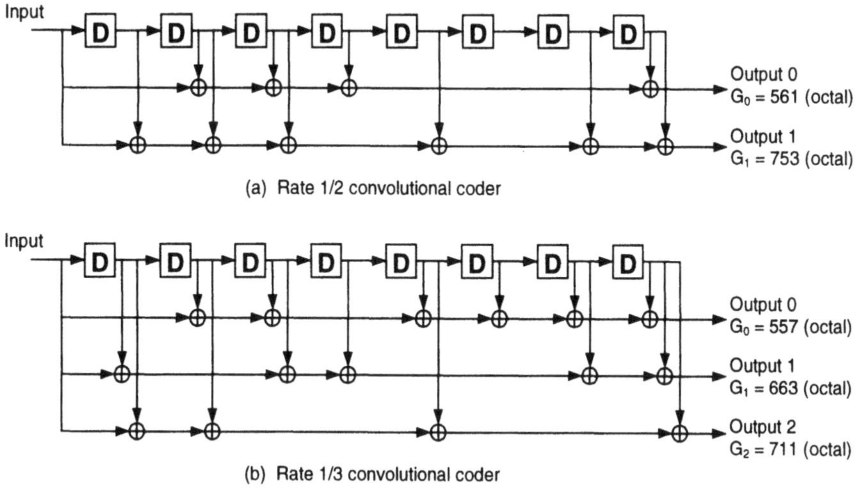

Figure 1.6 Rate 1/2 and rate 1/3 convolutional coders [4]

1.3.2 Turbo Coding

Turbo coding has relatively large coding gain with reasonable computation complexity. The main features of Turbo coding distinguishing it from conventional concatenated coding are (a) concatenating two or more recursive systematic codes, (b) using a pseudo-random interleaver, and (c) iterative decoding algorithm. The scheme of Turbo coder is a Parallel Concatenated Convolutional Code (PCCC) with two 8-state constituent encoders and one Turbo code internal interleaver. The coding rate of Turbo coder is 1/3. The structure of Turbo coder is illustrated in Figure 1.7. The encoder consists of two rate-1/2, constraint length of 3 recursive systematic convolutional encoders and a pseudo-random interleaver. The turbo-coded sequence $\{x_k, z_k, z'_k; k = 0, 1, .., N-1\}$, where N denotes the length of information sequence $\{x_k\}$, is transmitted over the radio channel. The transfer function of the 8-state constituent code for PCCC is:

$$G(D) = \left[1, \frac{g_1(D)}{g_0(D)}\right], \tag{1.1}$$

where

$$g_0(D) = 1 + D^2 + D^3, \tag{1.2}$$

$$g_1(D) = 1 + D + D^3. \tag{1.3}$$

The initial value of the shift registers of the 8-state constituent encoders shall be all zeros when starting to encode the input bits. Output from the Turbo coder is

$$x_1, z_1, z'_1, x_2, z_2, z'_2, \ldots, x_K, z_K, z'_K, \tag{1.4}$$

where $x_1, x_2, \ldots, x_K$ are the bits input to the Turbo coder i.e. both first 8-state constituent encoder and Turbo code internal interleaver, and K is the number of bits, and $z_1, z_2, \ldots, z_K$ and $z'_1, z'_2, \ldots, z'_K$ are the bits output from first and second 8-state constituent encoders, respectively.

The bits output from Turbo code internal interleaver are denoted by $x'_1, x'_2, \ldots, x'_K$, and these bits are to be input to the second 8-state constituent encoder.

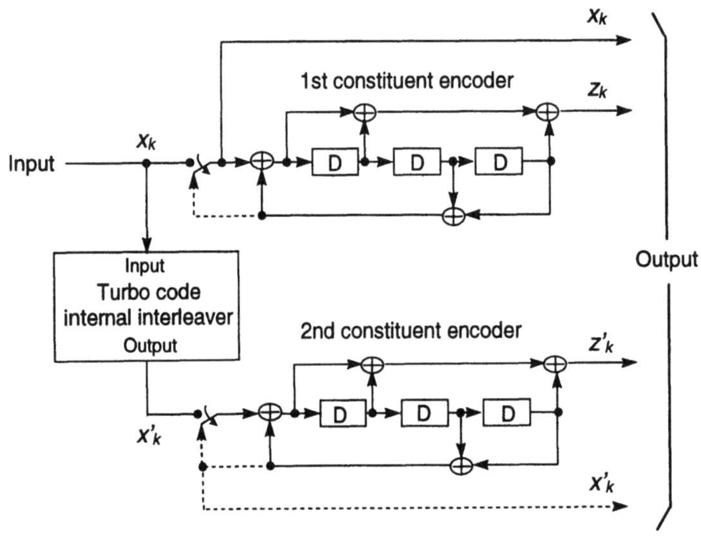

Figure 1.7 Structure of rate 1/3 Turbo coder (dotted lines apply for trellis termination only) [4]

Trellis termination is performed by taking the tail bits from the shift register feedback after all information bits are encoded. Tail bits are padded after the encoding of information bits. The first three tail bits shall be used to terminate the first constituent encoder (upper switch of Figure 1.7 in lower position) while the second constituent encoder is disabled. The last three tail bits shall be used to terminate the second constituent encoder (lower switch of Figure 1.7 in lower position) while the first constituent encoder is disabled. The transmitted bits for trellis termination shall then be:

$$x_{K+1}, z_{K+1}, x_{K+2}, z_{K+2}, x_{K+3}, z_{K+3}, x'_{K+1}, z'_{K+1}, x'_{K+2}, z'_{K+2}, x'_{K+3}, z'_{K+3}. \tag{1.5}$$

The Turbo code internal interleaver consists of bits-input to a rectangular matrix with padding, intra-row and inter-row permutations of the rectangular matrix, and bits-output from the rectangular matrix with pruning. The bits input to the Turbo code internal interleaver are denoted by $x_1, x_2, x_3, \ldots, x_K$, where K is the integer number of the bits and takes one value of $40 \leq K \leq 5114$. The relation between the bits input to the Turbo code internal interleaver and the bits input to the channel coding is defined by $x_k = o_{irk}$ and $K = K_i$.

The following is an example (Fig. 1.8) of Turbo coding for 64 kbps data, which can be applied to ISDN service.

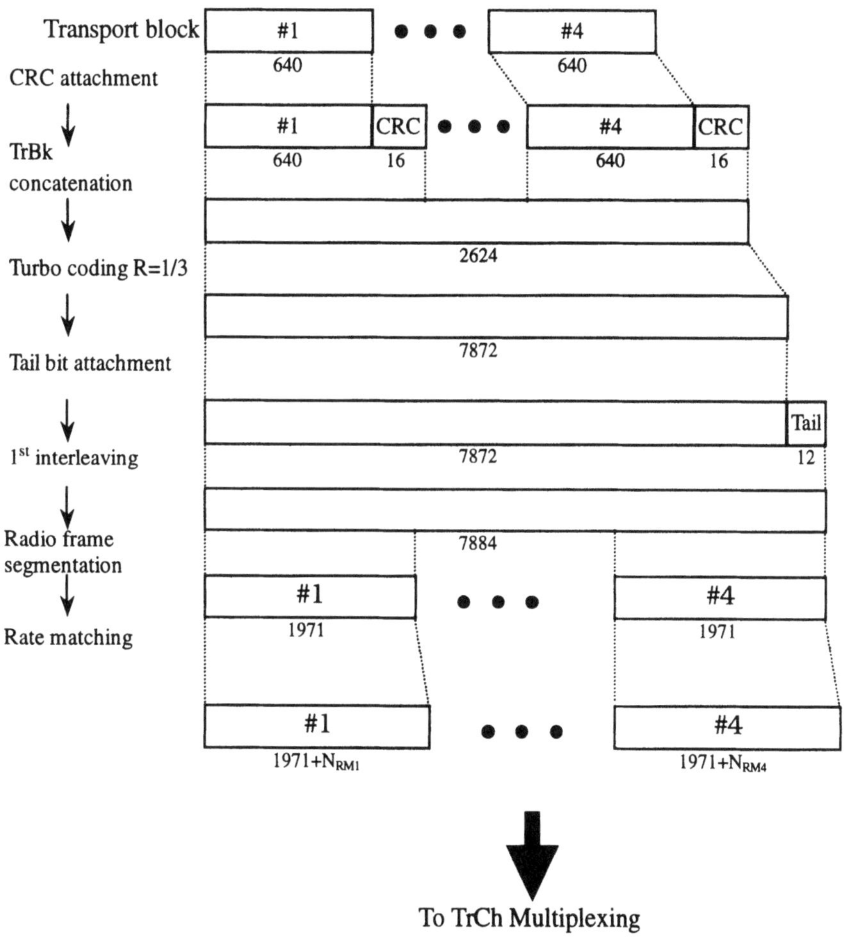

Figure 1.8 Channel coding and multiplexing example for 64 kbps data [4]

1.4 RATE MATCHING

One of the main features is the support of simultaneous services of different quality of service (QoS). However, when combining two or more services of different nature, it is very difficulty to achieve a common point of operation with optimum QoS. Using closed loop power control mechanism, one service having the highest SIR requirement would require the other service to be transmitted with unnecessary high QoS. This problem is alleviated by using rate matching technique. In rate matching algorithm, channel coded transport channel bit rate is adjusted by unequal puncturing or repetition to achieve QoS with minimum bit energy. The information bits with high QoS requirement or high bit energy are repeated, whereas information bits of other services with lower QoS requirement or lower bit energy are punctured. Additional functionality of rate matching algorithm is the matching of instantaneous bit rate that varies continuously.

Basically, rate matching is applied in order to match the bit rate of the CCTrCH to one of the limited set of bit rates of the uplink or downlink physical channels. As shown in Fig. 1.9, two different rate matching steps are carried out [2]:

(1) Static rate matching -- Static rate matching is carried out at the addition, removal, or redefinition of a transport channel (i.e., on a very slow basis). Static rate matching is applied after channel coding and uses code puncturing to adjust the channel coding rate of each transport channel so that the maximum bit rate of CCTrCH is matched to the bit rate of the physical channel. Static rate matching is applied on both uplink and downlink. On the downlink the static rate is used to, if possible, reduce the CCTrCH rate to the closest lower physical channel rate (next higher spreading factor), thus avoiding the over-allocation of orthogonal codes on the downlink and reducing the risk of a code-limited downlink capacity. Static rate matching should be distributed between the parallel transport channels is such a way that the transport channels fulfill their quality requirements at approximately the same channel signal-to-interference ratio (SIR); that is, static rate matching also performs "SIR matching,"

(2) Dynamic rate matching – Dynamic rate matching is carried out once every 10ms radio fram (i.e., on a very fast basis). Dynamic rate matching is applied after transport channel multiplexing and uses symbol repetition so that the instantaneous bit rate of the CCTrCH is exactly matched to the bit rate of the physical channel. Dynamic rate matching is only applied to the uplink. On the downlink, discontinuous transmission within each slot is used when the instantaneous rate of the CCTrCH doew not exactly match the bit rate of the physical channel.

Notice that although transport channel coding and multiplexing are carried out by the physical layer, the process is fully controlled by the radio resource controller, for example, in terms of choosing the appropriate coding scheme, interleaving parameters, and rate matching parameters.

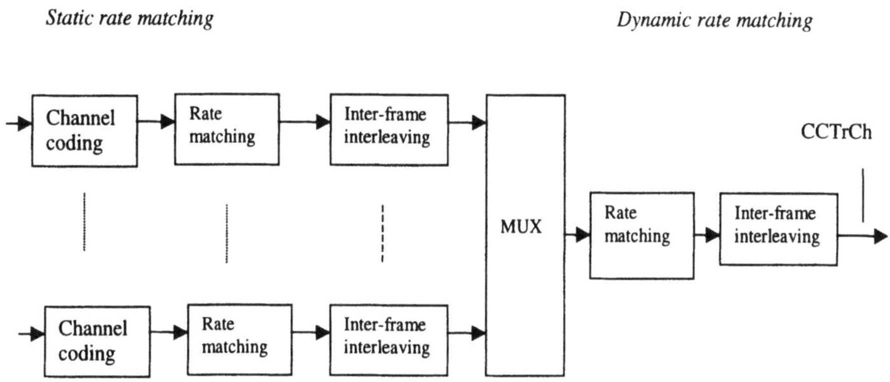

Figure 1.9 Transport channel coding and multiplexing in WCDMA

1.5 MODULATION AND SPREADING

The modulation scheme is QPSK. Pulse shaping square-root raised cosine with rolloff factor of α =0.22 in frequency domain is used. The impulse response of the chip impulse filter $RC_0(t)$ is:

$$RC_0(t)=\frac{\sin\left(\pi\frac{t}{T_C}(1-\alpha)\right)+4\alpha\frac{t}{T_C}\cos\left(\pi\frac{t}{T_C}(1+\alpha)\right)}{\pi\frac{t}{T_C}\left(1-\left(4\alpha\frac{t}{T_C}\right)^2\right)}$$

Where the roll-off factor α =0.22 and the chip duration (chip rate is 3.84Mcps) is

$$T=\frac{1}{chiprate}\approx 0.26042\mu s$$

The spreading (& scrambling) process is closely associated with modulation. Different spreading codes are used to spread the signal:

- For separating channels from same source or different mobiles, channelisation codes derived with the code tree structure [4].

- For separating different cells: Gold codes with 10 ms period (38400 chips at 3.84 Mcps) used, with the actual code itself length 2^{18}-1 chips.

1.5.1 Uplink Spreading and Modulation

Uplink spreading consists of two operations. The first is the channelization operation, which transforms every data symbol into a number of chips, thus increasing the bandwidth of the signal. The number of chips per data symbol is called the Spreading Factor (SF). The second operation is the scrambling operation, where a scrambling code is applied to the spread signal. With the channelization, data symbols on so-called I- and Q-branches are independently multiplied with an OVSF code. With the scrambling operation, the resultant signals on the I- and Q-branches are further multiplied by complex-valued scrambling code, where I and Q denote real and imaginary parts, respectively. In order to minimize the overall envelope variations and achieve high amplifier efficiency, a complex spreading technique is used. It has been shown in Chapter 2 that the normalized deviation of signal envelope with complex spreading is much stable than with dual channel spreading.

Figure 1.10 illustrates the principle of the uplink spreading of DPCCH and DPDCHs. The binary DPCCH and DPDCHs to be spread are represented by real-valued sequences, i.e. the binary value "0" is mapped to the real value +1, while the binary value "1" is mapped to the real value –1. The DPCCH is spread to the chip rate by the channelization code c_c, while the *n*:th DPDCH called $DPDCH_n$ is spread to the chip rate by the channelization code $c_{d,n}$. One DPCCH and up to six parallel DPDCHs can be transmitted simultaneously, i.e. $1 \leq n \leq 6$. For the case of a single DPDCH, the DPDCH and DPCCH are spread by two different channelization codes and transmitted on the I and Q branch, respectively. If more than one DPDCH is to be transmitted, the additional DPDCHs can be transmitted on either the I or Q branch using additional channelization codes (multicode transmission [13], see Chapter 3). The total spreading signal I+jQ is subsequently complex scrambled by a connection-specific complex scrambling code. The uplink channelization codes are the types of OVSF codes in order to ensure DPDCH/DPCCH transmit orthogonality.

After channelization, the real-valued spread signals are weighted by gain factors, β_c for DPCCH and β_d for all DPDCHs. At every instant in time, at

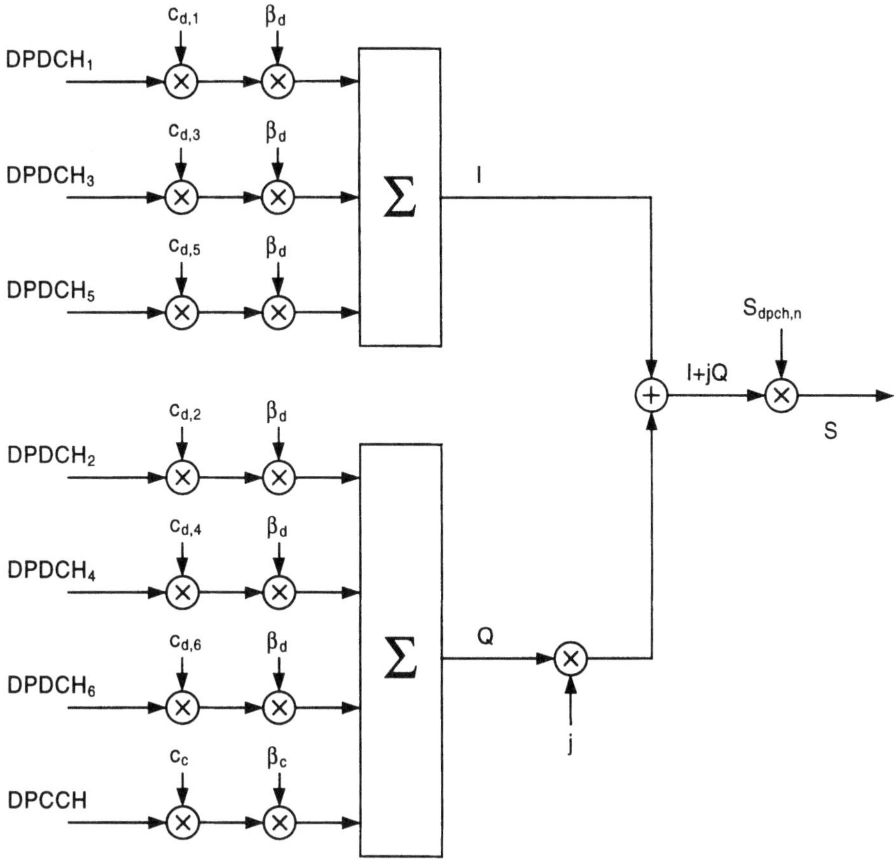

(a) Spreading for uplink DPCCH and DPDCHs

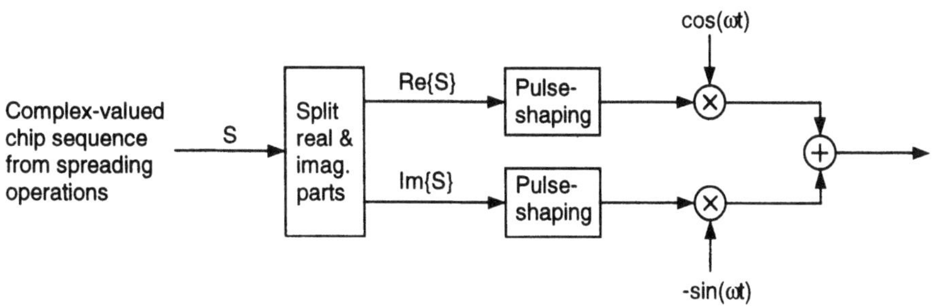

(b) uplink modulation

Figure 1.10 Uplink spreading and modulation [4]

least one of the values β_c and β_d has the amplitude 1.0. It has been shown in Chapter 2 that in order to achieve the maximum capacity of the WCDMA system the ratio of the DPCCH to DPDCH powers should range from –4 to –7dB. The β-values are quantized into 4 bit words. The quantization steps are given in Table 1.2.

Table 1.2 The quantization of the gain parameters [4]

Signalling values for β_c and β_d	Quantized amplitude ratios β_c and β_d
15	1.0
14	14/15
13	13/15
12	12/15
11	11/15
10	10/15
9	9/15
8	8/15
7	7/15
6	6/15
5	5/15
4	4/15
3	3/15
2	2/15
1	1/15
0	Switch off

After the weighting, the stream of real-valued chips on the I- and Q-branches are then summed and treated as a complex-valued stream of chips. This complex-valued signal is then scrambled by the complex-valued scrambling code $S_{dpch,n}$. The scrambling code is applied aligned with the radio frames, i.e. the first scrambling chip corresponds to the beginning of a radio frame.

Channelization codes

The channelization codes of Figure 1.10 are Orthogonal Variable Spreading Factor (OVSF) codes that preserve the orthogonality between a user's different physical channels. Improving the capability of multimedia communications is one of the targets for WCDMA mobile communications systems. WCDMA is designed to support a variety of data services from low to very high bit rates. Since the spread signal bandwidth is the same for all users, multiple rate transmissions need multiple spreading factors in the physical channels. Consider that each bit of the lowest bit-rate service (R_{min}) is spread by a code of length $N=2^n$. Since the bit duration for rate 2 R_{min} is half the duration of a bit in the minimum rate, we need a spreading code of length $N/2=2^{n-1}$ for spreading. Generally, a code length of 2^{n-k} is needed for bit rate $2^k R_{min}$. Depending on the maximum and minimum supported bit rates in the system and the spreading bandwidth, a range for the code length can be obtained. A method to obtain variable-length orthogonal codes that preserve orthogonality between different rates and SFs based on the modified Hadamard transformation is presented in the following [7]:

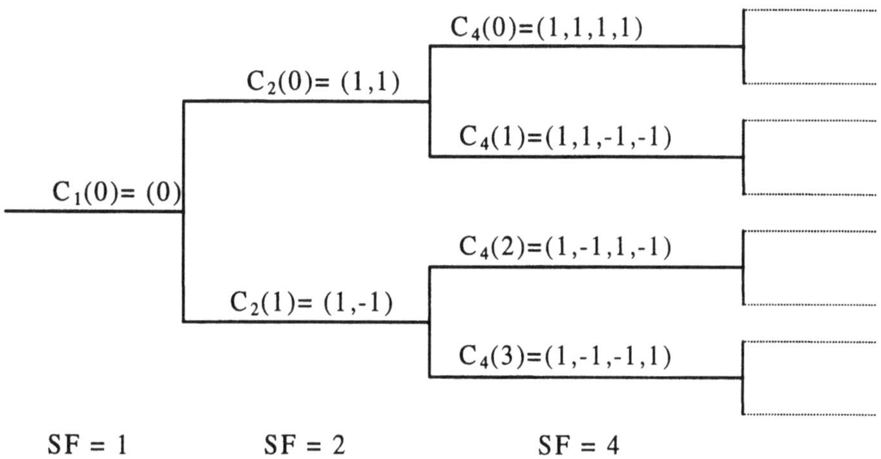

Figure 1.11 Code-tree for generation of Orthogonal Variable Spreading Factor (OVSF) codes

In Figure 1.11, the channelization codes are uniquely described as $C_{SF,}(k)$, where SF is the spreading factor of the code and k is the code number, $0 \leq k \leq$ SF-1. Each level in the code tree defines channelization codes of length SF, corresponding to a spreading factor of SF in Figure 1.10. The generation method for the channelization code is defined as:

$$C_1(0)=1,$$

$$\begin{bmatrix} C_2(0) \\ C_2(1) \end{bmatrix} = \begin{bmatrix} C_1(0) & C_1(0) \\ C_1(0) & -C_1(0) \end{bmatrix} = \begin{bmatrix} 1 & 1 \\ 1 & -1 \end{bmatrix}$$

$$\begin{bmatrix} C_{2^{(n+1)}}(0) \\ C_{2^{(n+1)}}(1) \\ C_{2^{(n+1)}}(2) \\ C_{2^{(n+1)}}(3) \\ \vdots \\ C_{2^{(n+1)}}(2^{(n+1)}-2) \\ C_{2^{(n+1)}}(2^{(n+1)}-1) \end{bmatrix} = \begin{bmatrix} C_{2^n}(0) & C_{2^n}(0) \\ C_{2^n}(0) & -C_{2^n}(0) \\ C_{2^n}(1) & C_{2^n}(1) \\ C_{2^n}(1) & -C_{2^n}(1) \\ \vdots & \vdots \\ C_{2^n}(2^n-1) & C_{2^n}(2^n-1) \\ C_{2^n}(2^n-1) & -C_{2^n}(2^n-1) \end{bmatrix}$$

The data channel may use the code from spreading factor 4 to 256. A mobile terminal is allowed to increase SF during the message transmission on a frame by frame basis. As s result, these variable length orthogonal codes can be generated recursively using a tree structure as shown in Fig. 1.11. Starting from $C_1(0)=1$, a set of 2^k spreading codes with the length of 2^k chips are generated at the kth layer. It can be understood that generated codes of the same layer constitute a set of Walsh functions and they are orthogonal, although the rows of C_N are not in the same order of H_N. Fortunately, any two codes of different layers are also orthogonal except for the case that one of the two codes is a mother code of the other; for example, all of $C_{16}(1)$, $C_8(0)$, $C_4(0)$ and $C_2(0)$ are mother codes of $C_{32}(2)$, and so are not orthogonal against $C_{32}(2)$. In other words, a code can be used in a channel if and only if no other code on the path from the specific code to the root of the tree or the sub-tree produced by the specific code is used in the same channel. From this observation, we can easily find that if $C_8(0)$ is assigned to a user, all the codes { $C_{16}(0)$, $C_{16}(1)$, $C_{32}(0)$, ..,$C_{32}(3)$, $C_{64}(0)$, ..,$C_{64}(7)$, $C_{128}(0)$,..} generated from this code cannot be assigned to other users requesting lower rates. In addition, mother codes {$C_4(1)$, C_2 (0)} of $C_8(0)$ cannot be assigned to users requesting higher rates. This means that the number of available codes is not fixed but depends on the rate and spreading factor of each physical channel. These restrictions are imposed in order to maintain orthogonality. *In conclusion, when a mother is used, its sons and grandsons cannot be used.*

Scrambling Codes

There are two types of scrambling codes used in WCDMA: long and short scrambling codes. All uplink physical channels are subjected to scrambling with a complex-valued scrambling code. The DPCCH/DPDCH may be scrambled by either long or short scrambling codes. In WCDMA, different base stations adopt different scrambling codes, unlike CDMA2000, which uses the same scrambling codes for all base stations. This means there is not requirement that each base station should be capable of reliable Global Positioning System (GPS) reception. This will significantly reduce the deployment efforts, especially in indoor environments. There are 2^{24} long and 2^{24} short uplink scrambling codes. The long scrambling code is built from constituent long sequences, while the constituent short sequences are used to build the short scrambling code.

The long scrambling sequences $c_{long,1,n}$ and $c_{long,2,n}$ (or Gold sequences) are constructed from position wise modulo 2 sum of 38400 chip segments of two binary *m*-sequences generated by means of two generator polynomials of degree 25 (see Fig. 1.12). Let *x* and *y* be the two *m*-sequences respectively. The *x* sequence is constructed using the primitive polynomial

$$X^{25}+X^{3}+1. \tag{1.6}$$

The *y* sequence is constructed using the polynomial

$$X^{25}+X^{3}+X^{2}+X+1. \tag{1.7}$$

The resulting sequences thus constitute segments of a set of Gold sequences. The sequence $c_{long,2,n}$ is a 16777232 chip shifted version of the sequence $c_{long,1,n}$. Therefore, the complex-valued long scrambling sequence $C_{long,\ n}$, is defined as:

$$C_{long,n}(i) = c_{long,1,n}(i)\left(1 + j(-1)^{i} c_{long,2,n}(2\lfloor i/2 \rfloor)\right) \tag{1.8}$$

where $i = 0, 1, \ldots, 2^{25} - 2$ and $\lfloor\ \rfloor$ denotes rounding to nearest lower integer.

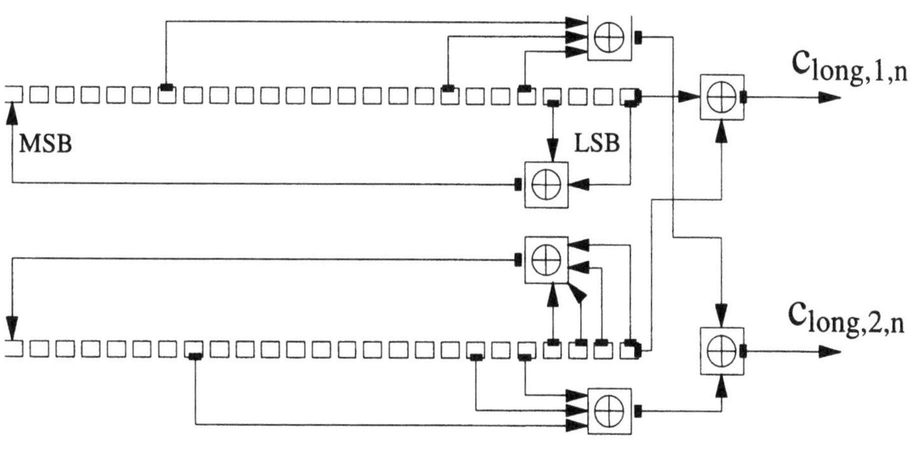

Figure 1.12 Configuration of uplink long scrambling sequence generator [4]

However, the short scrambling sequences $c_{short,1,n}(i)$ and $c_{short,2,n}(i)$ are defined from a sequence of the family of periodically extended S(2) codes. Let $n_{23}n_{22}\ldots n_0$ be the 24 bit binary representation of the code number n. The n:th quaternary sequence $z_n(i)$, $0 \le n \le 16777215$, is obtained by modulo 4 addition of three sequences, a quaternary sequence $a(i)$ and two binary sequences $b(i)$ and $d(i)$, where the initial loading of the three sequences is determined from the code number n. The sequence $z_n(i)$ of length 255 is generated according to the following relation:

- $z_n(i) = a(i) + 2b(i) + 2d(i)$ modulo 4, $i = 0, 1, \ldots, 254$; (1.9)

where the quaternary sequence $a(i)$ is generated recursively by the polynomial

$$g_0(x)= x^8+x^5+3x^3+x^2+2x+1 \text{ as:} \qquad (1.10)$$

and the binary sequence $b(i)$ is generated recursively by the polynomial

$$g_1(x)= x^8+x^7+x^5+x+1 \text{ as} \qquad (1.11)$$

and the binary sequence $d(i)$ is generated recursively by the polynomial

$$g_2(x)= x^8+x^7+x^5+x^4+1 \text{ as:} \qquad (1.12)$$

Finally, the complex-valued short scrambling sequence $C_{short,\,n}$, is defined as:

$$C_{short,n}(i) = c_{short,1,n}(i \bmod 256)\left(1 + j(-1)^i c_{short,2,n}\left(2\lfloor (i \bmod 256)/2 \rfloor\right)\right) \qquad (1.13)$$

An implementation of the short scrambling sequence generator for the 255 chip sequence to be extended by one chip is shown in Figure 1.13.

1.5.2 Downlink Spreading and Modulation

Figure 1.14 illustrates the spreading operation for downlink physical channels except synchronization channels. The non-spread physical channel consists of a sequence of real-valued symbols. For all channels the symbols can take the three values +1, -1, and 0, where 0 indicates discontinuous transmission. Each pair of two consecutive symbols is first serial-to-parallel converted and mapped to an I and Q branch. The mapping is such that even and odd numbered symbols are mapped to the I and Q branch respectively. For all channels, symbol number zero is defined as the first symbol in each frame. The I and Q branches are then spread to the chip rate by the same real-valued channelization code $C_{SF}(m)$. The sequences of real-valued chips on the I and Q branch are then treated as a single complex-valued sequence of chips. This sequence of chips is scrambled (complex chip-wise multiplication) by a complex-valued scrambling code $S_{dl,n}$.

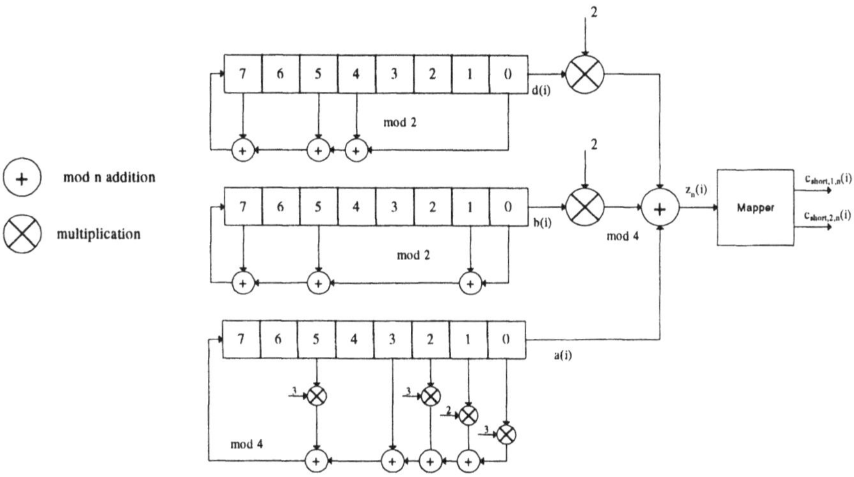

Figure 1.13 Uplink short scrambling sequence generator for 255 chip sequence [4]

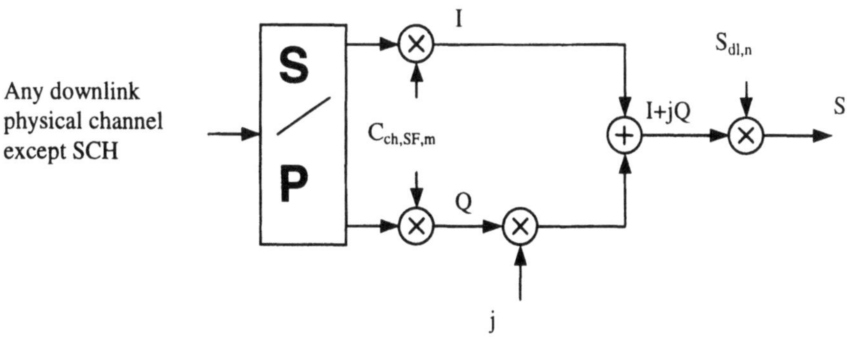

Figure 1.14 Spreading for all downlink physical channels except SCH

Figure 1.15 illustrates how different downlink channels are combined. Each complex-valued spread channel, corresponding to point S in Figure 1.14, is separately weighted by a weight factor G_i. The complex-valued P-SCH and S-SCH, are separately weighted by weight factors G_p and G_s. All downlink physical channels are then combined using complex addition.

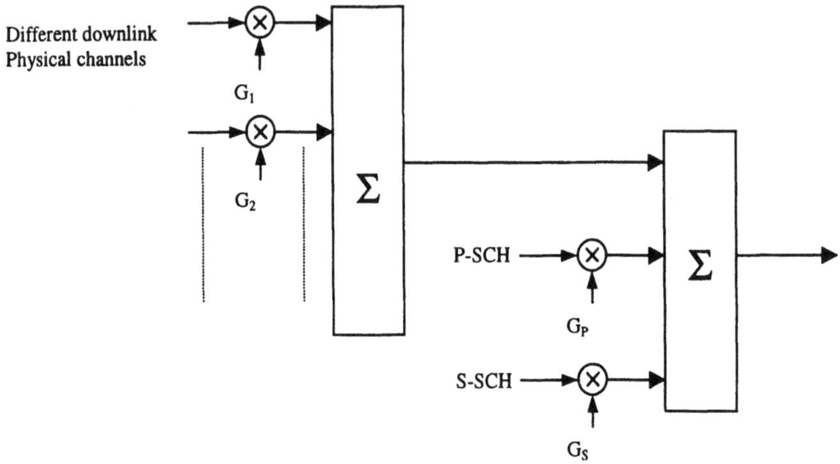

Figure 1.15 Spreading and modulation for SCH and P-CCPCH

The channelization codes of Figure 1.14 are the same codes as used in the uplink, namely Orthogonal Variable Spreading Factor (OVSF) codes that preserve the orthogonality between downlink channels of different rates and spreading factors. The channelization code for the Primary CPICH is fixed to $C_{256}(0)$and the channelization code for the Primary CCPCH is fixed to $C_{256}(1)$

Scrambling code

A total of 2^{18}-1 = 262,143 scrambling codes, numbered 0…262,142 can be generated. However not all the scrambling codes are used. The scrambling codes are divided into 512 sets each of a primary scrambling code and 15 secondary scrambling codes. The primary scrambling codes consist of scrambling codes n=16*i where i=0…511. The i:th set of secondary scrambling codes consists of scrambling codes 16*i+k, where k=1…15. There is a one-to-one mapping between each primary scrambling code and 15 secondary scrambling codes in a set such that i:th primary scrambling code corresponds to i:th set of secondary scrambling codes. Hence, according to the above, scrambling codes k = 0, 1, …, 8191 are used. Each of these codes are associated with a left alternative scrambling code and a right alternative scrambling code, that may be used for compressed frames. The left alternative scrambling code corresponding to scrambling code k is scrambling code number k + 8192, while the right alternative scrambling code corresponding to scrambling code k is scrambling code number k + 16384. The alternative scrambling codes can be used for compressed frames. In this case, the left alternative scrambling code is used if $n<SF/2$ and the right alternative scrambling code is used if $n\geq SF/2$, where $c_{ch,SF,n}$ is the

channelization code used for non-compressed frames. The usage of alternative scrambling code for compressed frames is signalled by higher layers for each physical channel respectively.

The set of primary scrambling codes is further divided into 64 scrambling code groups, each consisting of 8 primary scrambling codes. The j:th scrambling code group consists of primary scrambling codes 16*8*j+16*k, where j=0..63 and k=0..7. Each cell is allocated one and only one primary scrambling code. The primary CCPCH and S-CCPCH carrying paging channels are always transmitted using the primary scrambling code. The other downlink physical channels can be transmitted with either the primary scrambling code or a secondary scrambling code from the set associated with the primary scrambling code of the cell.

The scrambling code sequences are constructed by combining two real sequences into a complex sequence. Each of the two real sequences are constructed as the position wise modulo 2 sum of 38400 chip segments of two binary *m*-sequences generated by means of two generator polynomials of degree 18 (see Fig.1.16). The resulting sequences thus constitute segments of a set of Gold sequences. The scrambling codes are repeated for every 10 ms radio frame. Let x and y be the two sequences respectively. The x sequence is constructed using the primitive polynomial

$$1+X^7+X^{18}. \qquad (1.14)$$

The y sequence is constructed using the polynomial

$$1+X^5+X^7+X^{10}+X^{18}. \qquad (1.15)$$

The sequence depending on the chosen scrambling code number n is denoted z_n, in the sequel. Furthermore, let $x(i)$, $y(i)$ and $z_n(i)$ denote the i:th symbol of the sequence x, y, and z_n, respectively. The *m*-sequences x and y are constructed as:

Initial conditions:

- x is constructed with x (0)=1, x(1)= x(2)=...= x (16)= x (17)=0.
- y(0)=y(1)= … =y(16)= y(17)=1.

Recursive definition of subsequent symbols:

- $x(i+18) = x(i+7) + x(i)$ modulo 2, $i=0,\dots,2^{18}-20$. (1.16)
- $y(i+18) = y(i+10)+y(i+7)+y(i+5)+y(i)$ modulo 2, $i=0,\dots,2^{18}-20$. (1.17)

The n:th Gold code sequence z_n, $n=0,1,2,\dots,2^{18}-2$, is then defined as:

- $z_n(i) = x((i+n) \text{ modulo } (2^{18}-1)) + y(i)$ modulo 2, $i=0,\dots,2^{18}-2$. (1.18)

These binary sequences are converted to real valued sequences Z_n by the following transformation:

$$Z_n(i)=\begin{cases}+1 & if\ z_n(i)=0\\ -1 & if\ z_n(i)=1\end{cases} \quad for \quad i=0,1,\ldots,2^{18}-2. \qquad (1.19)$$

Finally, the n:th complex scrambling code sequence $S_{dl,n}$ is defined as:

- $S_{dl,n}(i) = Z_n(i) + j\, Z_n((i+131072) \text{ modulo } (2^{18}-1)),\ i=0,1,\ldots,38399.$ (1.20)

Note that the pattern from phase 0 up to the phase of 38399 is repeated.

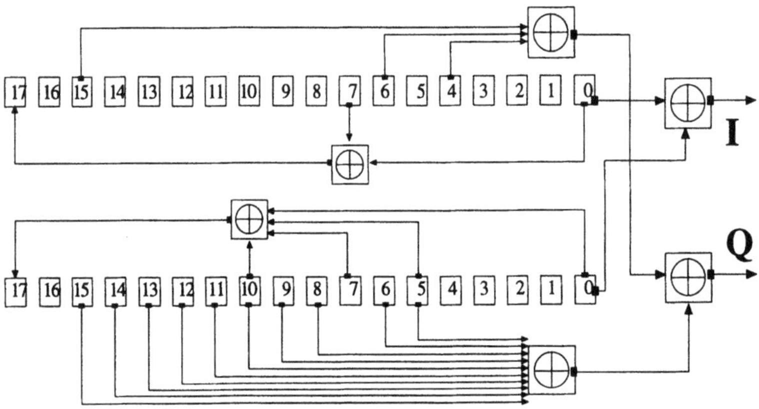

Figure 1.16 Configuration of downlink scrambling code generator [4]

1.6 PHYSICAL LAYER PROCEDURES

There are several physical layer procedures involved with WCDMA operation. Such procedures are cell search and the power control, with both inner loop and slow quality loop.

1.6.1 Cell Search

It is easier to realize continuous system deployment, from outdoors to indoors, with an inter-cell asynchronous system than with an inter-cell synchronous system. This is because the former does not require any external timing source such as Global Positioning System (GPS) as used in IS-95(or CDMA2000). Unlike IS95-based systems, a different cell site is assigned a unique scramble code on its downlink. In general, the use of different scramble codes at different cell sites increases the cell search time.

To support an efficient cell search with asynchronous operation, each WCDMA base station transmits a special synchronization signal. The synchronization signal consists of the following two signals in parallel:

(1) A repeatedly transmitted unmodulated orthogonal Gold code of length 256 chips, the primary synchronization code (PSC), with a period of one slot. The PSC is the same for every base station in the system and is transmitted time-aligned with the primary CCPCH (BCH) slot boundary. By detecting the PSC, the mobile station acquires slot synchronization to the target BS.

(2) A repeatedly transmitted length-16 sequence of unmodulated orthogonal Gold codes of length 256 chips, the secondary synchronization code (SSC), with a period on one frame. Each SSC is chosen from a set of 17 different orthogonal Gold colds of length 256. There are a total of 32 possible SSC sequences indicating to which of the 32 different code groups the base station downlink scrambling code belongs. The sequences are constructed in such a way that their cyclic-shifts are unique, that is a nonzero cyclic shift of one sequence is not equal to any of the other 31 sequences. Consequently, by detecting the SSC sequence the MS does not only determine the scrambling code group, but also the frame timing of the target BS.

After the detection of the scrambling code group, the MS searches all 16 downlink scrambling codes, typically using symbol-by-symbol correlation over the fixed-rate primary CCPCH.

The downlink control channels of all cell sites reuse the same short spreading code, e.g. $C_{64}(0)$-code, and the scrambling code sequence (whose length is the same as the data frame length) is periodically masked over one symbol duration so that the $C_{64}(0)$-code appears periodically during the scramble code period. During the masking period, the group identification (GI) code indicating the code group, to which the scramble code of each cell site belongs, is transmitted in parallel. The GI code can be chosen from the

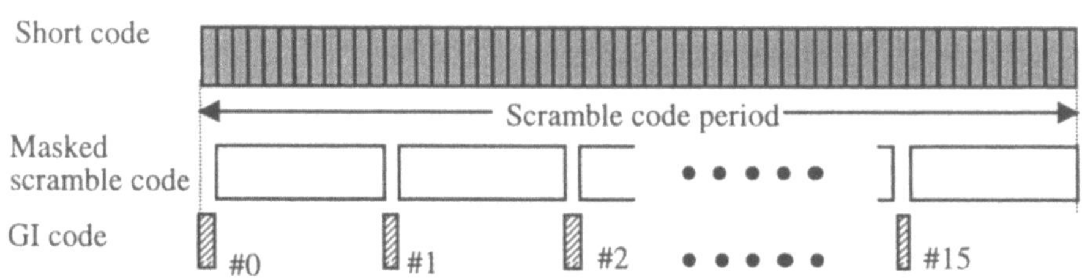

Fig. 1. 17 Scrambling code masking timing and GI code transmission.

set of orthogonal multi-SF codes, e.g. $\{C_{64}(j)\}_{j=1}^{63}$. The scramble code masking timing and GI code transmission are illustrated in Fig.1.17. A fast cell search algorithm is described in [9].

1.6.2 Power Control

Transmission power control (TPC) is a key to achieving a high capacity and reliable CDMA cellular system. Uplink TPC is required to combat the near-far problem because uplink signals are received at the base station from asynchronous and nonorthogonal MS's. Downlink TPC, on the other hand, is mainly used to minimize the excess use of transmission power by BS's, but it is unnecessary for coping with the near-far problem, which does not appear in the downlink. In both uplink and downlinks, TPC allows the reduction of inter-cell interference and system capacity can thus be enhanced. TPC enables the BS to maintain its output power at a minimum level and this greatly enhances system capacity in conjunction with the minimum path loss BS connection achieved by soft handoff.

DS-CDMA links are interference-limited. Fast TPC based on the measurement of SIR can always minimize the transmit power according to the traffic load and thus, interference to other users in the other cells can be reduced; thereby increasing the link capacity. SIR measurement is done after Rake combining. The frame structure using time-multiplexed (or code multiplexing) pilots well supports SIR measurement. Both pilot and data symbols are used to measure instantaneous received signal power, but only pilot symbols are used to measure instantaneous interference plus background noise power (followed by averaging using a first order filter). The channel estimation is first performed using the pilot symbols. The MF output signal $r_l(n, k)$ at the n-th symbol position of k-th slot associated with the l-th propagation path is weighted by the complex conjugate of channel estimate $\tilde{\xi}_l(k)$ to remove the fading-induced random phase. Assuming that the number of Rake fingers is L, the Rake combiner output is represented as $\tilde{r}(n,k) = \sum_{l=0}^{L-1} r_l(n,k)\,\tilde{\xi}_l^*(k),\ 0 \le n \le N_{slot} - 1$, where $\tilde{\xi}_l(k)$ is the simple average of N_p pilot symbols belonging to the k-th slot. $\tilde{r}(n,k)$ is used for measuring the SIR $\tilde{\lambda}(k)$ which is defined as $\tilde{\lambda}(k) = \tilde{S}(k)/\bar{I}(k)$, where $\tilde{S}(k)$ and $\bar{I}(k)$ are the instantaneous signal power and the average interference power, respectively.

The above fast TPC is applied to both up- and downlinks. The downlink is synchronous and all downlink channels are spread using orthogonal spreading codes (this is called orthogonal spreading). Orthogonal spreading with Rake combining provides larger capacity than random spreading even in frequency selective fading channels because of the absence of MAI within each path. In a cellular system, however, as the user moves away from the base station, the received signal power reduces due to increasing distance-dependence path loss and thus, the effects of background noise and the other-cell MAI become larger. In addition to distance-dependence path loss, shadowing exists. The instantaneous SIR on each downlink channel varies in a random manner. This suggests that the use of fast TPC can increase the link capacity of the downlink as well as that of the uplink.

The uplink inner-loop power control adjusts the MS transmit power in order to keep the received uplink signal-to-interference ratio (SIR) at a given SIR target, SIR_{target}. The serving cells (cells in the active set) should estimate signal-to-interference ratio SIR_{est} of the received uplink DPCH. The serving cells should then generate TPC commands and transmit the commands once per slot according to the following rule: if $SIR_{est} > SIR_{target}$ then the TPC command to transmit is "0", while if $SIR_{est} < SIR_{target}$ then the TPC command to transmit is "1". Upon reception of one or more TPC commands in a slot, the MS shall derive a single TPC command, TPC_cmd, for each slot, combining multiple TPC commands if more than one is received in a slot. Two algorithms shall be supported by the MS for deriving TPC bit.

The downlink inner-loop power control adjusts the network transmit power in order to keep the received downlink SIR at a given SIR target, SIR_{target}. A higher layer outer loop adjusts SIR_{target} independently for each connection. The MS should estimate the received downlink DPCCH/ DPDCH power of the connection to be power controlled. Simultaneously, the MS should estimate the received interference. The obtained SIR estimate SIR_{est} is then used by the MS to generate TPC commands according to the following rule: if $SIR_{est} > SIR_{target}$ then the TPC command to transmit is "0", requesting a transmit power decrease, while if $SIR_{est} < SIR_{target}$ then the TPC command to transmit is "1", requesting a transmit power increase.

SSDT Power Control

Although WCDMA base stations are generally mutually asynchronous, inter-base-station synchronization on a connection level is needed in case of soft handoff. Soft handoff synchronization is carried out as follows:

(1) From the cell search, the mobile station can estimate the timing offset between the downlink dedicated channel of the current base station and the primary CCPCH of the target base station.
(2) The estimated timing offset is transferred to the target base station using the current link with the old base station.
(3) The target base station uses the estimated timing offset to adjust the timing of the new downlink dedicated channel relative to that of the primary

CCPCH. The adjustment is done in steps of 256 chips to preserve the downlink transmit orthogonality of the target base station.

(4) Due to an approximately fixed offset between the downlink and uplink frame timing, the target base station can, from the estimated timing offset, estimate the approximate timing of the uplink dedicated physical channels to receive.

Site selection diversity transmit power control (SSDT) is a macro diversity method in soft handover mode. This method is optional WCDMA standard. Operation is summarised as follows. The MS selects one of the cells from its active set to be 'primary', all other cells are classed as 'non primary'. The main objective is to transmit on the downlink from the primary cell, thus reducing the interference caused by multiple transmissions in a soft handover mode. A second objective is to achieve fast site selection without network intervention, thus maintaining the advantage of the soft handover. In order to select a primary cell, each cell is assigned a temporary identification (ID) and MS periodically informs a primary cell ID to the connecting cells. The non-primary cells selected by MS switch off the transmission power. The primary cell ID is delivered by MS to the active cells via uplink FBI field. SSDT activation, SSDT termination and ID assignment are all carried out by higher layer signalling. The beneftis of SSDT can be summarized as follows [10].

(1) Since one of the BSs within an active set provides adequate power to the connected MS and the output power of the other BSs within the same active set is reduced, the increases in interference can be avoided.

(2) In conventional TPC, since multiple BSs transmit the same downlink signal with adequate power, a number of RAKE fingers must be equipped in order to collect many paths given by the active BS's. Since only one active BS serves the MS as a transmitting site in SSDT mode, the number of paths which should be collected by the RAKE receiver is smaller than that required for conventional TPC. This feature makes the path capturing efficiency of SSDT better than that of conventional TPC, given the same number of RAKE fingers.

(3) The power imbalance problem due to TPC command reception error never happens because only one BS transmits the downlink signal during soft handoff state.

1.7 TRANSMIT DIVERSITY

In WCDMA mobile radio, a transmitted signal is reflected by many obstacles between the transmitter and receiver, thus creating a multipath channel. Due to interference among the many paths, which have different time delays, the received signal suffers from frequency selective multipath fading, and transmission performance is significantly degraded. By using several spatially separated antennas at the receiver (using antenna diversity reception), transmission performance is improved. However, using antenna diversity reception at a MS increases the complexity and power consumption of the receiver. Therefore, it is preferable to use several transmit antennas at the BS, namely transmit diversity, in order to improve downlink transmission performance. Transmit diversity for the WCDMA downlink is divided into two categories: the open loop mode and closed loop mode [4,11]. In 3GPP, space time transmit diversity and time switched transmit diversity of the open loop mode will be applied to the common control physical channel and search channel, respectively [4]. However, the closed loop mode which controls the carrier phase and amplitude of the parallel signals transmitted from two antennas according to the feedback information (FBI) bit sent via the uplink will be adopted for a dedicated physical channel (DPCH) [4]. It has been shown in [11] that the improvement by the closed loop mode transmit diversity schemes compared to conventional single antenna transmission is heavily affected by the accuracy of channel estimation since the received signal power from each antenna becomes much lower.

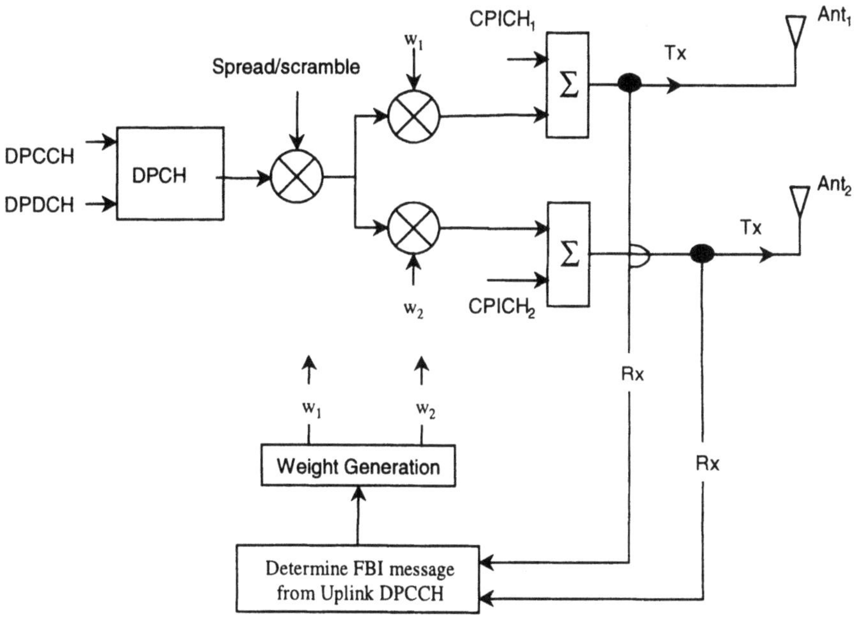

Figure 1.18 The generic downlink transmitter structure to support closed loop mode transmit diversity for DPCH transmission [4]

The general transmitter structure to support closed loop mode transmit diversity for DPCH transmission is shown in Figure 1.18 [4]. Channel coding, interleaving and spreading are done as in non-diversity mode. The spread complex valued signal is fed to both TX antenna branches, and weighted with antenna specific weight factors w_1 and w_2. The weight factors are complex valued signals (i.e., $w_{i,n} = a_{i,n} + jb_{i,n} = A_{i,n}e^{j\phi_{i,n}}$), in general. The weight factors (actually the corresponding phase phase/amplitude adjustments) are determined by the MS, and signalled to the BS transceiver using the D-bits of the FBI field of uplink DPCCH. The generation of FBI bit at the MS and the calculation of weighting factors at the BS for three odes are expressed as follows [11]:

(1) Mode I

$$\phi_{1,n} = 0, \phi_{2,n} = \{\pm\pi/4, \pm 3\pi/4\}, A_{1,n} = A_{2,n} = \sqrt{1/2} \tag{1.21}$$

The MS receiver estimates received carrier phases $\theta_{1,n}^{CP}$ and $\theta_{2,n}^{CP}$ of CPICHs transmitted from two antennas to generate FBI bit b_n, where n denotes the number of the slots. Let $\hat{\theta}_{1,n}^{CP}$ and $\hat{\theta}_{2,n}^{CP}$ be the estimated values of $\theta_{1,n}^{CP}$ and $\theta_{2,n}^{CP}$, respectively. When n is an even number

$$b_n = \begin{cases} 0, & -\pi/2 \le (\hat{\theta}_{1,n}^{CP} - \hat{\theta}_{2,n}^{CP} < \pi/2 \\ 1, & otherwise \end{cases} \tag{1.22}$$

When n is an odd number

$$b_n = \begin{cases} 0, & 0 \le (\hat{\theta}_{1,n}^{CP} - \hat{\theta}_{2,n}^{CP} < \pi \\ 1, & otherwise \end{cases} \tag{1.23}$$

Then, the BS determines the temporal phase adjustment of the (n+1)th slot for the second antenna (called the diversity antenna) $\varphi_{2,(n+1)}$ according to decoded FBI bit $\hat{b}_n$ ($\hat{b}_n = b_n$ if there is no FBI bit error). When n is an even number,

$$\varphi_{2,(n+1)} = \begin{cases} 0, & \hat{b}_n = 0 \\ \pi, & \hat{b}_n = 1 \end{cases} \tag{1.24}$$

When n is an odd number,

$$\varphi_{2,(n+1)} = \begin{cases} \pi/2, & \hat{b}_n = 0 \\ -\pi/2, & \hat{b}_n = 1 \end{cases} \tag{1.25}$$

Finally, the BS determines the final phase adjustment of the (n+1)th slot for the second antenna $\varphi_{2,(n+1)}$ as

$$\varphi_{2,(n+1)} = (\varphi_{2,n} + \varphi_{2,(n+1)})/2 \tag{1.26}$$

(2) Mode II

$$\phi_{1,n} = 0, \phi_{2,n} = \{0, \pi\}, A_{1,n} = A_{2,n} = \sqrt{1/2} \tag{1.27}$$

Similarly to Model I, the MS generates the FBI bit for the nth slot as

$$b_n = \begin{cases} 0, & -\pi/2 \le (\hat{\theta}_{1,n}^{CP} - \hat{\theta}_{2,n}^{CP} < \pi/2 \\ 1, & otherwise \end{cases} \tag{1.28}$$

Then, the BS determines the phase adjustment of the (n+1)th slot $\varphi_{2,(n+1)}$ according to the received FBI bit $\hat{b}_n$ as

$$\varphi_{2,(n+1)} = \begin{cases} 0, & \hat{b}_n = 0 \\ \pi, & \hat{b}_n = 1 \end{cases} \tag{1.29}$$

(3) Mode III

$$\phi_{1,n} = \phi_{2,n} = 0, (A_{1,n}, A_{2,n}) = \{(1,0), (0,1)\} \tag{1.30}$$

The MS receiver measures the received signal power of CPICH $P_{1,n}^{CP}$ and $P_{2,n}^{CP}$ for two transmit antennas. The MS generates the FBI bit for the nth slot as

$$b_n = \begin{cases} 0, & P_{1,n}^{CP} > P_{2,n}^{CP} \\ 1, & otherwise \end{cases} \tag{1.31}$$

Then, the BS determines the amplitude adjustment of the (n+1)th slot $A_{1,(n+1)}, A_{2,(n+1)}$ according to the received FBI bit $\hat{b}_n$ as

$$\{A_{1,(n+1)}, A_{2,(n+1)}\} = \begin{cases} \{1,0\}, & \hat{b}_n = 0 \\ \{0,1\}, & \hat{b}_n = 1 \end{cases} \tag{1.32}$$

This mode switches the transmit antenna according to the received FBI bit.

REFERENCES

[1] F. Adachi, M. Sawahashi and H. Suda, "Wideband DS-CDMA for Next-Generation Mobile Communications Systems", *IEEE Comm. Mag.*, pp56-69, Sept. 1998.

[2] E. Dahlman, B. Gudmundson, M. Nilsson and A. Skold, "UMTS/IMT-2000 Based on Wideband CDMA", *IEEE Comm. Mag.*, pp70-80, Sept. 1998.

[3] T. Ojanpera and R. Prasad, "An Overview of Air Interface Multiple Access for IMT-2000/UMTS", *IEEE Comm. Mag.*, pp82-95, Sept. 1998.

[4] 3GPP: http://www.3gpp.org

[5] Wideband CDMA I, IEEE Journal on Selected Areas in Communications, vol. 18, no. 8, August 2000.

[6] Wideband CDMA II, IEEE Journal on Selected Areas in Communications, vol. 19, no. 1, January 2001.

[7] F. Adachi, M. Sawahashi and K. Okawa, "Tree-structured generation of orthogonality spreading codes with different length for forward link of DS-CDMA mobile radio," Electronics Letters, pp. 27-28, Jan. 1997.

[8] E. Dinan and B. Jabbari, "Spreading codes for direct sequence CDMA and Wideband CDMA cellular networks", *IEEE Comm. Mag.*, pp48-54, Sept. 1998.

[9] K. Higuchi, M. Sawahashi, and F. Adachi, "Fast cell search algorithm in DS-CDMA mobile radio using long spreading codes," IEICE Trans.Commun., vol.E81-B, July 1998.

[10] H. Furukawa, K. Hanabe and A. Ushirokawa, "SSDT- site Selection diversity transmission power control for CDMA forward link," IEEE

Journal on Selected Areas in Communications, vol. 18, no. 8, pp. 1546-1554, August 2000.

[11] S. Fukumoto, K. Higuchi, M. Sawahashi and F. Adachi, "Field experiments on closed loop mode transmit diversity in WCDMA forward link," International Symposium on Spread Spectrum techniques and Applications 2000, Sept. 6-8, Virgina, USA

[12] F. Adachi, "The future of wireless communications," Presentation notes, Tohoku University, Japan Oct. 2000.

[13] L. B. Milstein, "Wideband code division multiple access," IEEE Journal on Selected Areas in Communications, August 2000.

Chapter 2

WIDEBAND CDMA

Abstract The coherent RAKE reception of wideband code division multiple access (Wideband CDMA) signals with complex spreading is considered in this chapter. A general multipath-fading channel model is assumed. A dedicated pilot channel, which is separate from data channels, is used for the purpose of channel estimation. Based on a digital implementation, the coherent demodulation scheme is presented. Pilot channel estimation error due to multiple access and multipath interference is studied. System performance is evaluated by means of bit error rate. The analysis shows that the error of channel estimation significantly degrades system performance and can be effectively suppressed by low pass filters (LPFs). A discussion on the envelope variation of complex spread signals is also included, which illustrates that the complex spread signal has more stable envelope than the dual-channel spread signal. The power ratio of pilot to data channels should be chosen in the range of 0.2 to 0.4 (-7 dB to –4 dB) in order to achieve maximum system capacity.

Keywords: Wideband CDMA, Complex Spreading, RAKE Reception, Channel Estimation, Envelope Variation

2.1 OVERVIEW

To provide higher data rates for end users, as well as to accommodate more users over wireless channels in the future mobile communication systems, wideband direct sequence code division multiple access (Wideband DS-CDMA) has become the focus of current research interests. Two of the important features of Wideband CDMA systems are the use of complex spreading and user dedicated pilot channel [1-4]. In the W-CDMA standard of 3GPP, time-multiplexed pilot is used in the downlink, whereas I/Q code-multiplexed pilot is used in the uplink [4]. In this chapter, I/Q code-multiplexed pilot is adopted. However, the results obtained can be applied to time-multiplexed pilot as well. In addition, the power ratio of pilot to data channels has been evaluated in order to achieve maximum system capacity.

Complex spreading is a term used in contrast to dual-channel spreading [5]. It can be implemented either by a complex valued sequence [6,7], e.g., polyphase sequence, or by two binary sequences. It is claimed that complex spreading can reduce the peak-to-average power ratio of modulated signals, thus improve RF power amplifier efficiency, but no explicit results have been given [3]. This declaration is studied in this chapter, and some useful results are presented.

Coherent reception outperforms non-coherent by 3dB in AWGN channels. In multipath fading channels, a coherent RAKE receiver with maximum ratio combining (MRC) has optimal performance. Eng and Milstein [8,9] and Efthymoglou etc [10] presented the performance of coherent DS-CDMA in Nakagami fading channels. Coherent reception requires knowledge of channel characteristics, which are time varying in fading environments. A conventional and effective method to accomplish this task is to use a separate pilot channel or insert pilot symbols in data symbols. Ling [11] described the optimal estimators for pilot symbols in a pilot channel scheme. The relative mean square error and optimal power allocation between data and pilot signals were also studied. Higuchi etc [12] and Abeta etc [13] presented a weighted multi-slot averaging (WMSA) channel estimation filter for time-multiplexed pilot channel. Choi [14] proposed an adaptive method to estimate channel parameters by jointly utilizing pilot and data channels. However, the analytical expression of the channel estimation error has been insufficient. In this chapter, the channel estimation error is analyzed in a multiple access and multipath environment, which leads to a better understanding of the nature of this estimation error and possible methods to reduce the error. The system performance degradation due to the estimation error is also investigated.

2.2 SYSTEM MODELS

In the uplink of Wideband CDMA systems, where I/Q code-multiplexed pilot is utilized, spreading consists of two operations. The first is channelization operation, in which pilot and data symbols on I- and Q-branches are independently multiplied with an orthogonal variable spreading factor (OVSF) code and transformed into a number of chips. The second operation is scrambling, where the resultant signals are further multiplied by a complex valued scrambling code. This spreading scheme is so called complex spreading as illustrated in Figure 2.1.

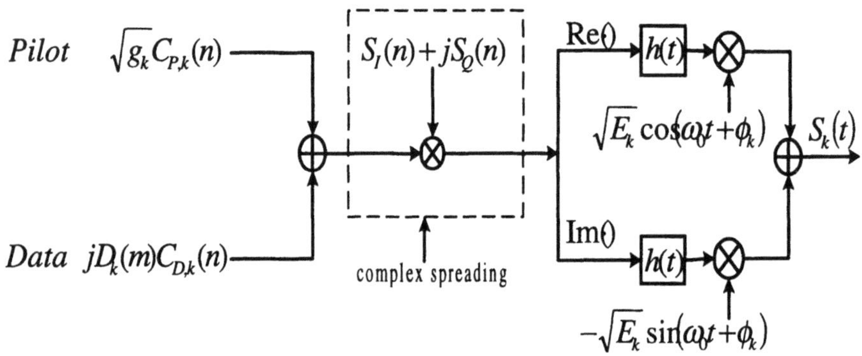

Figure 2.1 Transmitter block diagram of the k-th user

The transmitted signal of the k-th user can be written as:

$$S_k(t) = \sqrt{E_{D,k}} \sum_{n=-\infty}^{\infty} \left[\sqrt{g_k} C_{P,k}(n) S_I(n) - D_k(m) C_{D,k}(n) S_Q(n)\right] h(t - nT_c) \cos(\omega_0 t + \phi_k)$$

$$- \sqrt{E_{D,k}} \sum_{n=-\infty}^{\infty} \left[\sqrt{g_k} C_{P,k}(n) S_Q(n) + D_k(m) C_{D,k}(n) S_I(n)\right] h(t - nT_c) \sin(\omega_0 t + \phi_k) \quad (2.1)$$

where $E_{D,k}$ is the chip energy of data channel, g_k is the power ratio of pilot channel to the data channel, $D_k(m)$ are the data symbols for the k-th user, m stands for the integer part of n/N, i.e. $m = [n/N]$, N is the spreading factor, ϕ_k is the carrier phase, $C_{P,k}(n)$ and $C_{D,k}(n)$ are orthogonal channel codes for pilot and data symbols respectively (note that no data symbols are present in the pilot channel), $S_I(n)$ and $S_Q(n)$ represent the real and imaginary parts, respectively, of the cell-specific scrambling sequence, $h(t)$ stands for the impulse response of the pulse shaping filter and is truncated by the length of AT_c for practical systems, where $A \geq 1$ and T_c is the chip interval. Assuming that $\int_{-(A-1)Tc/2}^{(A+1)Tc/2} h^2(t)dt = 1$, the power of the transmitted signal is

$$P_k = (1+g_k)E_{D,k}/T_c = E_k/T_c \tag{2.2}$$

where E_k is the chip energy for both data channel and dedicate pilot channel of the k-th user.

The complex low-pass equivalent impulse response of a multipath fading channel can be written as

$$h_k(t) = \sum_{l=0}^{L-1} \alpha_{k,l}(t)\delta[t-\tau_{k,l}(t)]e^{j\theta_{k,l}(t)} \tag{2.3}$$

where L ($L \geq 1$) is the number of resolvable propagation paths. For the sake of simple notation, it is assumed that all users have the same number of multipaths. $\alpha_{k,l}(t)e^{j\theta_{k,l}(t)}$ and $\tau_{k,l}(t)$ are the complex fading factor and propagation delay of the l-th path of the k-th user, respectively. Note that $\alpha_{k,l}(t)$ can be Rayleigh, Rician or Nakagami distributed, depending on a specific channel model. All random variables in (2.3) are assumed independent for k and l. Assuming that there are K active users in the system, the received signal is given by:

$$r(t) = \eta(t) + \sum_{k=1}^{K}\sum_{l=0}^{L-1} \alpha_{k,l}(t)S_k[t-\tau_{k,l}(t)]e^{j\theta_{k,l}(t)} \tag{2.4}$$

where $\eta(t)$ is the background AWGN with double-side power spectrum density $\eta_0/2$.

In order to mitigate multipath effect, A RAKE receiver with pilot symbol aided coherent demodulation and maximum ratio combining (MRC) is employed. The RAKE receiver structure is shown in Figure 2.2, where the number of branches is less or equal to the number of resolvable paths. The received signal is multiplied by local carriers and passed through the pulse-matching filters, then sampled once per chip duration. The samples are fed to each branch of the RAKE receiver, where the pilot and data symbols are separately demodulated and pilot symbols are used to eliminate the phase error of different path and achieve MRC. Assuming that the i-th path delay $\tau_{1,i}$ can be accurately estimated for the reference user (k = 1), each path that corresponds to a RAKE branch gives an output component. The outputs of all branches are added together to form the decision statistic. The detailed demodulation block diagram of the i-th branch is shown in Figure 2.3.

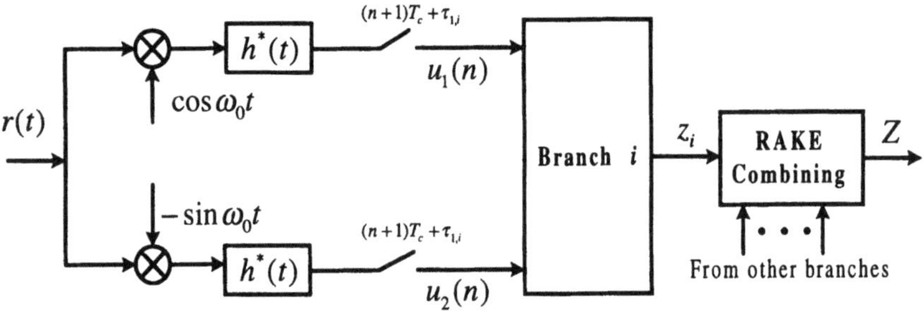

Figure 2.2 RAKE receiver block diagram

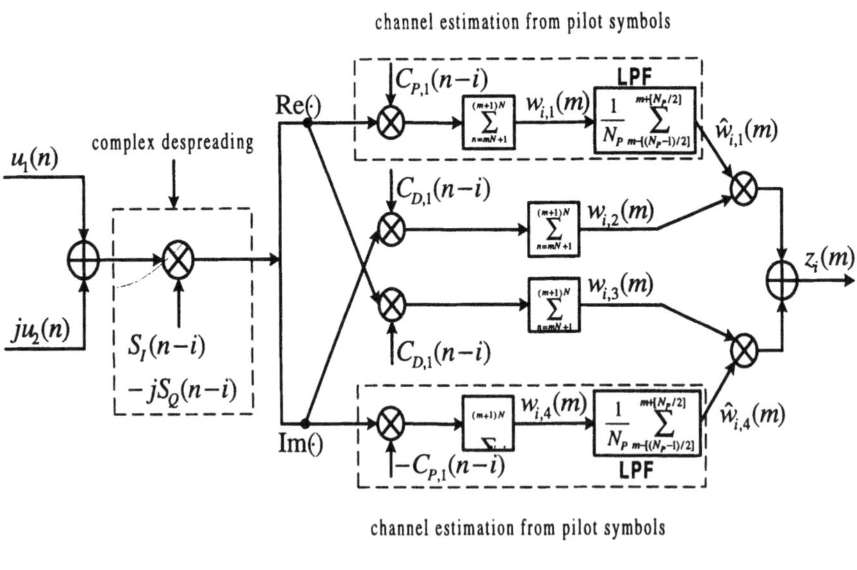

Figure 2.3 Detailed demodulation block diagram of the i-th branch [16]

2.3 CHANNEL ESTIMATION ERRORS

The two sampled outputs in Figure 2.2 are given by:

$$u_1(n) = \int_{nT_c+\tau_{1,i}}^{(n+1)T_c+\tau_{1,i}} r(t)\cos(\omega_0 t)h^*(t - nT_c - \tau_{1,i})dt \tag{2.5}$$

and

$$u_2(n) = -\int_{nT_c+\tau_{1,i}}^{(n+1)T_c+\tau_{1,i}} r(t)\sin(\omega_0 t)h^*(t - nT_c - \tau_{1,i})dt \tag{2.6}$$

where $h^*(t)$ is the complex conjugate of $h(t)$. For simple notation, nT_c is replaced by n. Neglecting the high frequency component, the two samples can be further written as:

$$u_1(n) = \sum_{k=1}^{K}\sum_{l=0}^{L-1} \beta_{k,l}^{(c)}(n)\left(\sqrt{g_k}\cdot S_I(n-l)C_{P,k}(n-l) - D_k(m_k)\cdot S_Q(n-l)C_{D,k}(n-l)\right)$$

$$-\sum_{k=1}^{K}\sum_{l=0}^{L-1} \beta_{k,l}^{(s)}(n)\left(\sqrt{g_k}\cdot S_Q(n-l)C_{P,k}(n-l) + D_k(m_k)\cdot S_I(n-l)C_{D,k}(n-l)\right) + \eta_1(n) \tag{2.7}$$

$$u_2(n) = \sum_{k=1}^{K}\sum_{l=0}^{L-1} \beta_{k,l}^{(s)}(n)\left(\sqrt{g_k}\cdot S_I(n-l)C_{P,k}(n-l) - D_k(m_k)\cdot S_Q(n-l)C_{D,k}(n-l)\right)$$

$$+\sum_{k=1}^{K}\sum_{l=0}^{L-1} \beta_{k,l}^{(c)}(n)\left(\sqrt{g_k}\cdot S_Q(n-l)C_{P,k}(n-l) + D_k(m_k)\cdot S_I(n-l)C_{D,k}(n-l)\right) + \eta_2(n) \tag{2.8}$$

where $m_k = [(n - n_k)/N]$ with $n_k = [\tau_{k,l}(nT_c)/T_c]$. Since data of interfering users have no effect on variance of multiple access interference, m_k is replaced by m for simple notation. $\eta_1(n)$ and $\eta_2(n)$ are the sampled noise components, and

$$\beta_{k,l}^{(c)}(n) = \frac{1}{2}\sqrt{E_{D,k}}\,\alpha_{k,l}(n)R[\overline{\tau}_{k,l}(n)]\cos[\varphi_{k,l}(n)] \tag{2.9}$$

$$\beta_{k,l}^{(s)}(n) = \frac{1}{2}\sqrt{E_{D,k}}\,\alpha_{k,l}(n)R[\overline{\tau}_{k,l}(n)]\sin[\varphi_{k,l}(n)] \tag{2.10}$$

where $\varphi_{k,l}(n) = \phi_k + \theta_{k,l}(nT_c) - \omega_0 \tau_{k,l}(nT_c)$ is the carrier phase offset, $\bar{\tau}_{k,l}(n)$ is the chip timing error given by $\bar{\tau}_{k,l}(n) = \tau_{k,l}(nT_c) - \tau_{1,i} - [(\tau_{k,l}(nT_c) - \tau_{1,i})/T_c]T_c$, and $|\bar{\tau}_{k,l}(n)| \leq T_c$. $R(\bar{\tau})$ represents the effect of timing error $\bar{\tau}$ on the sampled outputs. When the impulse response of the shaping filter is limited to T_c, the Appendix of [16] has proved that

$$R(\bar{\tau}) = \int_{-\infty}^{\infty} |H(f)|^2 \cos(2\pi f \bar{\tau}) df \tag{2.11}$$

where $H(f)$ is the frequency response of the pulse shaping filter.

Assuming that acquisition and synchronization have been accomplished for the reference user (k = 1), the local spreading codes in the i-th branch are aligned with the i-th path of the received signal. After descrambling by a complex scrambling code, the real and imaginary parts of the resultant signal are multiplied by separate channel codes respectively, then accumulated in one symbol period with symbol synchronization (N chips). The four outputs of the i-th branch are obtained at four sub-branches after this correlation, as show in Figure 2.3, which are given by

$$w_{i,1}(m) = \sum_{n=mN+1}^{(m+1)N} [u_1(n) \cdot S_I(n-i) + u_2(n) \cdot S_Q(n-i)] \cdot C_{P,1}(n-i) \tag{2.12}$$

$$w_{i,2}(m) = \sum_{n=mN+1}^{(m+1)N} [u_2(n) \cdot S_I(n-i) - u_1(n) \cdot S_Q(n-i)] \cdot C_{D,1}(n-i) \tag{2.13}$$

$$w_{i,3}(m) = \sum_{n=mN+1}^{(m+1)N} [u_1(n) \cdot S_I(n-i) + u_2(n) \cdot S_Q(n-i)] \cdot C_{D,1}(n-i) \tag{2.14}$$

$$w_{i,4}(m) = \sum_{n=mN+1}^{(m+1)N} [u_2(n) \cdot S_I(n-i) - u_1(n) \cdot S_Q(n-i)] \cdot [-C_{P,1}(n-i)] \tag{2.15}$$

Assuming that the channel parameters of (2.9) and (2.10) remain constant over one symbol duration, i.e. $\beta_{k,l}^{(c)}(n) = \beta_{k,l}^{(c)}(mN)$ and $\beta_{k,l}^{(s)}(n) = \beta_{k,l}^{(s)}(mN)$ for $mN < n \leq (m+1)N$, $w_{i,1}(m)$ can be rewritten as

$$w_{i,1}(m) = E_{i,1}(m) + I_{i,1}(m) + N_{i,1}(m) \tag{2.16}$$

$w_{i,1}(m)$ consists of three components: the desired signal component $E_{i,1}(m)$, the interference component $I_{i,1}(m)$ and the background AWGN component $N_{i,1}(m)$.

The desired signal component $E_{i,1}(m)$ is the pilot component of the reference user in the reference path (i.e. $k = 1$ and $l = i$), and is given by

$$E_{i,1}(m) = 2N\beta_{1,i}^{(c)}(mN)\sqrt{g_1} = N\alpha_{1,i}(mN)R[\overline{\tau}_{1,i}(mN)]\cos[\varphi_{1,i}(mN)]\sqrt{g_1 E_{D,1}} \tag{2.17}$$

For the reference path (i.e. $\overline{\tau}_{1,i}(mN) = 0$), $R[\overline{\tau}_{1,i}(mN)] = 1$. Therefore, a good estimate of the channel parameter $\{\alpha_{1,i}(t)\cos\varphi_{1,i}(t)\}$ can be derived from $E_{i,1}(m)$, which is approximated by

$$E_{i,1}(m) = N\alpha_{1,i}(mN)\cos[\varphi_{1,i}(mN)]\sqrt{g_1 E_{D,1}} \tag{2.18}$$

The interference component $I_{i,1}(m)$ is due to the cross-correlation with other users' spreading sequences, the auto-correlation of self-signal in other paths as well as the cross talk with quadrature components. Accurate evaluation of this interference component should involve all correlations in the specific sequence set. For simple analysis, it is assumed that all sequences are purely random binary sequences, except the case that $C_{p,1}(n-i)$ and $C_{D,1}(n-i)$ are orthogonal. $I_{i,1}(m)$ is a random variable conditioned on $\alpha_{1,i}(mN)$, with zero mean and variance given by [16]

$$\mathrm{var}\{I_{i,1}(m)\} = \Delta\frac{N}{2}\sum_{k=1}^{K} E_k \sum_{\substack{l=0 \\ if\ k=1, l\neq i}}^{L-1} \mathrm{E}\{\alpha_{k,l}^2(mN)\} + \Delta\frac{N}{4}E_1\alpha_{1,i}^2(mN) \tag{2.19}$$

where $\Delta = \mathrm{E}\{R^2[\overline{\tau}_{k,l}(mN)]\}$ is related to the frequency transfer function of pulse-shaping filter, which is given by

$$\Delta = \mathrm{E}\{R^2[\bar{\tau}_{k,l}(mN)]\} = \frac{2}{3}, \qquad \text{for retangular pulse shaping filter} \tag{2.20}$$

$\mathrm{var}\{I_{i,1}(m)\}$ in (2.19) is related to the instantaneous channel fading amplitude $\alpha_{1,i}(mN)$ of the i-th path of the reference user. However, the contribution of $\alpha_{1,i}(mN)$ to the variance is very small when *KL* is large even for small K. In the extreme case (K=1), the power ratio of the i-th path to other paths is $\frac{1}{2}\mathrm{E}\{\alpha_{1,i}^2(mN)\} \Big/ \sum_{l=0,l\neq i}^{L-1} \mathrm{E}\{\alpha_{k,l}^2(mN)\}$, which is very small when $L \gg 1$. Since adding $\alpha_{1,i}(mN)$ to $\mathrm{var}\{I_{i,1}(m)\}$ makes the analysis troublesome, for simple analysis the effect of $\alpha_{1,i}(mN)$ is averaged in the variance. Therefore, $\mathrm{var}\{I_{i,1}(m)\}$ is approximated by

$$\mathrm{var}\{I_{i,1}(m)\} \approx \Delta \frac{N}{2} \sum_{k=1}^{K} E_k \sum_{l=0}^{L-1} \mathrm{E}\{\alpha_{k,l}^2(mN)\} = \frac{\Delta}{2} NKE\Omega_T \tag{2.21}$$

where for the sake of simple analysis, it is assumed that $g_k = g_1 = g$, $E_k = E_1 = E$. Ω_T is the channel parameter given by

$$\Omega_T = \Omega_{T,k} = \sum_{l=0}^{L-1} \mathrm{E}\{\alpha_{k,l}^2(mN)\}, \qquad k = 1, 2, \cdots K \tag{2.22}$$

The AWGN component $N_{i,1}(m)$ is a zero mean Gaussian variable with variance:

$$\mathrm{var}\{N_{i,1}(m)\} = 2N \cdot \frac{\eta_0}{2} = N\eta_0 \tag{2.23}$$

Therefore, conditioned on $\alpha_{1,i}(mN)\cos[\varphi_{1,i}(mN)]$, $w_{i,1}(m)$ is a random variable with mean $\mathrm{E}_{i,1}(m)$, given by (2.18), and variance, $\mathrm{var}\{I_{i,1}(m)\} + \mathrm{var}\{N_{i,1}(m)\}$, i.e.

$$\mathrm{var}\{w_{i,1}(m)\} = N\left(\eta_0 + \frac{\Delta}{2} KE\Omega_T\right) \tag{2.24}$$

Similarly, $w_{i,2}(m)$, $w_{i,3}(m)$ and $w_{i,4}(m)$ can be approximated as conditional random variables, with same variance as $w_{i,1}(m)$, but different means:

$$\mathrm{E}\{w_{i,2}(m)\} = ND_1(m)\alpha_{1,i}(mN)\cos[\varphi_{1,i}(mN)]\sqrt{E_D} \tag{2.25}$$

$$\mathrm{E}\{w_{i,3}(m)\} = ND_1(m)\alpha_{1,i}(mN)\sin[\varphi_{1,i}(mN)]\sqrt{E_D} \tag{2.26}$$

$$\mathrm{E}\{w_{i,4}(m)\} = N\alpha_{1,i}(mN)\sin[\varphi_{1,i}(mN)]\sqrt{gE_D} \tag{2.27}$$

where $E_D = E_{D,1} = E_1/(1+g_1) = E/(1+g)$ is the chip energy for data channel of the reference user.

From analysis above, it has been seen that in such Wideband CDMA systems, the channel estimation using pilot symbols suffers from multi-access and multipath interference. In most publications on coherent demodulation, perfect channel estimation is assumed for the purpose of simplification. However, this assumption does not stand in our study and such estimation error is considered in the following performance analysis.

Because the variance (error) of this channel estimation is non-trivial and has direct contribution to the decision noise (as it will be shown in the following analysis), it must be suppressed. A simple way to reduce the estimation error is to use a low-pass filter (LPF) as shown in Figure 2.3. The interference from multi-access and multipath can be modeled as white noise with no correlation between successive symbols, and with a wide spectrum allocation of the chip rate, $R_c = 1/T_c$. Therefore, LPFs with a cut-off frequency much smaller than R_c should be able to reduce much noise on $w_{i,1}(m)$ and $w_{i,4}(m)$. In our study, a N_P-tap (N_P can be either even or odd numbers) FIR filter is adopted to suppress the noise, as shown in Figure 2.3. In order to assure that the channel parameters remain relatively constant over N_P symbol duration, the cutoff frequency of the filter should be much larger than the maximum Doppler frequency f_d, i.e. $R_b / N_p = R_c /(N_p N) >> f_d$, where R_b is the channel data rate. Higuchi etc [12] has shown by both simulations and experiments that system performance is very stable when the window size of the FIR filter is two time slots (around 1.3ms) for the Doppler frequency in the range of 0 to 200Hz.

Note that the cutoff frequency of its FIR filter is 1/1.3ms $\approx$ 800Hz, which is much larger than the maximum Doppler frequency of 200Hz.

It is worth noting here that in the uplink of W-CDMA standards in 3GPP [4], two different spreading factors (SF) are used for pilot and data channels, respectively. The SF, N_1, of pilot channel is fixed to be 256, while the SF, N_2, of data channel is varied from 4 to 256. The channel estimate that is derived from one long pilot symbol corresponds to N_1 / N_2 data symbols, depending on data rate. In this case, it is equivalent to say that the length of the FIR filter for 3GPP is $N_P /(N_1 / N_2)$.

The statistic of channel estimation after the averaging LPF can be written as follows

$$\hat{w}_{i,1}(m) = \frac{1}{N_P} \sum_{j=-[(N_p-1)/2]}^{[N_p/2]} w_{i,1}(m+j) \tag{2.28}$$

Assuming that the variances of successive $w_{i,1}(m)$ for different m are independent to each other, one obtains

$$\mathrm{E}\{\hat{w}_{i,1}(m)\} = \mathrm{E}\{w_{i,1}(m)\} \tag{2.29}$$

$$\mathrm{var}\{\hat{w}_{i,1}(m)\} = \frac{1}{N_P} \mathrm{var}\{w_{i,1}(m)\} \tag{2.30}$$

The estimation error is greatly reduced by the LPF when $N_p >> 1$. Similarly, the estimation error of $w_{i,4}(m)$ can be reduced significantly as well. On the other hand, because the variables $w_{i,2}(m)$ and $w_{i,3}(m)$ contain unknown data for successive m, LPF may not be used to reduce the variances of these two statistics.

The output of the *i*-th branch in RAKE receiver is given by:

$$z_i(m) = \hat{w}_{i,1}(m) \cdot w_{i,2}(m) + w_{i,3}(m) \cdot \hat{w}_{i,4}(m) \tag{2.31}$$

From (2.16), it is seen that the sub-branch outputs are actually the sums of a lot of independent random variables, thus $z_i(m)$ is in the form of $\left(\sum_i x_i\right)\cdot\left(\sum_j y_j\right)=\sum_i\sum_j x_i y_j$. That is, $z_i(m)$ consists of the sum of many independent random variables. Therefore, $z_i(m)$ can be approximated as a conditional Gaussian variable. Since Re(.) and Im(.) in Fig. 2.3 are independent, and $C_{P,1}$ and $C_{D,1}$ are orthogonal, $\hat{w}_{i,1}(m)$, $w_{i,2}(m)$, $w_{i,3}(m)$ and $\hat{w}_{i,4}(m)$ are mutually independent to each other, therefore, the mean of $z_i(m)$ is given by

$$\mathrm{E}\{z_i(m)\}=\mathrm{E}\{\hat{w}_{i,1}(m)\}\cdot\mathrm{E}\{w_{i,2}(m)\}+\mathrm{E}\{w_{i,3}(m)\}\cdot\mathrm{E}\{\hat{w}_{i,4}(m)\}$$

$$=N^2 D_1(m)\alpha_{1,i}^2(mN)E_D\sqrt{g} \tag{2.32}$$

The variance of $z_i(m)$ is given by

$$\begin{aligned}\mathrm{var}\{z_i(m)=(g+\frac{1}{N_P})\alpha_{1,i}^2(mN)N^3E_D\cdot\left(\eta_0+\frac{\Delta}{2}KE\Omega_T\right)\\ +\frac{2N^2}{N_P}\left(\eta_0+\frac{\Delta}{2}KE\Omega_T\right)^2\end{aligned} \tag{2.33}$$

In the case of perfect channel estimation, i.e. $\mathrm{var}\{\hat{w}_{i,1}(m)\}=\mathrm{var}\{\hat{w}_{i,4}(m)\}=0$ or $N_P\to\infty$ in (2.33), the mean of $z_i(m)$ remains unchanged, but the variance reduces to

$$\mathrm{var}\{z_i(m)\}\Big|_{N_P\to\infty}=g\alpha_{1,i}^2(mN)N^3E\cdot\left(\eta_0+\frac{\Delta}{2}KE\Omega_T\right) \tag{2.34}$$

Assuming that the fading of each path is independent of each other, and the output in each branch of the RAKE receiver is independent of each other, the summed output $Z(m)=\sum_{i=0}^{M-1}z_i(m)$, where M is the number of branches, is a Gaussian variable with mean and variance given by

$$\mathrm{E}\left\{Z(m)\,|_{\alpha_{1,i}(mN),\, i=0,1,\cdots,M-1}\right\} = \mathrm{E}\left\{Z(m)\,|_{\zeta}\right\} = \sum_{i=0}^{M-1} \mathrm{E}\{z_i(m)\}$$

$$= E_D N^2 \sqrt{g} \cdot D_1(m) \cdot \zeta \tag{2.35}$$

where

$$\zeta = \sum_{i=0}^{M-1} \alpha_{1,i}^2(mN) \tag{2.36}$$

and

$$\mathrm{var}\left\{Z(m)\big|_{\alpha_{1,i}(mN),\, i=0,1,\cdots,M-1}\right\} = \mathrm{var}\left\{Z(m)\big|_{\zeta}\right\} = \sum_{i=0}^{M-1} \mathrm{var}\{z_i(m)\}$$

$$= (g + \frac{1}{N_P}) N^3 E_D \zeta \left(\eta_0 + \frac{\Delta}{2} K E \Omega_T\right) + \frac{2MN^2}{N_P}\left(\eta_0 + \frac{\Delta}{2} K E \Omega_T\right)^2 \tag{2.37}$$

Therefore, conditioned on the instantaneous multipath fading amplitude ζ of the reference user, the bit error probability is given by

$$P_e(\zeta) = Q\left(\frac{\mathrm{E}\{\, Z(m)\,|_{\zeta}\}}{\sqrt{\mathrm{var}\{\, Z(m)\,|_{\zeta}\,\}}}\right) = Q\left(\frac{\zeta}{\sqrt{a\zeta + b}}\right) \tag{2.38}$$

where $Q(x) = \int_x^{\infty} \frac{1}{\sqrt{2\pi}} e^{-y^2/2} dy$, and

$$a = (1+g)\left(1 + \frac{1}{gN_P}\right)\left(\frac{1}{SNR} + \frac{\Delta}{2}\Omega_T \frac{K}{N}\right) \tag{2.39}$$

$$b = \frac{2M(1+g)^2}{gN_P}\left(\frac{1}{SNR} + \frac{\Delta}{2}\Omega_T \frac{K}{N}\right)^2 \tag{2.40}$$

$$SNR = NE/\eta_0 \tag{2.41}$$

For a specific channel model, ζ is a random variable with probability density function $p(\zeta)$. The average bit error probability P_e can be obtained by averaging $P_e(\zeta)$ over ζ, i.e.

$$P_e = \int_0^\infty P_e(\zeta) p(\zeta) d\zeta$$

$$= \underbrace{\int_0^\infty \int_0^\infty \cdots \int_0^\infty}_{M} P_e\left(\sum_{i=0}^{M-1} \alpha_{1,i}^2\right) \cdot p(\alpha_{1,0}, \alpha_{1,1} \cdots \alpha_{1,M-1}) d\alpha_{1,0} d\alpha_{1,1} \cdots d\alpha_{1,M-1} \tag{2.42}$$

where $p(\alpha_{1,0}, \alpha_{1,1} \cdots \alpha_{1,M-1})$ is the joint probability density function of multipath fading amplitude.

Equations (2.39) and (2.42) give a complete bit error rate (BER) performance evaluation of the Wideband CDMA system with user dedicated pilot symbols. However, (2.42) is generally too complicated to be computed, and the probability density function $p(\zeta)$ is unobtainable if the fading in each path is to be chosen arbitrarily. In order to simplify the numerical evaluation, it is assumed that each user has independent but the same fading characteristics. The amplitude fading in each path is assumed to be Nakagami distributed, since the Nakagami distribution is more versatile and more adequate to describe different fading situations. For example, Nakagami can closely represent Rayleigh and Rician fading under appropriate parameters [8,10]. The Nakagami probability density function (pdf) of $\alpha_{1,i} = \alpha_i$ is given by

$$p(\alpha_i) = 2\left(\frac{m_i}{\Omega_i}\right)^{m_i} \frac{\alpha_i^{2m_i-1}}{\Gamma(m_i)} \exp\left(-\frac{m_i}{\Omega_i}\alpha_i^2\right), \qquad \alpha_i \geq 0 \tag{2.43}$$

where $m_i = \dfrac{\Omega_i^2}{\mathrm{E}\{(\alpha_i^2 - \Omega_i)^2\}}$, $\Omega_i = \mathrm{E}\{\alpha_i^2\} = \Omega_0 e^{-i\delta}$, δ is the rate of exponential decay of the multipath intensity profile (MIP). $\Gamma(x) = \int_0^\infty t^{x-1} e^{-t} dt$ is the Gamma function. If the Nakagami distribution in each path is such that the ratio $\dfrac{m_i}{\Omega_i}$ is the same for all paths, it can be shown that ζ has the gamma pdf [15]:

$$p(\zeta)=\left(\frac{m_T}{\Omega_T}\right)^{m_T}\frac{\zeta^{m_T-1}}{\Gamma(m_T)}\exp\left(-\frac{m_T}{\Omega_T}\zeta\right),\qquad \zeta\ge 0 \tag{2.44}$$

where $\Omega_T=\sum_{i=0}^{M-1}\Omega_i=\sum_{i=0}^{M-1}\mathrm{E}\{\alpha_i^2\}$ and $m_T=\sum_{i=0}^{M-1}m_i$. Submitting (2.44) into (2.42), the average bit error probability P_e is given by

$$P_e=\int_0^\infty P_e(\zeta)p(\zeta)d\zeta$$

$$=\int_0^\infty Q\left(\frac{\zeta}{\sqrt{a\zeta+b}}\right)\cdot\left(\frac{m_T}{\Omega_T}\right)^{m_T}\frac{\zeta^{m_T-1}}{\Gamma(m_T)}\exp\left(-\frac{m_T}{\Omega_T}\zeta\right)d\zeta \tag{2.45}$$

In the case of perfect channel estimation, (2.45) has a known solution [8,10]. In general, the integral of (2.45) can be computed numerically using computer software like MATLAB.

It is seen from (2.38) – (2.41) that for a given SNR and number of active users, the change of the power ratio g has certain influence on the system performance. When g is very large or very small, at least one of a and b will be large, so that large error probability is caused. In other words, when g is very large, the desired component (see (2.18) or (2.27)) in pilot channel is strong (i.e. channel estimation quality is good), whereas the desired component (see (2.25) or (2.26)) in data channel is weak. This will cause large error probability. On the other hand, when g is very small, channel estimation quality is poor, so that large error probability is caused as well. Therefore, there exists an optimal value of g, which minimizes the error probability or maximizes the signal to interference plus noise ratio. The optimal g is obtainable by minimizing (2.38). That is by letting $\frac{d(a\zeta+b)}{dg}=0$, it is readily to have:

$$g_{opt}(\zeta)=\sqrt{\frac{2M\left(\frac{1}{SNR}+\frac{\Delta}{2}\Omega_T\frac{K}{N}\right)+\zeta}{2M\left(\frac{1}{SNR}+\frac{\Delta}{2}\Omega_T\frac{K}{N}\right)+\zeta\cdot N_p}} \tag{2.46}$$

Similar to the discussion of the average bit error probability, an average g_{opt} can be obtained by averaging $g_{opt}(\zeta)$ over ζ, i.e.

$$\begin{aligned} g_{opt} &= \int_0^\infty g_{opt}(\zeta)p(\zeta)d\zeta \\ &= \int_0^\infty g_{opt}(\zeta)\cdot\left(\frac{m_T}{\Omega_T}\right)^{m_T}\frac{\zeta^{m_T-1}}{\Gamma(m_T)}\exp\left(-\frac{m_T}{\Omega_T}\zeta\right)d\zeta \end{aligned} \tag{2.47}$$

2.4 NUMERICAL COMPUTATION

In this section, the effects of different system parameters on the BER performance of the Wideband CDMA system with complex spreading and imperfect channel estimation are investigated by numerical calculations. The fading is assumed to be Nakagami distributed as noted in Section 2.3. The exponential decay rate of the multipath intensity profile (MIP) δ is 0.2. The ratio m_l/Ω_l is 4 and Ω_0 is set as 1 for normalization. The number of multipaths L is 8, and the spreading factor N, i.e. number of chips per data symbol, is fixed to be 64. Unless noted otherwise, the number of RAKE fingers M is 4, The LPF length is N_P = 32, the number of active users in the system K is 10, the power ratio of in-phase and quadrature branches g is 0.3, the signal to noise ratio SNR is 15dB.

Figure 2.4 illustrates the bit error rate as a function of average received SNR for different number of RAKE fingers, i.e. M = 1, 2, 4, 8. It can be seen that the RAKE receiver takes advantage of the multipath and gives better performance when the number of fingers increases. For the case of M = 4, $BER \approx 10^{-2}$ can be obtained when SNR=8 dB. If channel coding is used, acceptable BER performance can be achieved.

Figure 2.5 shows the bit error rate versus average received SNR for different number of taps N_P of LPF. The last curve, where $N_P = \infty$, corresponds to the case of perfect channel estimation. From the analysis (2.38) -- (2.41), it is seen that the increase of N_P will increase the system performance. This is because the increase of the number of taps of LPFs decreases the estimation variance. It can be seen from the figure that using larger number of taps can improve performance significantly. Similar property can be found in Figure 2.6, which shows the BER as a function of the number of active users, K. It is clearly shown that the increment of N_P

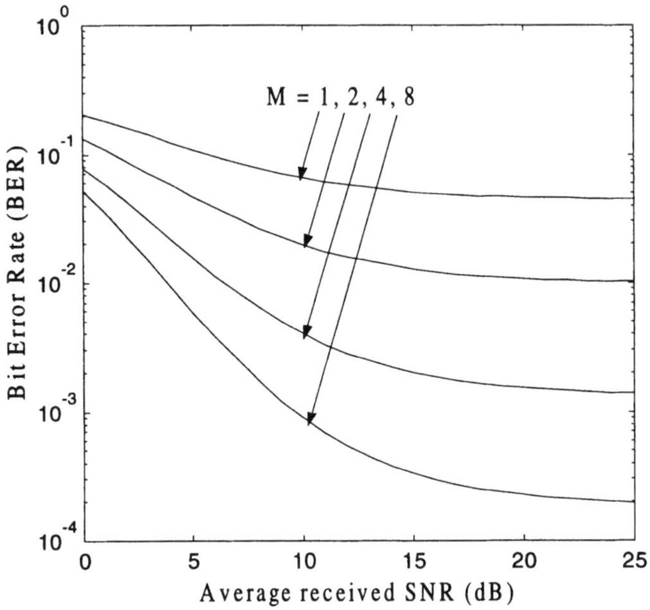

Figure 2.4. Bit error rate versus average received SNR with different number of RAKE fingers (N = 64, K = 10, L = 8, g = 0.3, δ = 0.2, m_l/Ω_l = 4, N_P = 32)

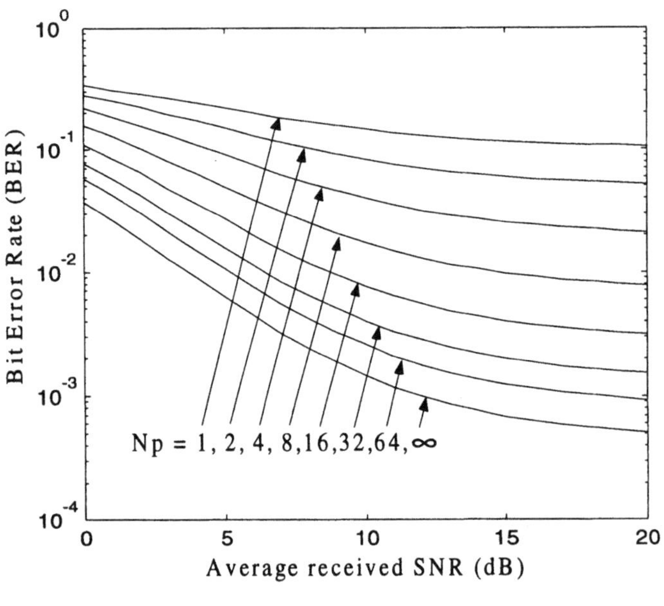

Figure 2.5 Bit error rate versus average received SNR with different number of taps of LPFs (N = 64, K = 10, L = 8, M = 4, g = 0.3, δ = 0.2, m_l/Ω_l = 4)

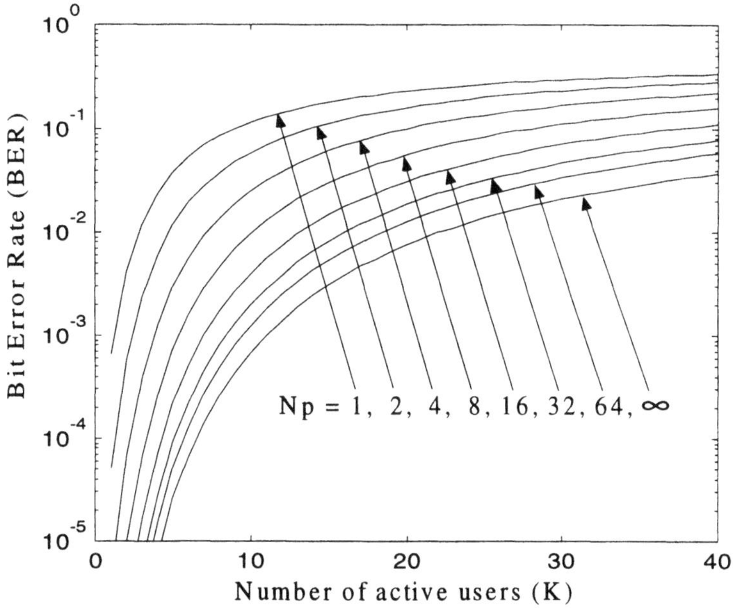

Figure 2.6 Bit error rate versus number of users with different number of taps of LPFs (SNR = 15dB, N = 64, L = 8, M = 4, g = 0.3, δ = 0.2, m_l/Ω_l = 4)

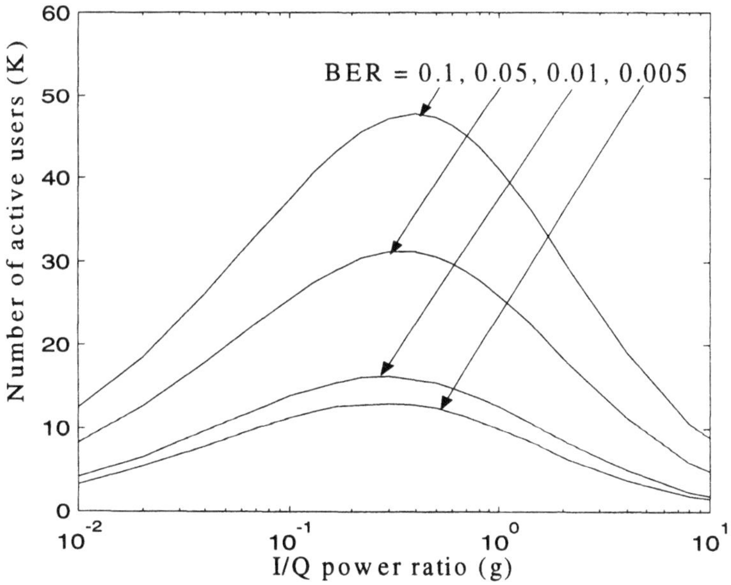

Figure 2.7 Number of users versus I/Q power ratio with different bit error rate (BER) (SNR = 15dB, N = 64, L = 8, M = 4, δ = 0.2, m_l/Ω_l = 4, N_P = 32)

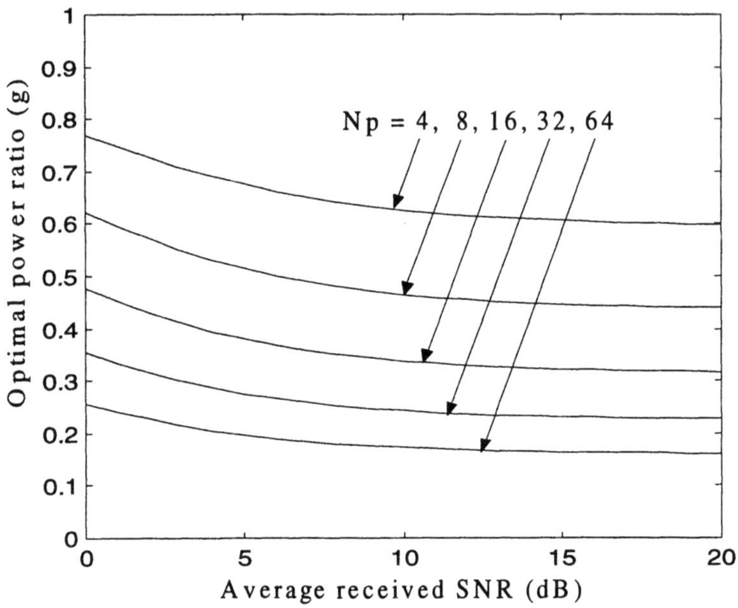

Figure 2.8 Optimal power ratio versus average received SNR with different number of taps of LPFs (K = 10, N = 64, L = 8, M = 4, $\delta = 0.2$, $m_l / \Omega_l = 4$)

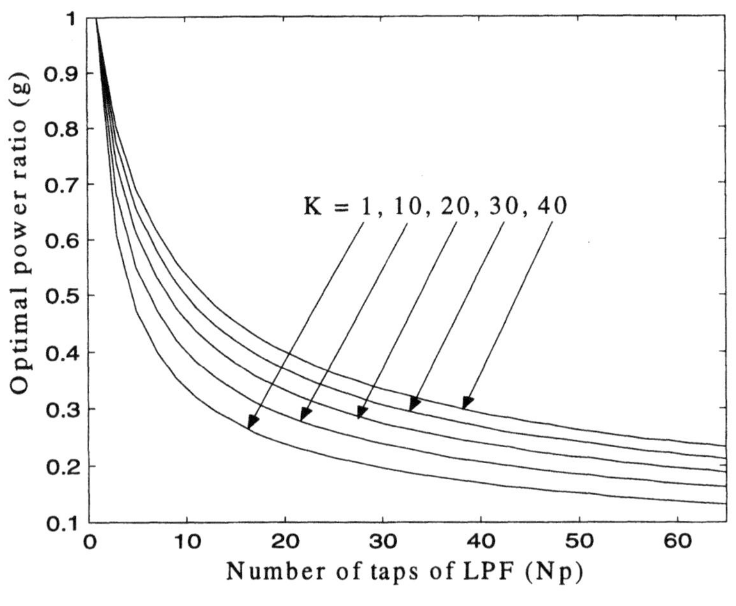

Figure 2.9 Optimal power ratio versus number of taps of LPFs with different number of active users (SNR = 15dB, N = 64, L = 8, M = 4, $\delta = 0.2$, $m_l / \Omega_l = 4$)

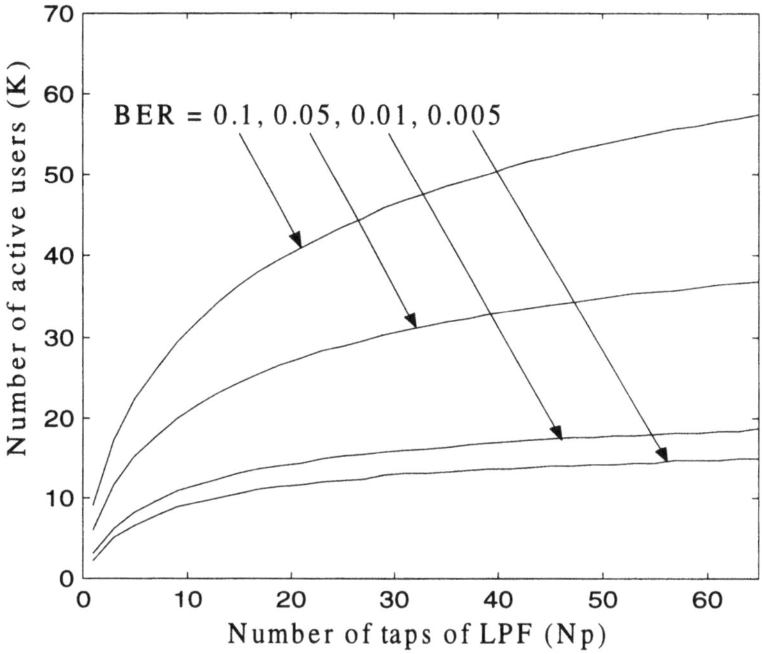

Figure 2.10 Number of active users versus number of taps of LPFs with different bit error rate (BER) (SNR = 15dB, N = 64, L = 8, M = 4, g = 0.3, $\delta = 0.2$, $m_l/\Omega_l = 4$)

always has positive effect to the system performance. For a given BER of 10^{-2}, doubling the number of taps means the increase of capacity by 3 or 4. However, the value of N_P is subject to the constraint that the channel parameters should be kept constant over N_P symbols time period. That is, N_P cannot be chosen too large.

Figure 2.7 illustrates the effects of I/Q power ratio g to the system capacity (number of active users, K). It is seen from the figure that for a given BER, when g is small, the increment of g increases the system capacity significantly, whereas the capacity decreases when g increases beyond its optimal value. One can find that the optimal values are slightly different for different BER level. Under the specified system parameters (i.e. SNR = 15 dB, N_P= 32, etc.) the optimal value for BER = 0.1 is about 0.4, while it is about 0.3 for BER = 0.005. The optimal power ratio g_{opt}, given by (2.47), is plotted in Figure 2.8 as a function of average SNR with different number of taps of LPF. As shown in the figure, the increase of SNR monotonously decreases the value of g_{opt}. However, the change is very slight when SNR>>0 dB, which implies that the *SNR* has little influence on g_{opt}. It can be also seen from the figure that increasing N_P means decreasing the optimal value. This is because increasing N_P improves the channel estimation quality so that the transmission power of the pilot can be reduced in order to maintain a given channel estimation error. This conclusion can also be obtained from Figure 2.9, where g_{opt} is shown as a function of number of taps of LPF, N_P, for different number of active users. Similarly, the increase of N_P monotonously decreases g_{opt}. The optimal value increases as the number of users increases. That is, when there are more users in the system, a larger portion of pilot power is preferred. It is worth noting that since *SNR* = NE/η_0=15 dB is fixed, a larger g means the increase of pilot power but the reduce of data channel power, thus the total power of the system remains unchanged, i.e. the total interference of this cell to other cells is not raised.

Finally, the system capacity (K) is shown in Fig. 2.10 as a function of the number of taps of the LPF for different values of BER. It can be clearly seen that the capacity can be significantly increased by the increase of N_P, especially when BER is small. Actually, the relative large BERs can be pulled down to acceptable levels using channel coding/decoding (e.g. convoluational or Turbo coding). That is, system performance can be significantly improved by use of LPFs with a large value of N_P in conjunction with channel coding.

2.5 ENVELOPE VARIATIONS

One of the reasons to use complex spreading instead of dual-channel spreading is that the complex spreading scheme can reduce the peak-to-average ratio (PAR) of the modulated signal, thus improve the efficiency of RF power amplifier [3] (see Fig. 2.11). It is expected that the peak power to average power ratio (or back-off) is very small in order for the power amplifier to work efficiently (see Fig. 2. 11(b)). The signal envelope is shown in Fig. 2. 11 (a) as a function of time. The PAR is a measurement of the signal envelope variation, which is defined as

$$PAR = \frac{\max\{|s(t)|\}}{\sqrt{E\{|s(t)|^2\}}} \tag{2.48}$$

where $|s(t)|$ is the envelope of $s(t)$. For a QPSK signal $s(t) = I(t)\cos\omega t - Q(t)\sin\omega t$, the envelope is effectively represented by $|s(t)| = \sqrt{I^2(t) + Q^2(t)}$. When PAR is evaluated, the max and mean operation in (2.48) should involve all kinds of symbol (chip) waveforms. The denominator of (2.48) is denoted as the root mean square (RMS) value of the signal, which corresponds to the average power transmitted. In practical measurements, the numerator of (2.48), i.e. the peak value of the envelope, is replaced by the 99.9 percentile of the amplitude distribution for reliable evaluation. Another convenient and useful parameter of the signal envelope variation is the normalized deviation (ND), which is defined as

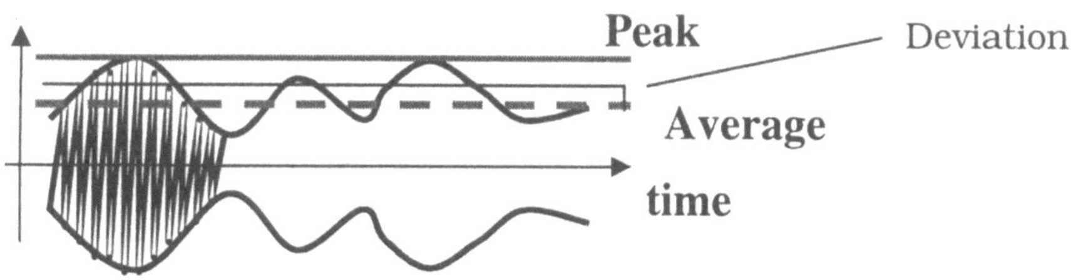

(a) Signal envelope

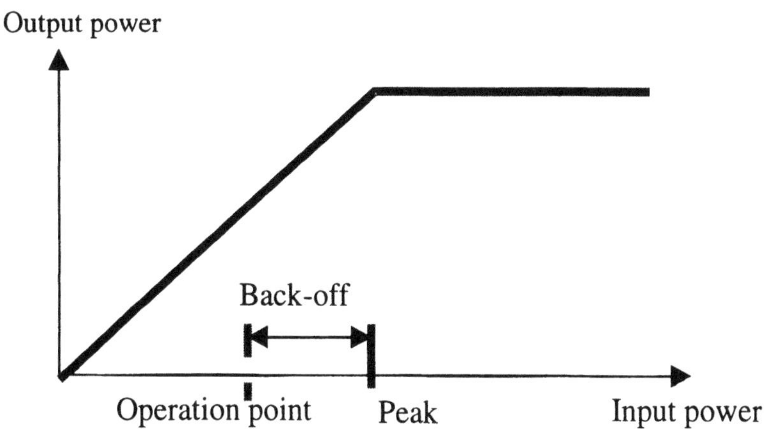

(b) Function of power amplifier

Figure 2.11 Signal envelope and efficiency of power amplifier

$$ND = \frac{\sqrt{\mathrm{var}\{|s(t)|\}}}{\sqrt{\mathrm{E}\{|s(t)|^2\}}} = \sqrt{1 - \frac{\mathrm{E}^2\{|s(t)|\}}{\mathrm{E}\{|s(t)|^2\}}} \tag{2.49}$$

This parameter effectively characterizes the deviation of signal envelope from its average value. The knowledge of the signal envelope variation helps to set the operating point as well as the back off of a RF power amplifier. Generally speaking, if the PAR and ND of a signal are small, the back off of the amplifier can be set smaller, which results in a higher power efficiency of the amplifier.

In (2.1), $S_k(t)$ can be rewritten as

$$S_k(t) = I(t)\cos(\omega_0 t + \phi_k) - Q(t)\sin(\omega_0 t + \phi_k) \tag{2.50}$$

where $I(t)$ and $Q(t)$ represent the baseband in-phase and quadrature components, and are given by

$$\begin{aligned} I(t) &= \sqrt{E_k}\sum_{n=-\infty}^{\infty}\left[\sqrt{g_k}C_{P,k}(n)S_I(n) - D_k(m)C_{D,k}(n)S_Q(n)\right]h(t-nT_c) \\ &= \sqrt{E_k}\sum_{n=-\infty}^{\infty} I_n h(t-nT_c) \end{aligned} \tag{2.51}$$

$$\begin{aligned} Q(t) &= \sqrt{E_k}\sum_{n=-\infty}^{\infty}\left[\sqrt{g_k}C_{P,k}(n)S_Q(n) + D_k(m)C_{D,k}(n)S_I(n)\right]h(t-nT_c) \\ &= \sqrt{E_k}\sum_{n=-\infty}^{\infty} Q_n h(t-nT_c) \end{aligned} \tag{2.52}$$

where for simple notation, the definition of h(t) is slightly different from (2.1). Here, h(t) is defined to be symmetric i.e. $\int_{-ATc}^{ATc} h^2(t)dt = 1$ where $A \geq 1$. Note that the values of I_n and Q_n are correlated to each other. For dual-channel spreading (see Figure 2.12), the transmitted signal is also given by (2.50), but with the baseband in-phase and quadrature components given by

$$
\begin{aligned}
I'(t) &= \sqrt{2E_k}\sum_{n=-\infty}^{\infty}\sqrt{g_k}C_{P,k}(n)S_I(n)h(t-nT_c) \\
&= \sqrt{2E_k}\sum_{n=-\infty}^{\infty}I'_n h(t-nT_c)
\end{aligned}
\tag{2.53}
$$

$$
\begin{aligned}
Q'(t) &= \sqrt{2E_k}\sum_{n=-\infty}^{\infty}D_k(m)C_{D,k}(n)S_Q(n)h(t-nT_c) \\
&= \sqrt{2E_k}\sum_{n=-\infty}^{\infty}Q'_n h(t-nT_c)
\end{aligned}
\tag{2.54}
$$

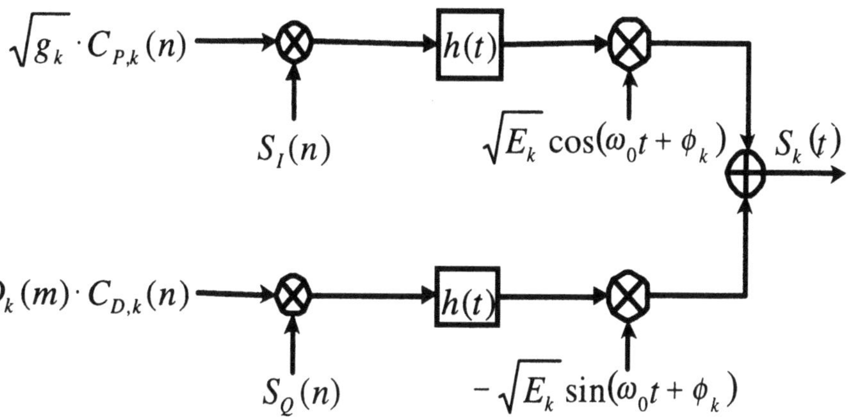

Figure 2.12 Transmitter block diagram of the *k*-th user

It can be seen that I'_n and Q'_n are independent to each other. Notice that both complex and dual-channel spread signals have the same average power. Given $D(m)=1$, all values of (I_n, Q_n) and (I'_n, Q'_n) are listed in Table 2.1 for all combinations of $C_{P,k}(n)$, $C_{D,k}(n)$, $S_I(n)$ and $S_Q(n)$.

As noted, the modulated signal envelope is effectively represented by $|s(t)|=\sqrt{I^2(t)+Q^2(t)}$. The relationships of the signal envelope with the modulating symbols (I_n, Q_n) or (I'_n, Q'_n) are very complicated if the baseband signals are pulse-shaped. This is because pulse-shaping does not only limit the bandwidth of the transmitted signal, but also extends the waveform in time domain. Therefore, the waveform is no longer determined by current chip or symbol, but subject to the effect of neighboring chips or symbols. In order to simplify the performance analysis, we have assumed that the chip waveform is limited to one chip interval, i.e. $A = 1$. However, to evaluate the signal envelope under difference spreading schemes, it is necessary to extend the chip waveform to multiple chips, which is the situation in any practical systems.

Table 2.1 Alphabet of complex spreading and dual-channel spreading

$S_I(n)$	$S_Q(n)$	$C_{P,k}(n)$	$C_{D,k}(n)$	I_n	Q_n	I'_n	Q'_n
1	1	1	1	$\sqrt{g_k}-1$	$\sqrt{g_k}+1$	$\sqrt{g_k}$	1
-1	-1	-1	-1				
1	1	1	-1	$\sqrt{g_k}+1$	$\sqrt{g_k}-1$	$\sqrt{g_k}$	-1
-1	-1	-1	1				
1	1	-1	1	$-\sqrt{g_k}-1$	$-\sqrt{g_k}+1$	$-\sqrt{g_k}$	1
-1	-1	1	-1				
1	1	-1	-1	$-\sqrt{g_k}+1$	$-\sqrt{g_k}-1$	$-\sqrt{g_k}$	-1
-1	-1	1	1				
1	-1	1	1	$\sqrt{g_k}+1$	$-\sqrt{g_k}+1$	$\sqrt{g_k}$	-1
-1	1	-1	-1				
1	-1	1	-1	$\sqrt{g_k}-1$	$-\sqrt{g_k}-1$	$\sqrt{g_k}$	1
-1	1	-1	1				
1	-1	-1	1	$-\sqrt{g_k}+1$	$\sqrt{g_k}+1$	$-\sqrt{g_k}$	-1
-1	1	1	-1				
1	-1	-1	-1	$-\sqrt{g_k}-1$	$\sqrt{g_k}-1$	$-\sqrt{g_k}$	1
-1	1	1	1				

As depicted in Figure 2.13, the impulse reponse of the square root raised cosine filter with rolloff factor of 0.22 is truncated by a length of AT_c for practical implementation. Actually, the chip waveform corresponding to each (I_n, Q_n) pair for different n is overlapped with each other. If $h(t)$ is truncated to the length AT_c, it can be observed that the current chip waveform is overlapped by its preceding $[A/2]$ chips as well as its subsequent $[A/2]$ chips. Thus, a complete evaluation of the maximum value will involve all different combinations of these continuous A chips.

According to Figure 2.13, the max occurs when the signs of I_n (or Q_n) are the same as that of the side lobes of the impulse response for corresponding n. The signs of the sidelops are defined as $+-++-+$. It is easy to verify that for a given g_k, the maximum values of the signal envelopes of two spreading schemes are the same, but with different probabilities, which are $2/8^A$ and $2/4^A$ for complex spreading and dual-channel spreading, respectively. Therefore, the occurrence of peak value is much more frequent for dual-channel spreading than for complex spreading. That is, when a nonlinear power amplifier is used, the nonlinear distortion caused by complex spreading is much less than by dual-channel spreading. In order to quantify this property, the variances and NDs of two spreading signals are evaluated by calculation, which are shown in Table 2.2.

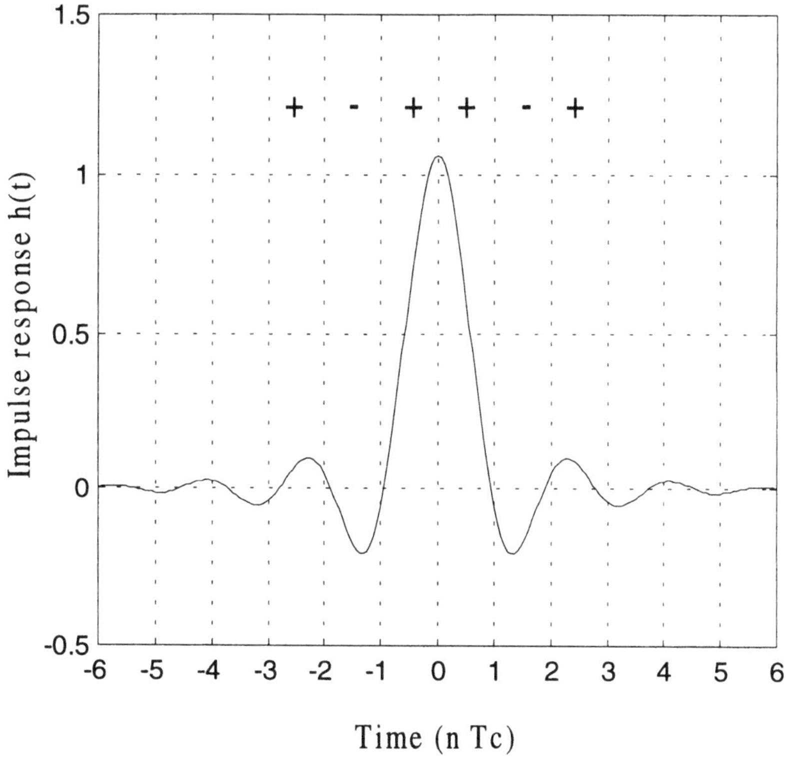

Figure 2.13 The impulse response of square RRC filter with rolloff factor 0.22

It can be observed that the theoretical PARs are the same for both complex and dual-channel spread signals, while the NDs are quite different. In Figure 2.14, the NDs are plotted versus the factor $g = g_k$ for all *k*, for both complex and dual-channel spread signals. It can be seen that when there is not much power difference between in-phase and quadrature branches, i.e. $g \approx 1$, the NDs are the same for both schemes. But when the power ratio g decreases from 1, the ND of the dual-channel spread signal increases significantly. That means the backoff of RF power amplifier becomes large when g is small, and results in less efficiency. The same situation occurs for dual-channel spreading when g increases from 1. On the other hand, the ND of the complex spread signal remains quite constant when the power ratio changes, which means the power efficiency can be kept as constant. This is obviously a benefit for RF power amplifier design.

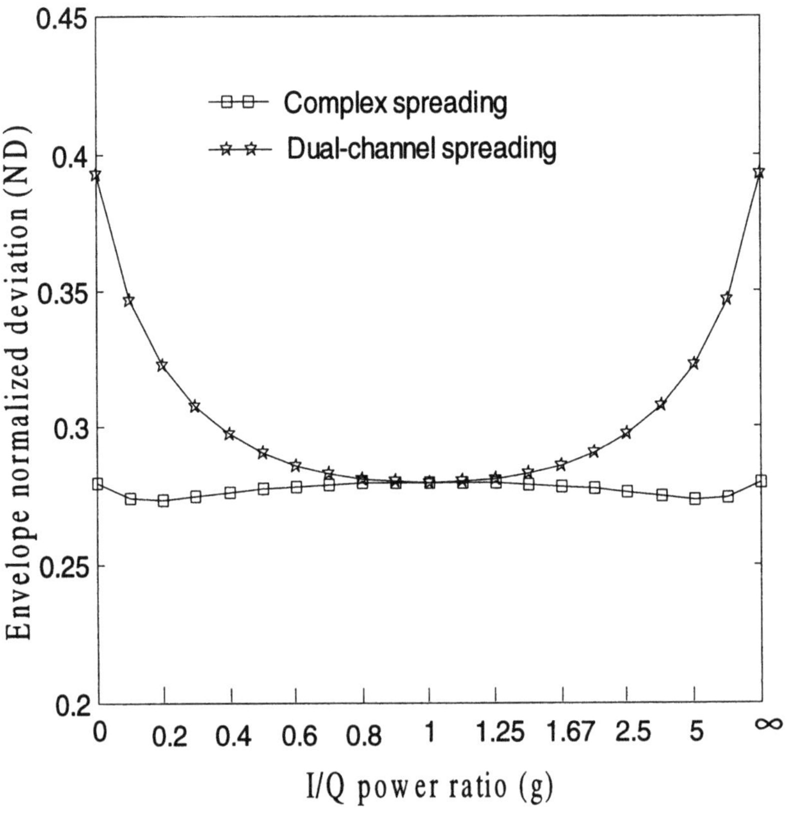

Figure 2.14. Envelope normalized deviation of complex and dual-channel spread signals

Table 2.2 Signal envelope of two spreading schemes (A = 7)

g_k	dual-channel spreading				complex spreading			
	RMS	MAX	VAR	ND	RMS	MAX	VAR	ND
1	1.4134	2.5358	0.1563	0.2797	1.4134	2.5358	0.1563	0.2797
0.9	1.3776	2.4716	0.1488	0.2800	1.3776	2.4716	0.1484	0.2796
0.8	1.3409	2.4057	0.1418	0.2809	1.3409	2.4057	0.1403	0.2794
0.7	1.3031	2.3379	0.1357	0.2826	1.3031	2.3379	0.1321	0.2789
0.6	1.2642	2.2681	0.1304	0.2856	1.2642	2.2681	0.1237	0.2782
0.5	1.2240	2.1961	0.1262	0.2902	1.2240	2.1961	0.1151	0.2771
0.4	1.1825	2.1216	0.1235	0.2971	1.1825	2.1216	0.1064	0.2759
0.3	1.1395	2.0425	0.1227	0.3074	1.1395	2.0425	0.0978	0.2745
0.2	1.0948	1.9642	0.1248	0.3226	1.0948	1.9642	0.0897	0.2735
0.1	1.0482	1.8806	0.1318	0.3463	1.0482	1.8806	0.0826	0.2741
0	0.9994	1.7931	0.1544	0.3932	0.9994	1.7931	0.0781	0.2797

2.6 SUMMARY

In this chapter, the performance evaluation of the Wideband CDMA system with complex spreading and imperfect channel estimation by dedicated pilot symbols is considered. The following conclusions have been drawn:

- In complex spreading, I- and Q- components of the modulated signal are correlated, whereas in dual-channel spreading, they are independent. Therefore, complex spreading keeps the normalized deviation of signal envelope invariant to the power ratio of in-phase and quadrature components, thus improves the efficiency of RF power amplifier.
- The system performance degrades significantly due to the channel estimation error, which results from the multiple access and multipath interference. However, this estimation error can be effectively suppressed by low pass filters (LPFs) with a large number of taps.
- The I/Q power ratio can be tuned to achieve the best system performance (the maximum capacity). The optimal value of this power ratio is mainly determined by the number of taps of the LPF and the number of active users in the system. The ratio can be chosen in the range of 0.2 to 0.4 (or –7 dB to –4 dB).

REFERENCES

[1] F. Adachi, M. Sawahashi and H. Suda, "Wideband DS-CDMA for Next-Generation Mobile Communications Systems", *IEEE Comm. Mag.*, pp56-69, Sept. 1998.

[2] E. Dahlman, B. Gudmundson, M. Nilsson and A. Skold, "UMTS/IMT-2000 Based on Wideband CDMA", *IEEE Comm. Mag.*, pp70-80, Sept. 1998.

[3] T. Ojanpera and R. Prasad, "An Overview of Air Interface Multiple Access for IMT-2000/UMTS", *IEEE Comm. Mag.*, pp82-95, Sept. 1998.

[4] 3G TS 25.213 version 3.3.0, "Spreading and Modulation (FDD)", 3GPP TSG-RAN, 2000-06.

[5] R. L. Peterson, R. E. Ziemer and D. E. Borth, *Introduction to Spread-Spectrum Communications*, pp63, Prentice Hall, 1995.

[6] L. Staphorst, M. Jamil and L. P. Linde, "Performance Evaluation of a QPSK System Employing Complex Spreading Sequences in a Fading Environment", *IEEE VTS 50th Veh. Tech. Conf. 1999,* pp2964-2968, Sept. 1999.

[7] T. G. Macdonald and M. B. Pursley, "Complex Processing in Quaternary Direct-Sequence Spread-Spectrum Receivers", *Proc. IEEE Mil. Comm. Conf. 1998*, pp494-498, Oct. 1998.

[8] T. Eng and L. B. Milstein, "Coherent DS-CDMA Performance in Nakagami Multipath Fading", *IEEE Trans. on Comm.,* vol. 43, No. 2/3/4, pp1134-1143, Feb./Mar./Apr. 1995.

[9] T. Eng and L. B. Milstein, "Partially Coherent DS-SS Performance in Frequency Selective Multipath Fading", *IEEE Trans. on Comm.,* vol. 45, No. 1, pp110-118, Jan. 1997.

[10] G. P. Efthymoglou, V. A. Aalo and H. Helmken, "Performance Analysis of Coherent DS-CDMA Systems in a Nakagami Fading Channel with Arbitrary Parameters", *IEEE Trans. on Vehicular Technology,* vol. VT-46, No. 2, pp289-297, May 1997.

[11] F. Ling, "Pilot Assisted Coherent DS-CDMA Reverse-link Communications with Optimal Robust Channel Estimation", *Proc. IEEE ICASSP'97*, pp163-166, Apr. 1997.

[12] K. Higuchi, H. Andoh, K. Okawa, M. Sawahashi and F. Adachi, "Experimental evaluation of combined effect of coherent Rake combining and SIR-based fast transmit power control for reverse link

of DS-CDMA mobile radio", *IEEE Journal on Selected Areas in Communications on Wideband CDMA I, vol. 18, no. 8, pp. 1526-1535, August 2000.*

[13] S. Abeta, M. Sawahashi and F. Adachi, "Performance Comparison between Time-Multiplexed Pilot Channel and Parallel Pilot Channel for Coherent RAKE Combining in DS-CDMA Mobile Radio", *IEICE Trans. Comm.,* vol. E81-B, No. 7, pp1417-1425, July 1998.

[14] J. Choi, "Multipath CDMA Channel Estimation by Jointly Utilizing Pilot and Traffic Channels", *IEE Proc. Comm.* vol. 146, No. 5, pp312-318, Oct. 1999.

[15] E. K. Al-Hussaini and A. A. M. al-Bassiouni, "Performance of MRC Diversity Systems for Detection of Signals with Nakagami Fading", *IEEE Trans. on Comm.,* vol. COM-33, No. 12, pp1315-1319, Dec. 1985.

[16] J. Wang and J. Chen, "Performance of Wideband CDMA systems with complex spreading and imperfect channel estimation", *IEEE J. Select. Areas Comm.,* vol. JSAC-19, pp. 152-163, January 2001.

Chapter 3

MULTICODE CDMA

Abstract Multicode CDMA is a new transmission scheme for flexible and high-speed data communications. The basic idea of multicode CDMA is to assign multiple channelization codes to any given user. If these channel codes are orthogonal, the self-interference among them can be eliminated in an AWGN channel. However, in a multipath environment, these intra-user signals from different delay paths no longer maintain orthogonality and thus cause interference, i.e. multicode interference (MCI), to each other. In high-speed data networks, where the number of users is much less than in the voice networks, the MCI may represent a large portion of the total interference and has great impact on the system performance. In this chapter, a complex spreading multicode wideband CDMA receiver with RAKE structure and MCI cancellation is studied. By pilot aided channel estimation, the MCI associated with the reference user is regenerated and subtracted from the received signal.

Keywords: Multicode CDMA, Interference cancellation, RAKE receiver, Channel estimation, Complex spreading

3.1 INTRODUCTION

Multicode Wideband CDMA has been introduced as a new transmission scheme for high speed and flexible data rate communications over wireless channels [1-3]. The basic idea of multicode CDMA is to split the user data into a number of streams and use parallel orthogonal channel codes to modulate. As described in the 3G standards of W-CDMA uplink physical layer, together with the orthogonal variable spreading factor (OVSF) method, multicode CDMA can provide data service up to the rate of 2Mbps [3]. The performance of multicode CDMA system has been addressed in several papers [4-8].

The introduction of multicode causes several problems. One is the high envelope variation, which results from a linear sum of signals over multicode channels. There have been a number of methods proposed to combat such envelope fluctuation, e.g. pre-coding [9] and pre-distortion [10-11]. However, as shown in Chapter 2, a convenient method is to use complex spreading, which can effectively suppress the envelope variation without introducing extra circuits or loss in information rate. In this chapter, complex spreading is adopted with the development of a unique representation of the complex signal. Another problem of multicode is the self-interference caused by the different delays of the intra-user signal in a multipath environment. These intra-user multicode signals have certain characteristics: synchronous, suffering the same fading for a given path and the channel codes associated with the reference user are known in both downlink and uplink. These features may be exploited to improve the receiver performance. That is, certain multicode detection techniques may be employed to achieve better performance.

Interference cancellation (IC) is to attempt the removal of the multiuser interference from each user's received signal before making final data decisions. IC has been vastly investigated [13-16]. The application of IC in multicode CDMA has also been suggested in some publications [17-21]. However, the following issues are not sufficiently addressed. First, in most publications, the system performance is examined by simulations rather than analysis, thus more insightful study is necessary. Second, the signal with very large bandwidth in the next generation communication system shall confront with a large number of multipaths. Most publications assume that the number of RAKE receiver fingers is equal to the number of resolvable paths, and full cancellation is achieved. However, in any practical system, the number of RAKE fingers is limited and cannot be chosen arbitrarily. That is, the multipath interference cannot be completely cancelled. Third, IC requires the knowledge of channel parameters, e.g. amplitudes and phases. However, the channel estimation using pilot symbols or data symbols suffers a lot from the multipath and multi-access interference and cannot be accurately obtained, which may greatly degrade the system performance. Fourth, since the tentative decision for MCI regeneration may be not completely reliable, thus it will be useful to use partial cancellation to mitigate the negative effects of incorrect decisions. [14-15] studied the effect of partial cancellation by experiments and simulations. In this chapter, the optimal partial MCI cancellation is analytically investigated and a multi-stage receiver with RAKE structure is presented. A consolidated theoretical analysis is given for system performance evaluation.

Basically, multi-user IC is used only in the uplink of the CDMA system, because in downlink, the mobile stations do not know the spreading codes of other users. However, in the case of multicode systems, the channel codes of the reference user are known in both uplink and downlink, thus it is possible to utilize MCI cancellation in mobile stations as well. The uplink transmission model is used in this chapter. However, the analysis and results can be applied to the downlink transmission as well.

Since multicode is only used for high-speed data transmission with high transmit power, the number of active users supported by the system is small. For instance, in 3GPP [3], when the number of data channels is 6 with a spreading factor (SF) of 4 in each code, only the multicode user is permitted for transmission in the whole system. That is, the number of active users in the system is one. Therefore, for simple analysis, only one user (i.e. the multicode user) is considered. However, even there are other interfering users in the system, they can be suppressed by other methods, such as antenna arrays [22].

3.2 MULTICODE CDMA MODELS

Consider the uplink of a single cell multicode CDMA system. In a general case, any user in the system may use multicode to transmit high-speed data or may use a single code to transmit voice and low-speed data. Here, only one multicode user is assumed. Figure 3.1 shows the transmitter block diagram of the multicode user, where $L+1$ denotes the number of code-channels. Note that one code (i.e. 0-th code) with a large SF is for pilot channel, whereas other L codes with small SF are for data channels. The ratio, g, is defined as the power ratio of pilot channel to one data channel. The user data is first split into L parallel streams, i.e. serial to parallel conversion, then all streams are multiplied by corresponding orthogonal codes. This is so called the spreading or channelization process and the data symbols are transformed into chip streams. The streams are divided into two groups. In one group, the data streams, together with the pilot channel, are added to form the Q-branch data (imaginary part of the spreading data), whereas data streams in another group are added to form the I-branch data (real part of the spreading data). It is worth noting that any pair of data channels on the I- and Q-branches may use the same channel code, as specified in the standard of W-CDMA uplink physical layer of 3GPP [3], and the I- and Q-data are distinguished by the following complex scrambling code. However, in this chapter, all the I- and Q-channels use different codes. But the system performance and models are identical for both cases.

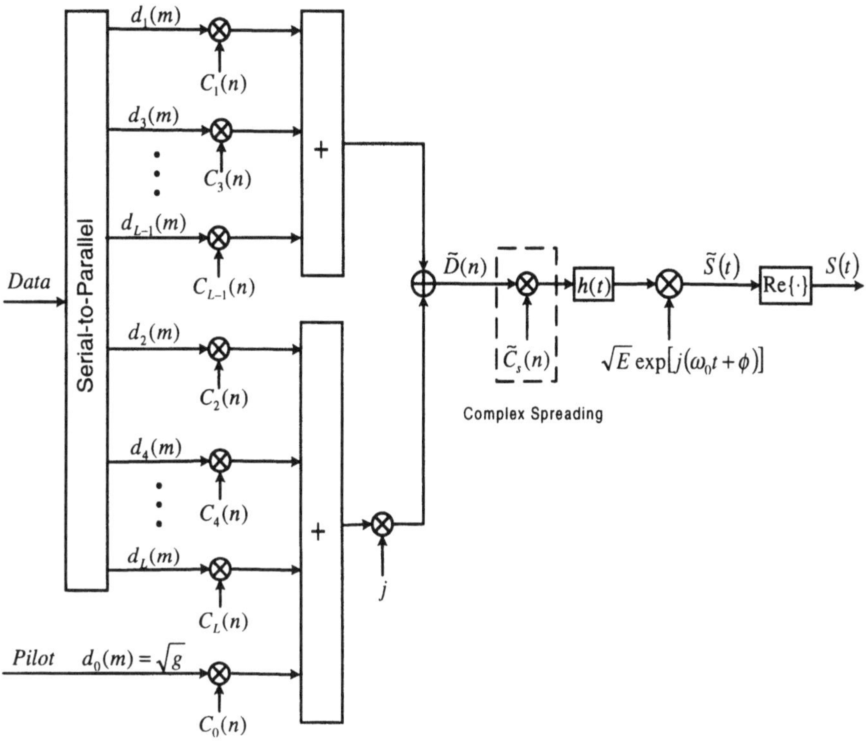

Figure 3.1 Transmitter block diagram of the multicode CDMA system [24]

The I- and Q-branches are treated as complex spreading data $\tilde{D}(n)$ and multiplied by a complex scrambling sequence $\tilde{C}_s(n)$, which consists of two binary sequences, i.e. $\tilde{C}_s(n) = \pm 1 \pm j$. Note that hereinafter, the symbol "~"

is used to denote a complex variable. After scrambling, the resultant complex signal is passed through the pulse-shaping filter with impulse response, $h(t)$, and then multiplied with a complex carrier $\sqrt{E}\exp[j(\omega_0 t+\phi)]$, where E is chip energy, ω_0 is angle frequency and ϕ is carrier phase. The real part of the modulated signal is transmitted and given by

$$\begin{aligned} S(t) &= \mathrm{Re}\left\{\tilde{S}(t)\right\} \\ &= \mathrm{Re}\left\{\sqrt{E}\sum_{n=-\infty}^{\infty}\tilde{D}(n)\cdot\tilde{C}_s(n)\cdot h(t-nT_c)\cdot\exp[j(\omega_0 t+\phi)]\right\} \end{aligned} \tag{3.1}$$

where $\mathrm{Re}\{\cdot\}$ stands for real part and $\tilde{D}(n)$ is the complex spreading data, of the form

$$\tilde{D}(n) = \sum_{l=1,3,5,\cdots} d_l(m)C_l(n) + j\sum_{l=0,2,4,\cdots} d_l(m)C_l(n)\ , \quad 0\le l\le L \tag{3.2}$$

where $d_l(m)$ is the data sequence in l-th sub-channel with $d_0(m)=\sqrt{g}$ and $d_l(m)=\pm 1$ for $l\neq 0$, and $m=\lfloor n/N\rfloor$, where $\lfloor\cdot\rfloor$ stands for the integer part and N is the spreading factor, $C_l(n)$ is the orthogonal channel code for l-th channel. Note that the pilot channel has a large SF $N=N_0$, while data channels take a small SF $N=N_1$.

One of the basic features of the next generation mobile communication system is the occupation of large bandwidth. When such a signal with large bandwidth is transmitted over wireless channels, it is very likely to demonstrate the characteristic of a large number of multipaths, especially in urban environment. A general multipath channel model is the tapped delay line model, with the complex low-pass equivalent impulse response that can be written as

$$h_{ch}(t) = \sum_{p=0}^{P-1}\alpha_p(t)\delta[t-\tau_p(t)]\,e^{j\theta_p(t)} \tag{3.3}$$

where P ($P\ge 1$) is the number of resolvable propagation paths. $\alpha_p(t)e^{j\theta_p(t)}$ and $\tau_p(t)$ are the complex fading factor and propagation delay, respectively, of the p-th path. It is assumed that all paths are independent. That is, all random variables in (3.3) are independent for p. An exponential multipath intensity profile is assumed, i.e. $\Omega_p = \mathrm{E}\{\alpha_p^2\} = e^{-p\delta}$, where δ is the rate of exponential decay of the path energy.

Therefore, the received signal is of the form

$$R(t) = \eta(t) + \mathrm{Re}\left\{\sum_{p=0}^{P-1} \alpha_p(t)\tilde{S}[t-\tau_p(t)]e^{j\theta_p(t)}\right\} \tag{3.4}$$

where $\eta(t)$ is the background AWGN with double-side power spectrum density $\eta_0/2$.

Note that $\alpha_p(t)$ can be Rayleigh, Rician or Nakagami distributed, depending on a specific channel model. $\alpha_p(t)$ is assumed to be Nakagami in this paper, because the Nakagami distribution is more versatile and more adequate to describe different fading situations. Accordingly, the probability density function (pdf) of $\{\alpha_p\}$ is given by

$$p(\alpha_p) = 2\left(\frac{m_p}{\Omega_p}\right)^{m_p} \frac{\alpha_p^{2m_p-1}}{\Gamma(m_p)} \exp\left(-\frac{m_p}{\Omega_p}\alpha_p^2\right), \qquad \alpha_p \geq 0 \tag{3.5}$$

where $m_p = \dfrac{\Omega_p^2}{\mathrm{E}\{(\alpha_p^2 - \Omega_p)^2\}}$, and $\Gamma(x) = \int_0^{\infty} t^{x-1}e^{-t}dt$ is the Gamma function. Furthermore, if the Nakagami distribution in each path is such that the ratio $\dfrac{m_p}{\Omega_p}$ is the same for all paths, it can be shown that $\zeta = \sum_{p=0}^{P-1}\alpha_p^2$ has the gamma pdf [23]:

$$p(\zeta) = \left(\frac{m_T}{\Omega_T}\right)^{m_T} \frac{\zeta^{m_T-1}}{\Gamma(m_T)} \exp\left(-\frac{m_T}{\Omega_T}\zeta\right), \qquad \zeta \geq 0 \tag{3.6}$$

where $\Omega_T = \sum_{p=0}^{P-1} \Omega_p = \sum_{p=0}^{P-1} \mathrm{E}\{\alpha_p^2\}$ and $m_T = \sum_{p=0}^{P-1} m_p$.

The receiver actually consists of several stages with similar structure. The overall structure is shown in Figure 3.2, and the detailed block diagram of each stage is shown in Figure 3.3. The basic idea is to make tentative decisions of the data symbols in the first stage using the conventional RAKE receiver. The tentative decisions are used to reconstruct MCI to be canceled in the second stage, and secondary decisions are used in third stage to further cancel the interference, and so forth. Note that the front-end circuit (before sampling) is shared for all stages. As in any digital processing systems, the received signal is first down-converted to baseband, then passed through a pulse-matching filter, whose impulse response is the conjugate of the impulse response of the transmitter filter. The output is sampled at the chip rate $1/T_c$, and these signal samples are fed to every stage correspondingly.

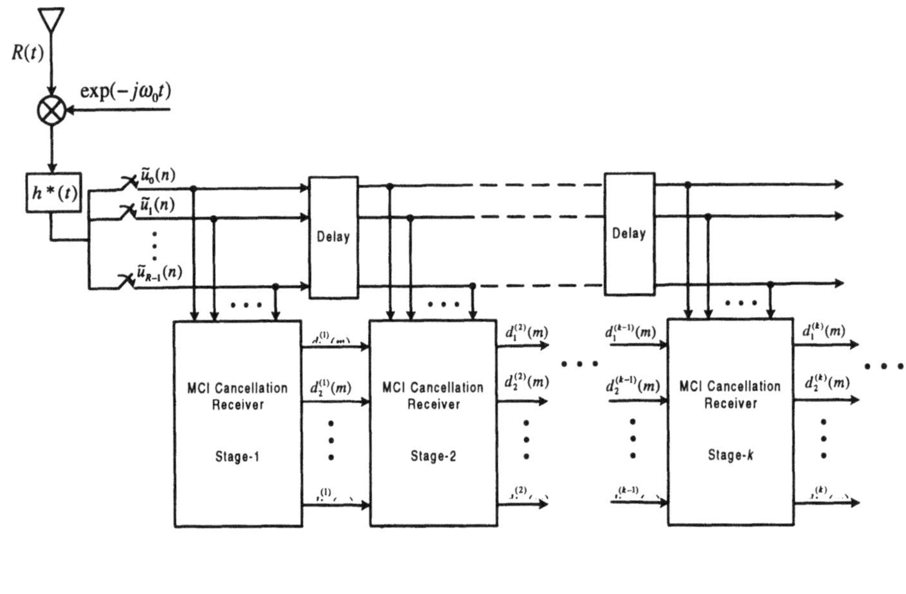

Figure 3.2 The overall structure of the multistage MCI cancellation receiver [24]

In all stages, a user-dedicated pilot channel is used to facilitate coherent reception. De-scrambled samples are multiplied by a known pilot channel code, $-jC_0(n-n_r)$, and summed over N_0 chips to obtain an unbiased channel estimate. This estimate is noisy due to the multipath interference and a low-pass-filter (LPF) is utilized to reduce the noise. In order to track the fading variation effectively, the cutoff frequency of the filter should be much larger than the maximum Doppler frequency f_d of the wireless channel. In this study, a N_P-tap FIR filter is adopted to suppress the noise with assumption $\dfrac{1}{N_0 N_P T_c} >> f_d$.

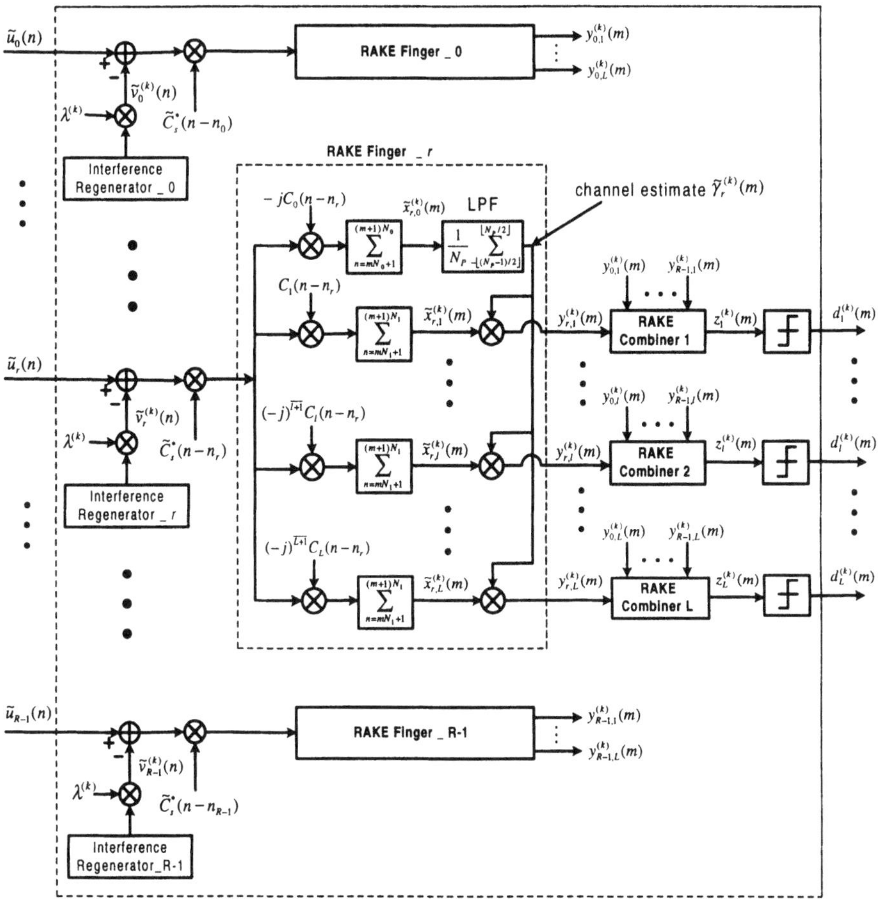

Figure 3.3 The detailed block diagram of the k-th stage [24]

In the first stage (without MCI cancellation), the sampled signal, $\tilde{u}_r(n)$, is first de-scrambled by multiplying conjugate of the scrambling code $\tilde{C}_s^*(n-n_r)$, then fed into a bank of RAKE processing units. There are P multipath components in the received signal, however, only the strongest R paths are processed, i.e., the number of RAKE fingers is $R(R \le P)$. In each RAKE finger, the de-scrambled samples are multiplied by individual channel code $(-j)^{\overline{l+1}} C_l(n-n_r)$, where $\overline{l+1} = (l+1) \bmod 2$, then integrated and dumped. After that, the outputs of all data channels are multiplied by channel estimate from pilot symbols to complete coherent demodulation and maximum ratio combining. Each finger produces L outputs and all outputs of R fingers are added correspondingly to form L decision statistics for the L parallel data symbols.

In the second stage ($\lambda^{(2)} \ne 0$), the MCI is first reconstructed for each RAKE finger, using previous data decisions and channel estimates. For each data channel, the total number of interference terms should be $(L+1)P-1$. However, since those multicode signals that arrive at the same path are orthogonal and do not cause interference, only $(L+1)(P-1)$ terms of interference need to be eliminated. Due to implementation limits of RAKE receiver, only the parameters of R strongest paths are estimated, so that only $(L+1)(R-1)$ terms of interference components could be regenerated and cancelled. The detailed block diagram of the interference regenerator in r-th finger is shown in Figure 3.4, where $\{d_l^{(k-1)}(m) = d_l^{(1)}(m), l = 1, \cdots, L\}$ are the tentative decisions from the first stage. These tentative data decisions $\{d_l^{(1)}(m)\}$ and scaled channel estimates $\{\tilde{\gamma}_p^{(1)}(m)/(2\sqrt{g}N_o)\}$ are multiplied by channel codes and scrambling sequences with different delay. Finally, a weighting factor $\lambda^{(k)}$, shown in Figure 3.3, is introduced to control the weight of interference being removed. Because when the interference estimate is poor, e.g. the signal-to-noise ratio (SNR) is low, it is preferable not to cancel the entire amount of estimated interference but to remove only a part of the interference. After subtraction, the resultant signal is processed as in the first stage and secondary decisions are made.

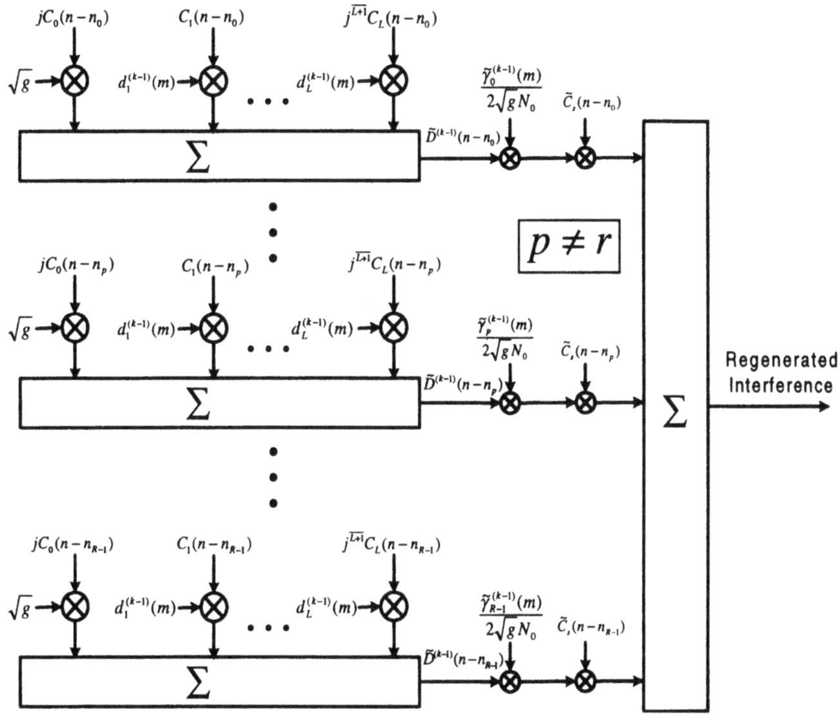

Figure 3.4 Interference regenerator of the *r*-th finger in the *k*-th stage [24]

In order to obtain acceptable performance, more stages with MCI cancellation are necessary. The operations of the *k*-th stages are like that of the second stage.

3.3 ANALYSIS OF MCI CANCELLATION

3.3.1 Decision Statistic in First Stage (Without MCI cancellation)

Assuming that acquisition and synchronization have been accomplished for the reference path, *r*, the path delay $\tau_r(t)$ is known, so that it is able to pick up the strongest R paths. Therefore, the local spreading codes in the *r*-th finger are aligned with the *r*-th path of the received signal. In Figure 3.3, the sampled output of the pulse-matching filter for the *r*-th finger is given by

$$\tilde{u}_r(n) = \int_{nT_c+\tau'_r(n)}^{(n+1)T_c+\tau'_r(n)} R(t)\exp(-j\omega_0 t)\, h^*[nT_c + \tau_r'(n) - t]dt$$

$$= \tilde{\eta}(n) + \sum_{p=0}^{P-1} \frac{1}{2}\sqrt{E}\alpha_p(n) R_w[\bar{\tau}_p(n)] \exp[j\varphi_p(n)] \cdot \tilde{D}(n-n_p) \cdot \tilde{C}_s(n-n_p)$$

$$= \tilde{\eta}(n) + \sum_{p=0}^{P-1} \frac{1}{2}\sqrt{E}\tilde{\beta}_p(n) \cdot \tilde{D}(n-n_p) \cdot \tilde{C}_s(n-n_p) \qquad (3.7)$$

where for the sake of simple notation, nT_c is replaced by n hereinafter, and $\tau'_r(n) = \tau_r(n) \bmod T_c$, $n_p = \lfloor [\tau_p(n) - \tau'_r(n)]/T_c \rfloor$, $\tilde{\eta}(n)$ is the sampled noise component, and $\tilde{\beta}_p(n)$ is of the form

$$\tilde{\beta}_p(n) = \alpha_p(n) R_w[\bar{\tau}_p(n)] \exp[j\varphi_p(n)] \qquad (3.8)$$

where $\varphi_p(n) = \phi + \theta_p(n) - \omega_0 \tau_p(n)$ is the overall carrier phase offset. $\bar{\tau}_p(n)$ is the chip timing error given by $\bar{\tau}_p(n) = \tau_p(n) - \tau'_r(n) - n_p T_c$. $R_w(\bar{\tau})$ represents the effect of timing error $\bar{\tau}$ on the sampled output. If the impulse response of the shaping filter is limited to one T_c interval, then we have [12]

$$R_w(\bar{\tau}) = \int_{-\infty}^{\infty} |H(f)|^2 \cos(2\pi f \bar{\tau}) df \qquad (3.9)$$

where $H(f)$ is the frequency response of the pulse shaping filter.

It is assumed that for different path *p*, the fading amplitudes $\alpha_p(n)$ are independent, chip timing errors $\bar{\tau}_p(n)$ are i.i.d random variables uniformly distributed over $(0, T_c]$ except for $\bar{\tau}_r(n) = 0$, carrier phase offsets $\varphi_p(n)$

are i.i.d random variables uniform over $(0,2\pi]$. Accordingly, $\tilde{\beta}_p(n)$ in (3.8) can be modeled as a zero mean variables with variance

$$\mathrm{var}\{\tilde{\beta}_p(n)\} = \mathrm{E}\{\tilde{\beta}_p^2(n)\} = (\frac{1}{2} + j\frac{1}{2})\Delta\Omega_p \tag{3.10}$$

where $\Omega_p = \mathrm{E}\{\alpha_p^2(n)\}$ is the channel parameter given in the channel model, $\Delta = \mathrm{E}\{R_w^2[\tau_p(n)]\}$ is related to the frequency transform function of pulse-shaping filter, and $\Delta = 2/3$ for rectangle pulse shaping filter, $\mathrm{E}\{\exp^2[j\varphi_p(n)]\}$ is defined as $\mathrm{E}\{\cos^2\varphi_p(n)\} + j\mathrm{E}\{\sin^2\varphi_p(n)\}$ $= \frac{1}{2} + j\frac{1}{2}$. Notice that the complex variables used in this paper are for the purpose of avoiding ponderous notations. In most cases, the real and imaginary parts are corresponding to the I- and Q- components in processing. Hence in mathematical operations, the real and imaginary parts will be dealt individually.

The output of the I&D correlator in the data branch l ($l \geq 1$) of the finger r at the first stage is

$$\tilde{x}_{r,l}^{(1)}(m) = \sum_{n=mN_1+1}^{(m+1)N_1} \tilde{u}_r(n) \cdot \tilde{C}_s^*(n-n_r) \cdot (-j)^{\overline{l+1}} C_l(n-n_r) \tag{3.11}$$

where $(-j)^{\overline{l+1}} = 1$ and $-j$ for odd and even l, respectively. Assuming that the channel parameters remain constant over one symbol duration, i.e. $\tilde{\beta}_p(n) = \tilde{\beta}_p(mN_1)$ for $mN_1 < n \leq (m+1)N_1$, it can be shown that the output, $\tilde{x}_{r,l}^{(1)}(m)$, consists of three distinct components: the desired signal component $\tilde{E}_{r,l}^{(1)}(m)$, the intra-user multicode interference $\tilde{I}_{r,l}^{(1)}(m)$ and the background AWGN component $\tilde{\eta}_{r,l}^{(1)}(m)$. That is,

$$\tilde{x}_{r,l}^{(1)}(m) = \tilde{E}_{r,l}^{(1)}(m) + \tilde{I}_{r,l}^{(1)}(m) + \tilde{\eta}_{r,l}^{(1)}(m) \tag{3.12}$$

By assuming that the products of scrambling sequence and channel codes are pure random sequence, one can have

- $\widetilde{E}_{r,l}^{(1)}(m)$ is a deterministic term, conditioned on the instantaneous channel parameters,

$$\widetilde{E}_{r,l}^{(1)}(m) = N_1\sqrt{E}\alpha_r(mN_1)R_w[\bar{\tau}_r(mN_1)]\exp[j\varphi_r(mN_1)]d_l(m) \qquad (3.13)$$

Since $\bar{\tau}_r(mN_1) = 0$ and $R_w[\bar{\tau}_r(mN_1)] = 1$, $\widetilde{E}_{r,l}^{(1)}(m)$ is to be further written as

$$\widetilde{E}_{r,l}^{(1)}(m) = N_1\sqrt{E}\alpha_r(mN_1)\exp[j\varphi_r(mN_1)]d_l(m) \qquad (3.14)$$

- $\widetilde{I}_{r,l}^{(1)}(m)$ is a zero mean random variable given by

$$\widetilde{I}_{r,l}^{(1)}(m) = \sum_{\substack{p=0\\p\neq r}}^{P-1}\frac{1}{2}\sqrt{E}\widetilde{\beta}_p(mN_1)\sum_{n=mN_1+1}^{(m+1)N_1}\widetilde{D}(n-n_p)\,\widetilde{C}_s(n-n_p)\cdot\widetilde{C}_s^*(n-n_r)\cdot(-j)^{\overline{l+1}}C_l(n-n_r)$$

(3.15)

with variance

$$\mathrm{var}\left\{\widetilde{I}_{r,l}^{(1)}(m)\right\} = (1+j)\Delta\frac{N_1}{2}\sum_{\substack{p=0\\p\neq r}}^{P-1}\Omega_p(L+g)E \qquad (3.16)$$

- $\widetilde{\eta}_{r,l}^{(1)}(m)$ is a zero mean Gaussian variable with variance

$$\mathrm{var}\left\{\widetilde{\eta}_{r,l}^{(1)}(m)\right\} = (1+j)2N_1\cdot\frac{\eta_0}{2} = (1+j)N_1\eta_0 \qquad (3.17)$$

Therefore, $\widetilde{x}_{r,l}^{(1)}(m)$ is a random variable with mean given by (3.14) and variance

$$\mathrm{var}\left\{\widetilde{x}_{r,l}^{(1)}(m)\right\} = \mathrm{var}\left\{\widetilde{I}_{r,l}^{(1)}(m)\right\} + \mathrm{var}\left\{\widetilde{\eta}_{r,l}^{(1)}(m)\right\} = (1+j)N_1\Lambda^{(1)} \qquad (3.18)$$

Notice that $\Lambda^{(1)} = \Delta \frac{1}{2} \sum_{p=0}^{P-1} \Omega_p (L+g)E + \eta_0$ is a statistic that gives the measurement of the overall interference in the first stage. The approximation in (3.18) is valid when $P >> 1$. Actually, the interference in (3.18) is slightly enlarged, so that the BER performance to be derived may represent an upper bound.

In any finger of the RAKE receiver, the output of the top branch, namely $\tilde{x}_{r,0}^{(1)}(m)$, is the channel estimate from the pilot symbol. The mean and variance of this estimate are also given by (3.14) and (3.18), respectively, with the replacement of N_1 with N_0. It can be seen from (3.18) that the variance of this estimate is nontrivial and may result in very inaccurate estimation of channel parameters. In order to reduce the error of channel estimate, $\tilde{x}_{r,0}^{(1)}(m)$ is passed through a low pass FIR filter with N_P taps. Hence the refined channel estimate is given by

$$\tilde{\gamma}_r^{(1)}(m) = \frac{1}{N_P} \sum_{m'=-\lfloor (N_P-1)/2 \rfloor}^{\lfloor N_P/2 \rfloor} \tilde{x}_{r,0}^{(1)}(m+m') \qquad (3.19)$$

The pdf of successive $\tilde{x}_{r,0}^{(1)}(m)$ for different m are independent to each other, because for different *m*, the scrambling codes are uncorrelated. Assuming that the channel parameters are unchanged over N_P symbols, one obtains

$$\mathrm{E}\{\tilde{\gamma}_r^{(1)}(m)\} = \mathrm{E}\{\tilde{x}_{r,0}^{(1)}(m)\} = N_0\sqrt{gE}\alpha_r(mN_0)\exp[j\varphi_r(mN_0)] \qquad (3.20)$$

$$\mathrm{var}\{\tilde{\gamma}_r^{(1)}(m)\} = \frac{1}{N_P}\mathrm{var}\{\tilde{x}_{r,0}^{(1)}(m)\} = (1+j)\frac{N_0}{N_P}\Lambda^{(1)} \qquad (3.21)$$

For coherent detection and maximum ratio combining, the data estimate is multiplied by channel estimate, that is

$$y_{r,l}^{(1)}(m) = \mathrm{Re}\left\{\tilde{x}_{r,l}^{(1)}(m) \cdot \tilde{\gamma}_r^{(1)*}(m)\right\}, \qquad l \neq 0 \qquad (3.22)$$

where $\tilde{\gamma}_r^{(1)*}(m)$ is the conjugate of $\tilde{\gamma}_r^{(1)}(m)$. The mean and variance of $y_{r,l}^{(1)}(m)$ are

$$\mathrm{E}\{y_{r,l}^{(1)}(m)\} = N_0 N_1 \sqrt{g} E \alpha_r^2(mN_1) d_l(m) \tag{3.23}$$

where $\alpha_r^2(mN_1) = \alpha_r^2(mN_0)$, and

$$\mathrm{var}\{y_{r,l}^{(1)}(m)\} = \left(gN_0 + \frac{N_1}{N_P} \right) N_0 N_1 E \alpha_r^2(mN_1) \Lambda^{(1)} + \frac{2N_0 N_1}{N_P} [\Lambda^{(1)}]^2 \tag{3.24}$$

After RAKE combining, the decision statistic of data symbol in the l-th code channel is the sum of outputs in R corresponding fingers

$$z_l^{(1)}(m) = \sum_{r=0}^{R-1} y_{r,l}^{(1)}(m) \tag{3.25}$$

Because the variables in different paths are independent, the mean and variance of $z_l(m)$ are

$$\mathrm{E}\{z_l^{(1)}(m)\} = N_1 N_0 \sqrt{g} E \zeta_R d_l(m) \tag{3.26}$$

$$\mathrm{var}\{z_l^{(1)}(m)\} = \left(gN_0 + \frac{N_1}{N_P} \right) N_0 N_1 E \Lambda^{(1)} \zeta_R + \frac{2RN_0 N_1}{N_P} [\Lambda^{(1)}]^2 \tag{3.27}$$

where $\zeta_R = \sum_{r=0}^{R-1} \alpha_r^2(mN_1)$ is a random variable determined by the instantaneous amplitude fading.

In order to regenerate the multicode interference, a tentative decision of each data symbol is to be made in the first stage. For hard decision, the estimated data symbol in the first stage is

$$d_l^{(1)}(m) = \mathrm{sgn}[z_l^{(1)}(m)] \tag{3.28}$$

Notice that the number of independent interference components in $y_{r,l}^{(1)}(m)$ is P, consequently the number of interference terms in $z_l^{(1)}(m)$ is

PR. When PR is very large (which is true for general cases), $z_l^{(1)}(m)$ can be modeled as a Gaussian variable. Therefore an instantaneous error probability of this tentative decision is given by

$$P_e^{(1)} = P_{e,l}^{(1)}(\zeta_R) = Q\left(\frac{|\mathrm{E}\{z_l(m)\}|}{\sqrt{\mathrm{var}\{z_l(m)\}}}\right)$$

$$= Q\left(\frac{\zeta_R}{\sqrt{\zeta_R\left(1+\frac{1}{gN'}\right)\frac{\Lambda^{(1)}}{N_1 E}+\frac{2R}{gN'}\cdot\left(\frac{\Lambda^{(1)}}{N_1 E}\right)^2}}\right) \tag{3.29}$$

where $N' = N_P \cdot N_0/N_1$, $|x|$ denotes the absolute value, and $Q(x) = \int_x^\infty \frac{1}{\sqrt{2\pi}} e^{-y^2/2} dy$ is the Q-function. Note that the overall error probability is equal to that of each data channel.

3.3.2 Decision Statistic in Second Stage (with MCI Cancellation)

The regenerated interference of the r-th finger in the second stage is given by

$$\tilde{v}_r^{(2)}(n) = \lambda^{(2)} \cdot \sum_{\substack{p=0 \\ p\neq r}}^{R-1} \frac{\tilde{\gamma}_p^{(1)}(m)}{2\sqrt{g N_0}} \tilde{D}^{(1)}(n-n_p)\cdot \tilde{C}_s(n-n_p), \tag{3.30}$$

where $\tilde{D}^{(1)}(n) = \sum_{l=1,3,5,\cdots} d_l^{(1)}(m) C_l(n) + j \sum_{l=0,2,4,\cdots} d_l^{(1)}(m) C_l(n)$, $0 \le l \le L$, is the reconstructed complex spreading data using tentative decisions in the first stage. It is seen that the reconstructed interference is irrelevant to different sub-channel l, and distinct only for different RAKE finger. This is because the multicode signals that arrive at the same delay are orthogonal to each other and the resultant interference is zero. Therefore, there are only (L+1)(R-1) terms of interference components involved in cancellation.

The correlator output in the l-th data channel of the second stage is

$$\tilde{x}_{r,l}^{(2)}(m) = \sum_{n=mN_1+1}^{(m+1)N_1} \left[\tilde{u}_r(n) - \tilde{v}_r^{(2)}(n)\right] \cdot \tilde{C}_s^*(n-n_r) \cdot (-j)^{\overline{l+1}} C_l(n-n_r)$$

$$= \tilde{E}_{r,l}^{(2)}(m) + \tilde{I}_{r,l}^{(2)}(m) + \tilde{\eta}_{r,l}^{(2)}(m)$$

$$= \tilde{E}_{r,l}^{(1)}(m) + \tilde{I}_{r,l}^{(2)}(m) + \tilde{\eta}_{r,l}^{(1)}(m) \tag{3.31}$$

where the desired component $\tilde{E}_{r,l}^{(2)}(m) = \tilde{E}_{r,l}^{(1)}(m)$, and the noise component $\tilde{\eta}_{r,l}^{(2)}(m) = \tilde{\eta}_{r,l}^{(1)}(m)$, the only component different from that in the first stage is the MCI, i.e. $\tilde{I}_{r,l}^{(1)}(m) \to \tilde{I}_{r,l}^{(2)}(m)$. $\tilde{I}_{r,l}^{(2)}(m)$ is of the form

$$\tilde{I}_{r,l}^{(2)}(m) = \sum_{p=R}^{P-1} \frac{1}{2}\sqrt{E}\tilde{\beta}_p(mN_1) \sum_{n=mN_1+1}^{(m+1)N_1} \tilde{D}(n-n_p)\tilde{C}_s(n-n_p)\tilde{C}_s^*(n-n_r)(-j)^{\overline{l+1}} C_l(n-n_r)$$

$$+\sum_{\substack{p=0 \\ p\neq r}}^{R-1} \sum_{n=mN_1+1}^{(m+1)N_1} \left[\frac{1}{2}\sqrt{E}\tilde{\beta}_p(mN_1)\tilde{D}(n-n_p) - \lambda^{(2)} \cdot \frac{\tilde{\gamma}_p^{(1)}(m)}{2\sqrt{g}N_0}\tilde{D}^{(1)}(n-n_p)\right]$$

$$\cdot \tilde{C}_s(n-n_p) \cdot \tilde{C}_s^*(n-n_r) \cdot (-j)^{\overline{l+1}} C_l(n-n_r)$$

$$= \tilde{I}'^{(2)}_{r,l}(m) + \tilde{I}''^{(2)}_{r,l}(m) \tag{3.32}$$

Notice that after cancellation, the MCI still consists of two residual parts. The first part is the interference that is beyond the R strongest paths, which is not being cancelled at all. The variance of this part does not change with the increase of stages, and is given by

$$\mathrm{var}\left\{\tilde{I}'^{(2)}_{r,l}(m)\right\} = (1+j)\Delta\frac{N_1}{2}\sum_{p=R}^{P-1}\Omega_p(L+g)E \tag{3.33}$$

The second part is the residual interference within the R paths, which is due to two factors. First is the error of scaled channel estimate, that is,

$\frac{\tilde{\gamma}_p^{(1)}(m)}{2\sqrt{g N_0}} \neq \frac{\sqrt{E}}{2} \tilde{\beta}_p(mN_1)$, $0 \le p \le R-1$. The second factor is the tentative data decision errors from the first stage, that is, $d_l^{(1)}(m) \neq d_l(m), l = 1, 2, \ldots, L$. Therefore, the interference can be approximated as the summation of the interferences in two cases. First is the interference where the channel estimation is error free, while the tentative data decisions have errors with error rate $P_e^{(1)}$ given by (3.29). Second is the interference where the tentative data decisions are error free, while the scaled channel estimate has deviation (variance)

$$\mathrm{var}\left\{ \frac{\tilde{\gamma}_p^{(1)}(m)}{2\sqrt{g N_0}} \right\} = (1+j)\frac{\Lambda^{(1)}}{4 g N_P N_0} = \left(\frac{1}{8} + j\frac{1}{8}\right)\Delta E \left(\sigma_p^{(1)}\right)^2 \tag{3.34}$$

where

$$\left(\sigma_p^{(1)}\right)^2 = \frac{L+g}{g N_P N_0} \sum_{p=0}^{P-1} \Omega_p + \frac{\eta_0}{g E N_P N_0} \tag{3.35}$$

Notice that $\left(\sigma_p^{(1)}\right)^2$ is approximately irrelevant to the subscript p, although it represents the variance of channel estimate of the p-th path.

When $B^{(1)}$ out of L data channel decisions are incorrect at the first stage, the MCI caused by the error decisions is increased by $(1+\lambda^{(2)})^2$ times. For the rest correct $(L - B^{(1)})$ decisions, the interference is reduced by $(1-\lambda^{(2)})^2$ times. Moreover, the channel estimation error is additive to this interference, no matter whether the data decision are correct or not. Note that there is no tentative decision error in the pilot channel. Therefore, the variance of this residual interference is given by [24]

$$\begin{aligned}
\mathrm{var}\left\{\tilde{I}''^{(2)}_{r,l}(m)\right\} &= (1+j)\Delta \frac{N_1 E}{2}\left(\sum_{\substack{p=0\\ p\neq r}}^{R-1} \Omega_p \left[(1+\lambda^{(2)})^2 \overline{B}^{(1)} + (1-\lambda^{(2)})^2 (L - \overline{B}^{(1)})\right] + \sum_{\substack{p=0\\ p\neq r}}^{R-1} \left(\lambda^{(2)}\right)^2 \left(\sigma_p^{(1)}\right)^2 (L+g) \right) \\
&= (1+j)\Delta \frac{N_1 E}{2}\left(\sum_{\substack{p=0\\ p\neq r}}^{R-1} \Omega_p \left[4\lambda^{(2)} \overline{B}^{(1)} + (1-\lambda^{(2)})^2 L\right] + \sum_{\substack{p=0\\ p\neq r}}^{R-1} \left(\lambda^{(2)}\right)^2 \left(\sigma_p^{(1)}\right)^2 (L+g) \right)
\end{aligned} \tag{3.36}$$

where $\overline{B}^{(1)} = \mathrm{E}\{B^{(1)}\} = LP_e^{(1)}$ is the average number of error bits at the first stage. Strictly speaking, $B^{(1)}$ should not be averaged here. Rather, subsequent conditional error probability should be averaged over $B^{(1)}$. However, since the tentative data decision errors are not independent to each other, it is very difficult to obtain the distribution of $B^{(1)}$. Nevertheless, the simulation results in Section 3.5 show that the alternative is quite accurate.

The total variance of MCI after cancellation in the second stage is given by

$$\mathrm{var}\{\tilde{I}_{r,l}^{(2)}(m)\} = \mathrm{var}\{\tilde{I}'^{(2)}_{r,l}(m)\} + \mathrm{var}\{\tilde{I}''^{(2)}_{r,l}(m)\} \tag{3.37}$$

Similar to $\tilde{x}_{r,l}^{(1)}(m)$, the variance of $\tilde{x}_{r,l}^{(2)}(m)$ can be written as

$$\begin{aligned} \mathrm{var}\{\tilde{x}_{r,l}^{(2)}(m)\} &= \mathrm{var}\{\tilde{I}_{r,l}^{(2)}(m)\} + \mathrm{var}\{\tilde{\eta}_{r,l}^{(2)}(m)\} \\ &= (1+j)N_1\Lambda^{(2)} \end{aligned} \tag{3.38}$$

where $\Lambda^{(2)}$ is given by

$$\Lambda^{(2)} \approx \frac{\Delta E}{2}\left(\sum_{p=R}^{P-1}\Omega_p(L+g) + \sum_{p=0}^{R-1}\Omega_p\left[4\lambda^{(2)}\overline{B}^{(1)} + (1-\lambda^{(2)})^2 L\right] + \sum_{p=0}^{R-1}\left(\lambda^{(2)}\right)^2\left(\sigma_p^{(1)}\right)^2(L+g)\right) + \eta_0 \tag{3.39}$$

$\Lambda^{(2)}$ is the measurement of total interference after MCI cancellation in the second stage. Similar to the first stage, the mean and variance of the decision statistic in the second stage can also be written in the forms of (3.26) and (3.27), but with $\Lambda^{(1)} \rightarrow \Lambda^{(2)}$. Therefore, it is possible to calculate the error performance of the second stage on the basis of the first stage.

3.3.3 Optimal Partial MCI Cancellation

In view of the fact that the tentative decisions are not completely reliable, a possible way to improve the performance is not to attempt to cancel interference completely, but partially. This can be accomplished by the weighting factor $\lambda^{(2)}$, shown in Figure 3.3. The second part of the residual interference after cancellation is re-written as

$$\mathrm{var}\left\{\tilde{I}''^{(2)}_{r,l}(m)\right\} = (1+j)\Delta\frac{N_1 E}{2}\left(\sum_{\substack{p=0\\p\neq r}}^{R-1}\Omega_p\left[4\lambda^{(2)}\overline{B}^{(1)} + (1-\lambda^{(2)})^2 L\right] + \sum_{\substack{p=0\\p\neq r}}^{R-1}\left(\lambda^{(2)}\right)^2\left(\sigma_p^{(1)}\right)^2(L+g)\right) \tag{3.40}$$

It can be seen from the first term of the right hand side of (3.40) that in the presence of the weighting factor ($\lambda^{(2)} \neq 0$), interference cannot be completely removed even though tentative data decisions are correct. When $\lambda^{(2)}$ is small, interference caused by both incorrect tentative data decisions and imperfect channel estimation is small, whereas the interference introduced by correct tentative decision is large. Therefore, an optimal value of $\lambda^{(2)}$ should exist according to different channel conditions. That is, when channel condition is good (the error rate of tentative decision is low), a large value of $\lambda^{(2)}$ is desirable (close to 1). However, when the channel condition is poor (the noise and interference level is high), a smaller value of $\lambda^{(2)}$ is preferable.

The optimal $\lambda^{(2)}$ is to minimize the variance given by (3.40). Thus by letting $\dfrac{d\,\mathrm{var}\left\{\tilde{I}''^{(2)}_{r,l}(m)\right\}}{d\lambda^{(2)}} = 0$, it is readily to have [24]

$$\lambda_{opt}^{(2)} = \frac{\sum\limits_{\substack{p=0\\p\neq r}}^{R-1}\Omega_p(L-2\overline{B}^{(1)})}{\sum\limits_{\substack{p=0\\p\neq r}}^{R-1}\Omega_p L + \sum\limits_{\substack{p=0\\p\neq r}}^{R-1}\left(\sigma_p^{(1)}\right)^2(L+g)} \tag{3.41}$$

This result agrees with the above discussion on how to choose the optimal value of $\lambda^{(2)}$, which should be in the range $\lambda^{(2)} < 1$. It is seen from (3.41) that when $2\overline{B} > L$, $\lambda_{opt}^{(2)} < 0$. This can be instinctively explained as following. The regenerated interference component from an error decision is a negative term compared to that from a correct data decision. If more than half of the data decisions in the first stage are incorrect ($\overline{B} > L/2$), then the overall regenerated interference would be a negative one. In this case, to subtract the regenerated interference cannot reduce the interference at all, but increase the interference. Thus the optimal $\lambda^{(2)}$ goes to negative to make the subtraction meaningful. The optimal $\lambda^{(2)}$ given by (3.41) depends on the particular channel parameters as well as the average bit error rate of previous stage. However, (3.41) can be approximated by

$$\lambda_{opt}^{(2)} = \frac{\sum_{\substack{p=0\\p\neq r}}^{R-1}\Omega_p(L-2\overline{B}^{(1)})}{\sum_{\substack{p=0\\p\neq r}}^{R-1}\Omega_p L+\sum_{\substack{p=0\\p\neq r}}^{R-1}\left(\sigma_p^{(1)}\right)^2(L+g)} \approx \frac{\sum_{p=0}^{R-1}\Omega_p(L-2LP_e^{(1)})}{\sum_{p=0}^{R-1}\Omega_p L+\sum_{p=0}^{R-1}\left(\sigma_p^{(1)}\right)^2(L+g)}$$

$$= \frac{1-2P_e^{(1)}}{1+\delta^{(1)}} \tag{3.42}$$

where

$$\delta^{(1)} = (1+g/L)\sum_{p=0}^{R-1}\left(\sigma_p^{(1)}\right)^2 \Big/ \sum_{p=0}^{R-1}\Omega_p \tag{3.43}$$

In the case of perfect channel estimation, i.e. $\left(\sigma_p^{(1)}\right)^2 = 0$, or if the channel estimation error $\left(\sigma_p^{(1)}\right)^2$, given by (3.35), is much smaller than Ω_p, $\delta^{(1)}$ is very small, and (3.42) can be approximated by [24]

$$\lambda_{opt}^{(2)} \leq 1-2P_e^{(1)} \tag{3.44}$$

3.3.4 Decision Statistics in more stages

The analytical results of the second stage can be extended to more stages without difficulty. The regenerated interference of the *r*-th finger in the *k*-th stage is given by

$$\tilde{v}_r^{(k)}(n) = \lambda^{(k)}\cdot\sum_{\substack{p=0\\p\neq r}}^{R-1}\frac{\tilde{\gamma}_p^{(k-1)}(m)}{2\sqrt{g}N_0}\tilde{D}^{(k-1)}(n-n_p)\cdot\tilde{C}_s(n-n_p), \qquad 0\leq r\leq R-1 \tag{3.45}$$

where $\tilde{D}^{(k-1)}(n) = \sum_{l=1,3,5,\cdots} d_l^{(k-1)}(m)C_l(n) + j \sum_{l=0,2,4,\cdots} d_l^{(k-1)}(m)C_l(n)$, $0 \le l \le L$.

The correlator output in the l-th data channel of the k-th stage is

$$\tilde{x}_{r,l}^{(k)}(m) = \sum_{n=mN_1+1}^{(m+1)N_1} \left[\tilde{u}_r(n) - \tilde{v}_r^{(k)}(n)\right] \cdot \tilde{C}_s^*(n-n_r) \cdot (-j)^{\overline{l+1}} C_l(n-n_r)$$

$$= \tilde{E}_{r,l}^{(1)}(m) + \tilde{I}_{r,l}^{(k)}(m) + \tilde{\eta}_{r,l}^{(1)}(m) \tag{3.46}$$

The variance of $\tilde{x}_{r,l}^{(k)}(m)$ is of the form

$$\text{var}\left\{\tilde{x}_{r,l}^{(k)}(m)\right\} = (1+j)N_1\Lambda^{(k)} \tag{3.47}$$

where $\Lambda^{(k)}$ is given by

$$\Lambda^{(k)} \approx \frac{\Delta E}{2}\left(\sum_{p=R}^{P-1}\Omega_p(L+g) + \sum_{p=0}^{R-1}\Omega_p\left[4\lambda^{(k)}B + (1-\lambda^{(k)})^2 L\right] + \sum_{p=0}^{R-1}\left(\lambda^{(k)}\right)^2\left(\sigma_p^{(k-1)}\right)^2(L+g)\right) + \eta_0 \tag{3.48}$$

and

$$\left(\sigma_p^{(k-1)}\right)^2 = \frac{2\Lambda^{(k-1)}}{\Delta E g N_p N_0}, \qquad k > 2 \tag{3.49}$$

The optimal cancellation factor of the k-th stage, $\lambda_{opt}^{(k)}$, is of the form [24]

$$\lambda_{opt}^{(k)} = \frac{1-2P_e^{(k-1)}}{1+\delta^{(k-1)}}$$

$$\le 1 - 2P_e^{(k-1)} \tag{3.50}$$

where

$$\delta^{(k-1)} = (1+g/L)\sum_{p=0}^{R-1}\left(\sigma_p^{(k-1)}\right)^2 \Big/ \sum_{p=0}^{R-1}\Omega_p \tag{3.51}$$

3.4 SYSTEM PERFORMANCE

When hard decision is applied and no channel coding is used, the instantaneous bit error probability of l-th branch data in the k-th stage is in similar form as that of the first stage, with substitution of $\Lambda^{(1)}$ by $\Lambda^{(k)}$. That is

$$P_e^{(k)} = P_{e,l}^{(k)}(\zeta_R) = Q\left(\frac{\left|\mathrm{E}\{z_l^{(k)}(m)\}\right|}{\sqrt{\mathrm{var}\{z_l^{(k)}(m)\}}}\right)$$

$$= Q\left(\frac{\zeta_R}{\sqrt{\zeta_R\left(1+\frac{1}{gN'}\right)\frac{\Lambda^{(k)}}{N_1E}+\frac{2R}{gN'}\left(\frac{\Lambda^{(k)}}{N_1E}\right)^2}}\right) \tag{3.52}$$

where $\Lambda^{(k)}$ is given by (3.48). The error probability is conditioned on instantaneous multipath fading parameter ζ_R. In order to get an average bit error rate (BER) of the final decisions, the conditioned error probability should be averaged over the distributions of ζ_R. The final average BER of the proposed system is given by

$$P_e = \int_0^\infty P_e^{(k)}(\zeta_R)\cdot p(\zeta_R)\,d\zeta_R \tag{3.53}$$

where $P_e^{(k)}$ is given by (3.52), and $p(\zeta_R)$ is of the form

$$p(\zeta_R) = \left(\frac{m_R}{\Omega_R}\right)^{m_R} \frac{\zeta_R^{m_R-1}}{\Gamma(m_R)} \exp\left(-\frac{m_R}{\Omega_R}\zeta_R\right) \tag{3.54}$$

where $\Omega_R = \sum_{r=0}^{R-1} \Omega_r = \sum_{r=0}^{R-1} \mathrm{E}\{\alpha_r^2\}$ and $m_R = \sum_{r=0}^{R-1} m_r$.

3.5 NUMERICAL RESULTS

In this section, the effects of different system parameters on the BER performance of the multicode Wideband CDMA system are investigated by numerical calculations. Unless noted otherwise, the exponential decay rate of the multipath intensity profile (MIP) $\delta = 0.2$ and the ratio $m_l/\Omega_l = 4$ for all paths. The number of multipaths $P = 8$, and the number of RAKE fingers $R = 4$. The spreading factor for pilot channel $N_0 = 256$, and for data channel $N_1 = 4$. The number of data channels for the reference user $L = 6$. The number of LPF taps $N_P = 2$ and the power ratio of pilot channel $g = 0.3$. The signal-to-noise ratio (SNR) is defined as $SNR = NE/\eta_0$, which is actually the average received SNR in the first path.

In Figure 3.5, the BER performance is plotted versus the received SNR for different number of stages. Simulation results are also provided for comparison. The first stage ($k = 1$) is the BER without MCI cancellation. It is seen that the MCI cancellation effectively reduces the bit error rate. The improvement is quite significant at stage-2 and stage-3, but little improvement is observed beyond stage-4. Therefore, in practical system, a receiver with 3 or 4 stages would be sufficient. Also note that the cancellation factor in each stage is roughly chosen in this plotting according to the error rate. The values are 0, 0.5, 0.7, 0.85 and 0.95 for $k = 1, 2, 3, 4, 5$ respectively. They should be considered as typical values rather than optimal values. As it is shown in the following figures, these values are slightly shifted from the optimal ones. The simulation results (curves with marks only) demonstrate that the analytical results are quite accurate.

Figure 3.6 shows the system BER performance in terms of received SNR with different number of data channels. The system utilizes a 4-stage MCI cancellation receiver with cancellation factor set to be 0, 0.5, 0.75 and 0.95, respectively. The performance degradation due to the increasing number of

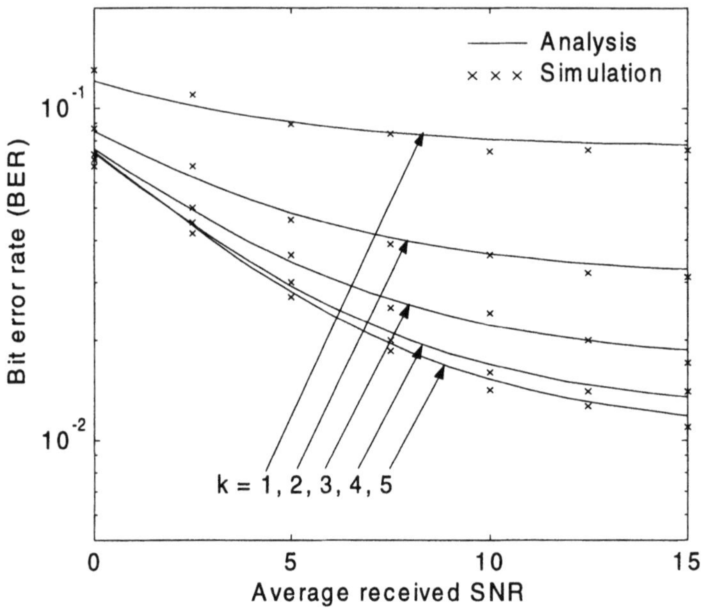

Figure 3.5 BER versus SNR with increase of number of stages in system

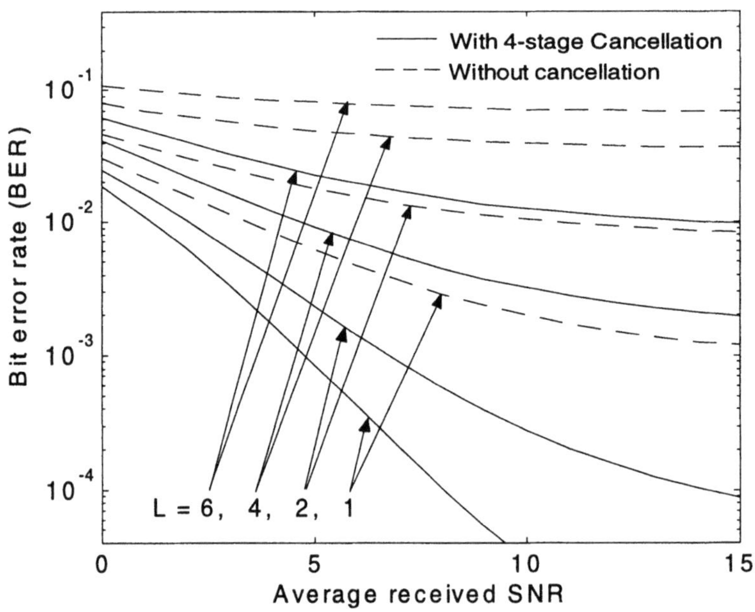

Figure 3.6 BER versus SNR with different number of data channels

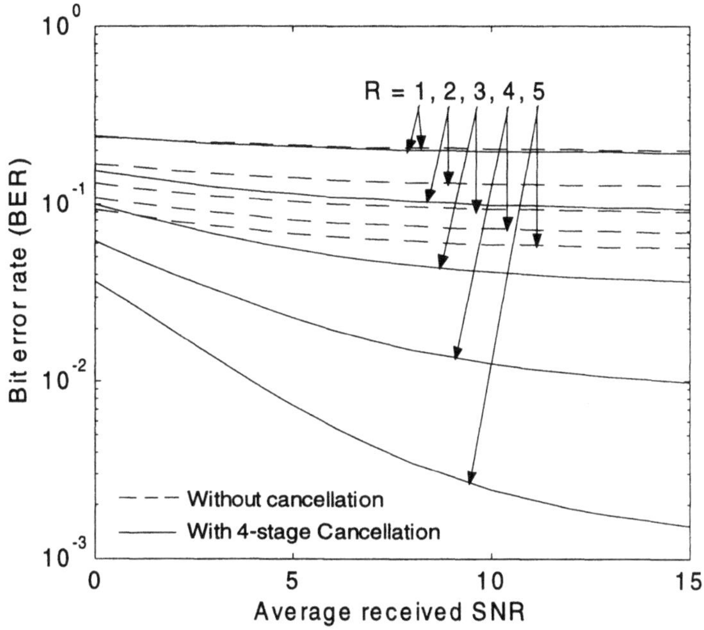

Figure 3.7 BER versus SNR with different number of RAKE fingers

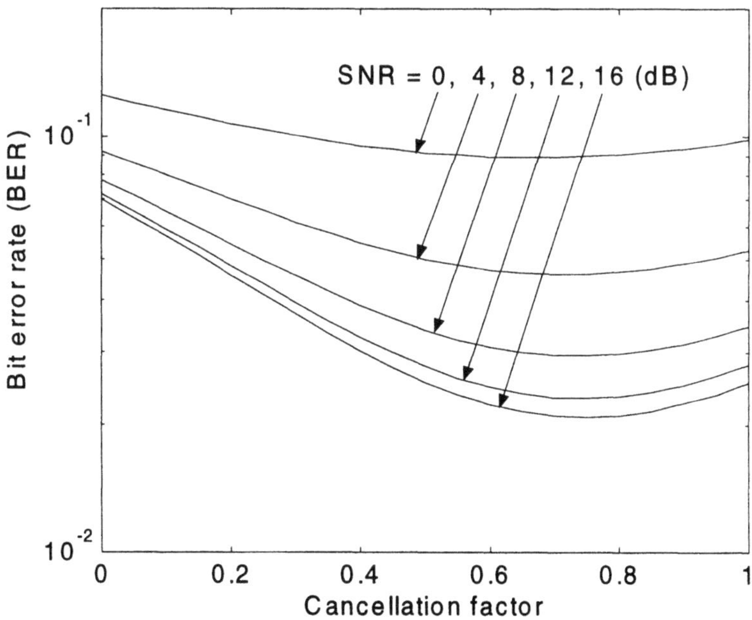

Figure 3.8 BER versus cancellation factor with different SNR level

data channels is obvious. However, acceptable performance ($BER \approx 10^{-2}$) can be achieved even if total 6 data channels are in use.

In Figure 3.7, the BER is plotted versus SNR with different number of RAKE fingers *R*. The receiver is also of 4 stages with cancellation factor set as 0, 0.5, 0.75 and 0.95, respectively. It is clearly shown that the more RAKE finger is utilized, the more advantage we can get from MCI cancellation. This is because when more RAKE fingers are used, more multipaths are estimated and the MCI in these multipaths can be cancelled and energy in more paths can be employed for data detection.

In Figure 3.8, the BER performance is plotted versus the cancellation factor. The curves are differed from each other in SNR level. In order to obtain simple conclusion, only a 2-stage MCI cancellation receiver is used when plotting this figure. There is no cancellation in the first stage as it should be, and the cancellation factor at the second stage is varied from 0 to 1 to show the distinct results. As expected, the increase of cancellation factor gradually decreases the BER until it reaches an optimal value, then the BER increases again. The situations are the same for different SNR level, but the optimal value may slightly shift from each other.

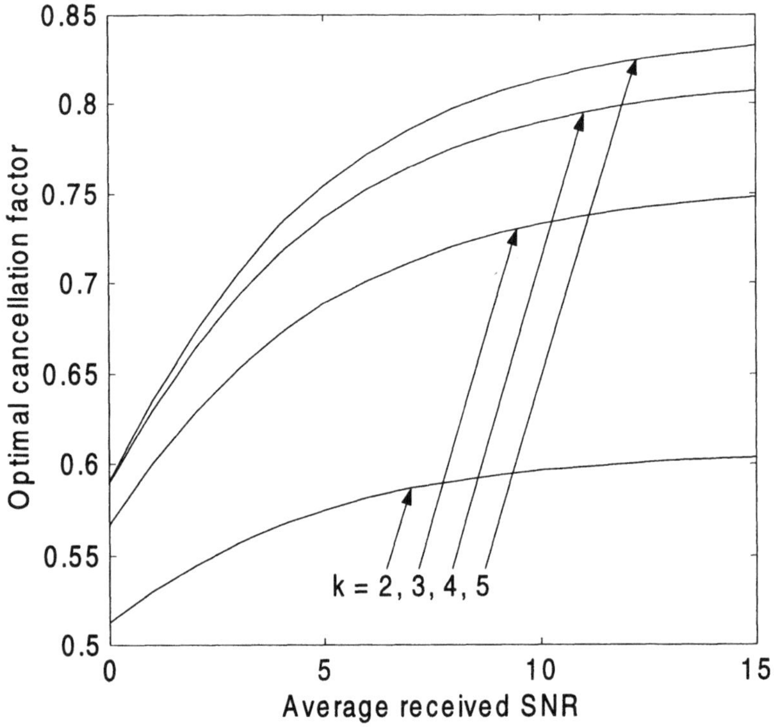

Figure 3.9 Optimal cancellation factor versus SNR in different stage

Figure 3.9 illustrates the optimal cancellation factor $\lambda_{opt}^{(k)}$, given by (50), as the function of received SNR level for different stages (i.e. $k = 2,3,4,5$). As it is seen from the figure, when the SNR is low, the optimal cancellation factor is small, while a higher SNR requires a higher cancellation factor. It is also shown that with the increase of number of stages, the optimal cancellation factor goes larger. The reason is that the BER is reduced with the increase of stages. However, same as what we found in Figure 4.5, there is little performance improvement between stage-4 and stage-5, which result in little changes of the optimal cancellation factor. The optimal value of the cancellation factor should be chosen from 0.5 to 0.85 for a wide range of signal to noise ratio, as observed from the figure.

3.6 CONCLUSIONS

This chapter presents a multicode RAKE receiver with MCI cancellation for high and flexible rate data transmission over multipath fading channels. The following conclusions are drawn:

- MCI cancellation can effectively reduce the BER in multipath environment. With a 4-stage receiver, the BER can be lowered by approximately an order ($BER \approx 10^{-1} \rightarrow BER \approx 10^{-2}$).
- The optimal cancellation factor in the k-th stage is approximated by $\lambda_{opt}^{(k)} \leq 1 - 2P_e^{(k-1)}$, where $P_e^{(k-1)}$ is the error probability of the (k-1)-th stage.
- The optimal cancellation factor of each stage is gradually increased with the decreasing of BER. The values of these factors can be roughly chosen from 0.5 to 0.85 for different signal to noise ratios.

REFERENCES

[1] Chih-Lin I and R. D. Gitlin, "Multi-code CDMA wireless personal communications networks", *Proc. IEEE ICC'95*, Vol. 2, pp1060 – 1064, June 1995.

[2] F. Adachi, K. Ohno, A. Higashi, T. Dohi and Y. Okumura, "Coherent multicode DS-CDMA mobile radio access", *IEICE Trans. on Comm.*, Vol. E79-B, No. 9, pp1316-1325, Sept. 1996

[3] 3G TS 25.213 version 3.3.0, "Spreading and Modulation (FDD)", 3GPP TSG-RAN, 2000-06.

[4] N. Guo and L. B. Milstein, " On rate-variable multidimensional DS/SSMA with sequence sharing", *IEEE J. Select. Areas Comm.,* Vol. 17, No. 5, pp. 902-917, May 1999.

[5] Z. Liu, M. J. Karol, M. E. Zarki and K. Y. Eng, "Interference issues in multi-code CDMA networks", *Proc. IEEE PIMRC'96,* Vol. 1, pp98-102, Oct. 1996.

[6] H. D. Schotten, H. Elders-Boll and A. Busboom, "Adaptive multi-code CDMA systems for variable data rates", *Proc. IEEE ICPWC'97,* pp334-337, Dec. 1997.

[7] D. N. Kalofonos, E. Kurtas and J. G. Proakis, "Performance of multi-code CDMA systems with concatenated coding and iterative decoding in multipath Rayleigh fading channels", *Proc. IEEE VTC'99-Fall,* Vol. 5, pp2566 –2570, Sept. 1999.

[8] D. Ayyagari and A. Ephremides, "Cellular multicode CDMA capacity for integrated (voice and data) services", *IEEE J. Select. Areas Comm.,* Vol. 17, No. 5, pp928-938, May 1999.

[9] T. Ottosson, "Precoding in multicode DS-CDMA systems", *Proc. IEEE ISIT'97,* pp351, June 1997.

[10] N. Guo and L. B. Milstein, "Uplink performance evaluation of multicode DS/CDMA systems in the presence of nonlinear distortions", *IEEE J. Select. Areas Comm.,* Vol. 18, No. 8, pp1418-1428, Aug. 2000.

[11] D. I. Kim and K. H. Park, "A hybrid nonlinear distortion compensator for multicode DS/CDMA systems", *Proc. IEEE ICUPC'98*, Vol. 2, pp1087-1091, Oct. 1998

[12] J. Wang and J. Chen, "Performance of Wideband CDMA systems with complex spreading and imperfect channel estimation", *IEEE J. Select. Areas Comm.,* January 2001.

[13] A. C. K. Soong and W. A. Krzymien, "A novel CDMA multiuser interference cancellation receiver with reference symbol aided estimation of channel parameters", *IEEE J. Select. Areas Comm.,* Vol. 14, No. 8, pp1536-1547, Oct. 1996

[14] M. Sawahashi, H. Andoh, K. Higuchi, "Interference rejection weight control for pilot symbol-assisted coherent multistage interference canceller using recursive channel estimation in DS-CDMA mobile radio", *IEICE Trans. Fundamentals*, Vol. E81-A, No. 5, pp957-972. May 1998.

[15] N. S. Correal, R. M. Buehrer and B. D. Woerner, "Real-time DSP implementation of a coherent partial interference cancellation multiuser receiver for DS-CDMA", *Proc. IEEE ICC'98*, Vol. 3, pp1536-1540, June 1998

[16] P. K. Frenger, P. Orten and T. Ottosson, "Code-spread CDMA with interference cancellation", *IEEE J. Select. Areas Comm.,* Vol. 17, No.12, pp2090-2095, Dec. 1999.

[17] M. C. -C Chan and J. C. -I Chuang "Multicode high-speed transmission with interference cancellation for wireless personal communications", *Proc. IEEE VTC'96,* Vol. 2, pp661-665, May 1996.

[18] R. Herzog, "Interference cancellation for a high data rate user in coded CDMA systems", *Proc. IEEE ICC'98,* Vol. 2, pp709-713, June 1998.

[19] D. Koulakiotis and A. H. Aghvami, "Performance enhancement of multi-code CDMA using interference cancellation", *Proc. IEEE ISSTA'98,* Vol. 1, pp130-134, Sept. 1998.

[20] Y. Suzuki and K. Kobayashi, "Interference cancellation method for DS-CDMA multicode-packet transmission", *Proc. IEEE GLOBECOM'98,* Vol. 6, pp3578 –3583, Nov. 1998.

[21] S. I. Kim and H. S. Lee, "Performance improvement of RAKE receiver for a multicode DS-CDMA system with multistage interference cancellation detectors", *Proc. IEEE TENCON'99,* Vol.1, pp573 –576, Sept. 1999.

[22] S. Tanaka, H. Harada, M. Sawahashi and F. Adachi, "Experiments on coherent antenna array diversity for wideband DS-CDMA mobile radio", *IEEE J. Select. Areas Comm.,* Vol. , No. , pp, August 2000.

[23] E. K. Al-Hussaini and A. A. M. al-Bassiouni, "Performance of MRC Diversity Systems for Detection of Signals with Nakagami Fading", *IEEE Trans. on Comm.,* vol. COM-33, No. 12, pp1315-1319, Dec. 1985.

[24] J. Chen, J. Wang and M. Sawahashi, "MCI cancellation for multicode wideband CDMA systems," Submitted to IEEE Transactions on Communications.

Chapter 4

ADVANCED TRACKING LOOP

ABSTRACT In this chapter, half-sine and triangular chip waveforms are proposed for the locally generated early and late despreading codes in a coherent DLL for a single user in a DS/SS systems. Both ideal and band-limited received signals are investigated. These non-rectangular waveforms reduce the effective loop noise power with only minor effect on the DLL sensitivity. Analysis and computer simulation conclude that the proposed chip waveforms reduce the RMS tracking error, increase the MTLL compared to that of the conventional rectangular chip waveform. It is also shown that the proposed chip waveform reduces the tracking offset introduced by the multipath signals.

Keywords Chip Waveforms, Delay-Locked Loop, PN Code Tracking, Synchronization

4.1 OVERVIEW

4.1.1 Background

All applications of direct sequence spread-spectrum (DS/SS) systems require that the received and the locally generated spreading waveforms to be synchronized. The synchronization process generally consists of two steps: acquisition (course synchronization) and tracking (fine synchronization). During the acquisition process, the locally generated despreading signal is brought to one chip time accuracy relative to the incoming spreading signal. When this required accuracy is achieved, acquisition unit hands over to the tracking unit for fine synchronization.

The Delay-Locked-Loop (DLL) technique has been widely used for tracking purpose. The performance of the DLL is generally measured by tracking error variance when the received signal has an adequate signal to noise ratio. To ensure low bit error rate transmission during normal operations, small tracking error in the DLL should be maintained. Even though mean time to loose lock (MTLL) is much larger than the average connection time under normal circumstances, sporadic signal fading during the transmission may cause sudden reduction in the received SNR and may cause the tracking loop to loose lock. The MTLL is therefore an important parameter used to measure system performance. Multipath signals introduce a tracking offset in the DLL and a higher signal-to-noise ratio may be required to achieve the desired bit error rate. Therefore, the tracking offset of the DLL should also be kept as small as possible in multipath environments. Hence, to ensure good system performance, the root-mean-square (RMS) tracking error and tracking offset should be as small as possible whereas the MTLL should be as large as possible.

The operation of the DLL can be categorized into two modes: coherent and non-coherent. The coherent operation mode requires a good carrier recovery system and decision feedback in order to cancel the data modulation effect. On the other hand, non-coherent mode operations do not require carrier recovery and decision feedback due to the squaring operation. The squaring operation increases the noise power level in the DLL and therefore the performance of the coherent DLL is always better (by approximately 3dB signal power) compared to that of non-coherent DLL for the same operating conditions. In this chapter, we are only concerned with coherent mode operations. However all techniques proposed here can be extended to non-coherent DLL also.

4.1.2 The Development of DLL Performance Analysis

In early years, the analysis and optimization of the DLL performance have been performed based on the carrier phase-locked loops (PLL). The performance of the DLL was measured in terms of the tracking error variance. The linear analysis was used for preliminary study and design purposes under reasonable SNR conditions. In linear analysis, the discriminator characteristic function is approximated by a linear curve at its origin. The tracking error variance obtained by linear analysis is still used as the performance measure of the DLL since the DLL is considered in high SNR conditions. However, results obtained by linear analysis differ from the actual performance of the DLL when the system is operating around threshold SNR or below. Thus non-linear analysis becomes necessary. The computation of MTLL also requires non-linear analysis.

Initially, the use of non-linear analysis to evaluate the DLL performance was based on the analysis of the carrier PLL. In the carrier tracking loop, the non-linear phase error characteristic is periodic. Meyr [1-3] analyzed the tracking loop performance assuming that the DLL characteristic function is approximately periodical. He also suggested that the transition probability density of the phase error in the tracking loop depends on the present value and not on the past values (markov process). In fact, the characteristic function of the code tracking loop is aperiodic and has a finite range. Polydoros and Weber [4-5] proposed a method of analysis for the DLL with aperiodic finite characteristic function. This method yields more accurate results and has been commonly used for the DLL performance analysis. In this chapter, the DLL analysis is performed using this method.

4.1.3 A Review of DLL Techniques

The delay-locked loop has been used in DS/SS receivers from the primitive period of spread spectrum communications. Many DLL techniques have been suggested in past years to improve the tracking performance. Most of them focused on the DLL structure and improved the performance of the DLL under some predefined operating conditions.

The first technique suggested in the DLL was the tau-dither loop proposed by Hartman [6]. In the conventional DLL, the incoming signal is correlated with the early and late versions of the locally generated spreading signals. Thus, it requires two correlators and the gain imbalance of these two correlator branches causes a tracking offset, which increases the bit-error-rate (BER). The dithering technique uses a single correlator in which the incoming signal is alternatively correlated with the early and late local spreading signal. Hartman compared the performances of the conventional DLL and dithering loop and showed that the dithering loop requires an additional SNR (about 3 dB) in order to have the same tracking error [6]. Following this, Hopkins [7] proposed a double-dither non-coherent DLL structure. This scheme solved the gain imbalance problem at the expense of increased hardware complexity. This structure has the same loop noise power as that of the classical non-coherent DLL and hence there is no performance degradation.

The tracking performance of the DLL can be improved by reducing the loop noise power. Based on this phenomena, another non-coherent DLL technique was proposed by Yost and Boyd [10]. They suggested an equivalent structure to replace the squaring operations in the non-coherent DLL. In the non-coherent DLL, the received signal is correlated with the local early and late despreading signal, squared and then subtracted to produce the error signal which is used to correct the code phase error. In this structure [10], the early and late correlated signals are multiplied by the despreaded incoming signal which is already available in the receiver. Thus, the squaring operation is avoided and the noise power that goes into the DLL is reduced. They also showed that the tracking error of this scheme is considerably less than that of the conventional non-coherent DLL or tau-dither loop.

Even though the tracking error is the prime performance measure of the DLL, the MTLL is also an imperative measure under transient operating conditions. Larger tracking range of the DLL reduces the possibilities of loss of lock. Bowles [11], Wakabayashi et al [12] proposed a structure called "Extended Range DLL". The tracking range of the coherent DLL was extended by adding an additional correlator branches. Note that only two correlator branches (early and late) are used in the conventional DLL. They also showed that the tracking range of the DLL is proportional to the number of correlator branches used. However, the noise power that affects the DLL is also increased as the number of correlators increases.

This problem was solved by Wilde [13-15]. He proposed a technique to reduce the loop noise power of the "Extended Range DLL". He compares all the correlator branch outputs and select only two largest outputs to generate the error signal. Thus, this scheme has the same loop noise power as the conventional system by selecting only two correlator outputs at a time. This results in the same tracking error for both conventional and the scheme described in [14-15] while the later one has a wider tracking range. Wilde [27-28] also showed that the "Extended Range DLL" has larger MTLL compared to that of the conventional DLL because of its wider tracking range. Another version of this technique was described in [15]. In the proposed structure, all correlator outputs are linearly combined to generate the error signal. Wilde obtained different S-curves by using different linear combinations. Improved performances were observed as the number of correlator branches increases at the expense of the hardware complexity. This work also demonstrated the dependency of tracking performance on the DLL characteristic function.

Wilde [29] also proposed another modified DLL which uses a single correlator to generate the error signal. A DLL characteristic function must have a positive slope in its origin in order to drive the phase error to zero. The auto-correlation function of PN codes also has a positive slope on the left side of its peak. He used this phenomena to generate the error signal by shifting the auto-correlation function appropriately. This proposed structure reduces the hardware complexity at the expense of reduced MTLL and increased tracking error variance compared to that of the conventional DLL with one chip time early-late spacing.

The DLL techniques discussed so far have fixed S-curves. However, Doppler effects may cause the DLL to loose lock frequently. The loss of lock requires reacquisition which is a time consuming process and results in the loss of data. Welti and Bernhard [30] proposed a modified DLL technique with dynamic S-curve to track signals subject to Doppler effects, when the classical DLL failed. The proposed DLL has three correlator arms that can be switched adaptively to other code positions according to predefined criteria based on the estimate of the power level in each correlator arm. The early, on-line and late versions of the local PN codes are correlated with received signals in each arm and the switching of shifted codes are done in such a way that the correlator arm with on-line PN code gives the maximum power level. Other two correlator outputs are used to generate the loop error signal.

All DLL techniques described so far suffer severe performance degradation in a multipath environment. Multipath signal causes a tracking offset in the DLL which requires additional signal power to maintain the required BER. Note that the tracking offset caused by multipath signals are independent of transmitted signal power. Van Nee [25-26] showed that the DLL with smaller early-late spacing is more robust to tracking offset in multipath environments. However the power in the multipath signals have not been utilized in this technique. Sheen and Stuber [22] proposed a DLL structure which utilizes the inherent multipath diversity in frequency-selective fading channels. The channel is modeled by a tapped delay line (chip time spaced) with complex tap gains. The tap gain and phase of each individual path were locally estimated. The proposed scheme performs better than the conventional DLL.

4.1.4 Motivation

The quality of CDMA receivers depends on the tracking performance of the DLL [8-9]. The performance of the DLL has been analyzed in many papers. Substantial work has been done in the past to enhance the performance by modifying the DLL structure, for example [10-16]. However most of these modifications focus on the structure of the DLL, which changes the S-curve, for example [15]. The S-curve of the DLL is highly dependent on the locally generated chip waveform since it is obtained as the difference of the early and late correlation functions. These papers analyse the DLL performance by considering the normalized (chip time) cross-correlation function $R(\tau)$ of the incoming and locally generated spreading codes as a triangular function, where $R(\tau) = 1 - |\tau|$ for $|\tau| \leq 1$ and $R(\tau) = 0$, otherwise. In practice, this happens only if the received and locally generated chip waveforms are rectangular in shape.

The employment of non-rectangular spreading chip waveforms for DS/CDMA transmitted signals has been discussed in [17-18]. Recently, weighted despreading chip waveforms, while the spreading chip waveform is the conventional rectangular, have been proposed [19] for DS/CDMA systems. The purpose is to optimize the shape of the chip waveforms in the receiver to reduce the multiple access interference (MAI) and to increase the system capacity. In this Chapter, the effect of non-rectangular despreading chip waveforms for locally generated early and late despreading codes in a coherent DLL for DS/SS systems is investigated. The transmitted signal is assumed to be conventional with a rectangular spreading chip waveform. The waveforms considered include half sine and triangular and are assumed to be time limited to the chip duration which causes no inter-chip interference. The code tracking loop considered here is a first order DLL with an arbitrary early-late spacing.

4.2 SYSTEM MODELS

In the following, a coherent DLL is considered and signals are analyzed in the baseband. The block diagram of a DLL in the baseband is given in Fig. 4.1. The baseband received signal $r(t)$ in a DS/SS system can be expressed as

$$r(t) = \sqrt{P}d(t)c(t)\cos(\eta) + n(t) \tag{4.1}$$

where P is the received signal power, $d(t)$ is a binary data stream, $c(t)$ is the data-synchronous PN code sequence, η is the phase difference between the locally generated and the incoming carrier signals, and $n(t)$ is the additive white noise with double sided power spectral density of $N_o/2$. The spreading modulation is assumed to be BPSK (binary phase-shift keying) with conventional rectangular chip spreading waveform. The early-late spacing of the DLL is denoted by $2\Delta T_c$ and the parameter Δ $(0 < \Delta \leq 1)$ is considered as an arbitrary value. Thus, the early and late despreading sequences of the DLL are separated by Δ chip time from the despreading sequence which is used for data demodulation.

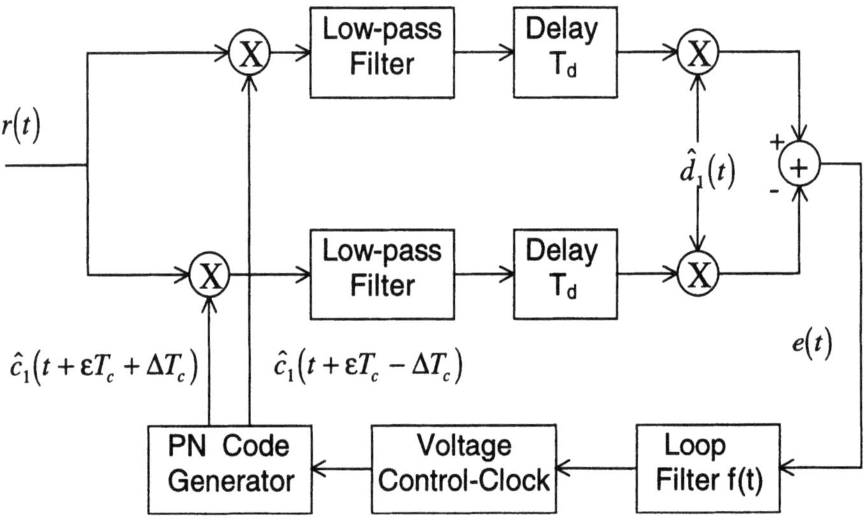

Fig. 4.1 The block diagram of the coherent DLL in baseband.

In the conventional DLL, the early and late despreading sequences are obtained by the shifted versions of the locally generated $c(t)$, denoted by $\hat{c}(t)$. Conventionally, the chip waveform of $\hat{c}(t)$ is $p(t)$ where $p(t) = 1$ for $0 \leq t < T_c$ and $p(t) = 0$ elsewhere. Now consider (i) a half sinusoidal pulse $\hat{p}_s(t)$ and (ii) a triangular pulse $\hat{p}_t(t)$, instead of the commonly used rectangular pulse $p(t)$. Fig. 4.2 shows the chip arrangements of single early and late PN pulses with an incoming PN pulse for $\Delta = 1/2$ when the received and the locally generated PN sequences are aligned with negligible phase error. We use the subscript s and t to denote the corresponding parameter of half-sine and triangular chip waveforms respectively, i.e.,

$$\hat{p}_s(t) = \begin{cases} \sin\left(\dfrac{\pi t}{T_c}\right), & 0 \le t < T_c \\ 0, & \text{otherwise} \end{cases} \tag{4.2}$$

and

$$\hat{p}_t(t) = \begin{cases} 1 - \left|\dfrac{2t}{T_c} - 1\right|, & 0 \le t < T_c \\ 0, & \text{otherwise} \end{cases} \tag{4.3}$$

For the triangular chip waveform case, the maximum value of $\hat{p}_t(t)$ is fixed at the middle $(t = T_c/2)$ of the chip waveform. Otherwise, two different chip wave forms for the early and the late despreading signals will

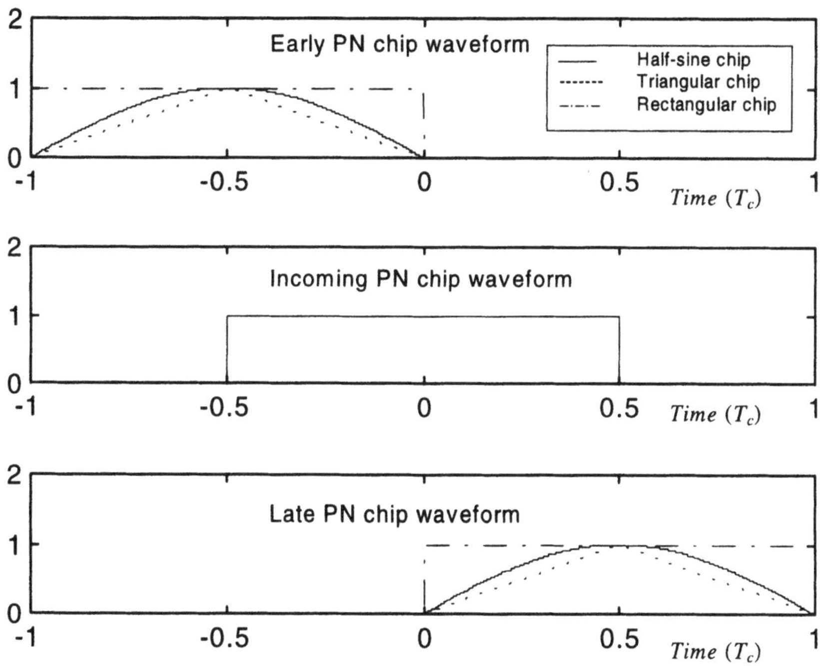

Figure 4.2 The early, late and incoming PN pulses.

be needed and the reason will be given in Section 4.3. Since the chip waveforms are restricted to the duration T_c, there is no inter-chip interference in the local PN signals. These chip waveforms, $\hat{p}_s(t)$ and $\hat{p}_t(t)$, are normalized by limiting their maximum value to 1. We assume that the interval of each data bit in $d(t)$ is T_d and is equal to the period of the despreading sequence NT_c where N is the PN sequence length.

4.3 PERFORMANCE ANALYSIS

The effect of data signal $d(t)$ in the coherent code tracking loop is removed by multiplying (decision-feedback) the early and late signals with the estimated binary data stream $\hat{d}(t)$ as shown in Fig. 4.1. The correlated signals at the early and late branches of the DLL are delayed by a data-bit time T_d in order to cancel the delay introduced by $\hat{d}(t)$ due to the demodulation process. A similar technique, which cancels the effect of data signal in a coherent DLL, can be found in [20]. In perfect conditions, the average value of $d(t)\hat{d}(t)$ denoted by $\overline{d(t)\hat{d}(t)}$ is 1. In practical situations, the value of $\overline{d(t)\hat{d}(t)}$ will be less than 1 due to both the estimation error in $\hat{d}(t)$ and the effects of filtering. It is clear from (4.1) that the parameters $\overline{d(t)\hat{d}(t)}$ and $\cos(\eta)$ can be grouped with the received signal power P. Note that both terms $\overline{d(t)\hat{d}(t)} \le 1$ and $\cos(\eta) \le 1$. Therefore, the term $\sqrt{P}\left\{\overline{d(t)\hat{d}(t)}\right\}\cos(\eta)$ can be considered as a degradation on the received signal power P of the desired user. This degradation is independent of the local despreading chip waveform used in the tracking loop. Thus the performance degradation of the coherent DLL due to $\overline{d(t)\hat{d}(t)}$ and $\cos(\eta)$ can be separately expressed in terms of the reduction in SNR. From this point onwards, we shall consider the resultant received signal power $P_o = P\left\{\overline{d(t)\hat{d}(t)}\right\}^2 \cos^2(\eta)$.

4.3.1 Stochastic Differential Equation

The cross-correlation function $R(\tau)$ of the incoming and the locally generated PN sequences is given by

$$R(\tau) \triangleq \frac{1}{NT_c} \int_0^{NT_c} c(t)\hat{c}(t + \tau T_c)dt. \tag{4.4}$$

We assume that the PN sequence is very long so that $R(\tau)$ vanishes when $|\tau| \geq 1$. Combining (4.2) and (4.3) with (4.4), the respective cross-correlation functions are obtained as

$$R_s(\tau) = \begin{cases} \frac{1}{\pi}\{1 + \cos(\pi\tau)\}, & |\tau| < 1 \\ 0, & \text{otherwise} \end{cases} \tag{4.5}$$

and

$$R_t(\tau) = \begin{cases} \frac{1}{2} - |\tau|^2, & |\tau| < \frac{1}{2} \\ (1 - |\tau|)^2, & \frac{1}{2} \leq \tau < 1 \\ 0, & \text{otherwise} \end{cases} \tag{4.6}$$

The cross-correlation functions $R_s(\tau)$ and $R_t(\tau)$ are symmetric due to the symmetric chip waveforms $\hat{p}_s(t)$ and $\hat{p}_t(t)$. If the triangular chip waveform $\hat{p}_t(t)$ is not symmetric, the cross-correlation function $R_t(\tau)$ will not be symmetric. It follows that the discriminator characteristic function of the DLL, generated by the shifted versions of $R_t(\tau)$ used as the early and late cross-correlation functions, will introduce a tracking offset. Alternatively, we can use two different chip waveforms $\hat{p}_{t+}(t)$ and $\hat{p}_{t-}(t)$, which are mirror images of each other for the early and late despreading sequences. We also observe that the slope of $R(\tau)$ is maximum for both triangular and half sine chip waveforms at $\tau = \pm 0.5$.

The delay-locked discriminator output in the absence of white noise is obtained by the difference of the early and late cross-correlation functions and it is given by $\sqrt{P_o}S(\varepsilon)$, where $S(\varepsilon)$ is the discriminator S-curve [9] and ε is the phase error normalized to T_c.

$$S(\varepsilon) \;=\; R(\varepsilon-\Delta)-R(\varepsilon+\Delta). \tag{4.7}$$

Let $S'(\varepsilon)=dS(\varepsilon)/d\varepsilon$. The slope at the origin given by $S'(0)$ plays a major role in the performance of the DLL as we shall see in Section 4.3.2. Better performance of the DLL can be achieved by obtaining larger value of $S'(0)$. From (4.7), the slope $S'(0)$ can be expressed as $R'(-\Delta)-R'(+\Delta)$ where $R'(\tau)=dR(\tau)/d\tau$. It is now clear from (4.5) and (4.6) that the maximum values of the chip waveforms, $\hat{p}_s(t)$ and $\hat{p}_t(t)$, in (4.2) and (4.3) at $t=T_c/2$ achieves the maximum slope $S'(0)$ for the early-late spacing $\Delta=1/2$. The plots of S-curves $S(\varepsilon)$ for the three chip waveforms, rectangular, half sine, and triangular, are presented in Fig. 4.3 for $\Delta=1/2$. We note that the slope of $R(\tau)$ at $\tau=\pm 1/2$ is the same for all three chip waveforms, as is that of $S(\varepsilon)$ at $\varepsilon=0$. We also observe that the linear region of the S-curves $S(\varepsilon)$ for half-sine and triangular chip waveforms are substantially reduced compared to that of rectangular pulses. Therefore, a change in threshold values, where the results of linear and non-linear analyses differs, can be expected. This also makes the use of nonlinear analysis to evaluate the DLL performances more appropriate. Referring to Fig. 4.1, the error signal $e(t)$ can be expressed as:

$$e(t) \;=\sqrt{P_o}S(\varepsilon) \;+\; n_T(t) \tag{4.8}$$

where the total noise $n_T(t)$ is given by

$$n_T(t) \;=\; \hat{d}(t)\left\{\hat{c}(t+\varepsilon T_c-\Delta T_c)-\hat{c}(t+\varepsilon T_c+\Delta T_c)\right\}n(t) \tag{4.9}$$

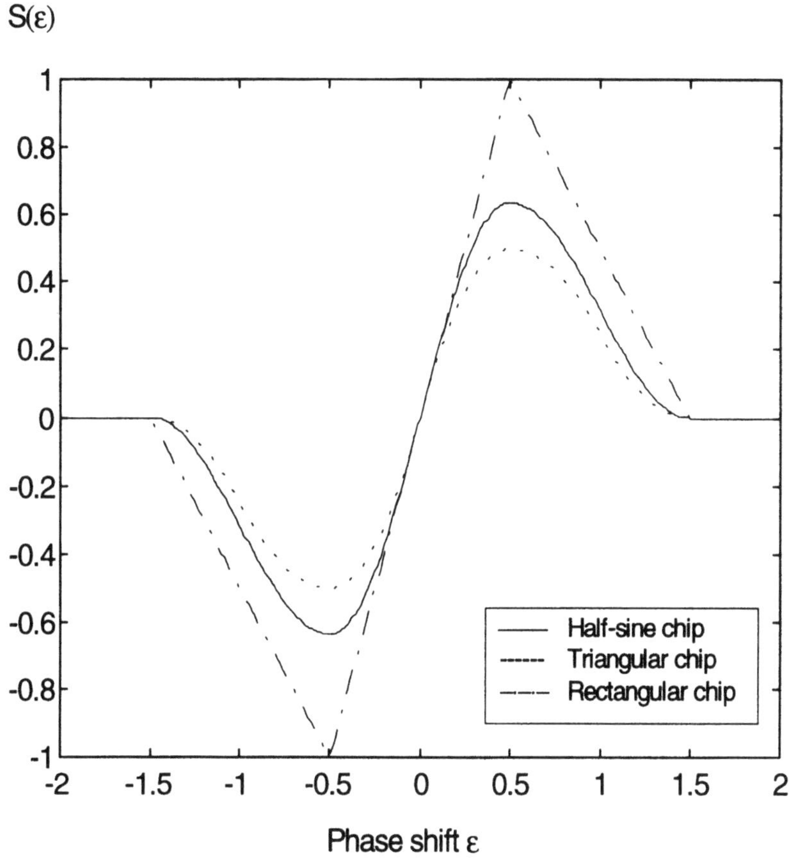

Fig. 4.3 The plots of S-curves for chip waveforms.

The code self-noise term has been neglected in deriving (4.8). As shown in Fig. 4.1, the error signal at the output of the loop filter drives the voltage controlled oscillator (VCO), shown as PN code generator, to correct the code phase error. The operation of the VCO is described by the following equation [9]

$$\varepsilon(t) \quad = \quad -K\int_0^t f(z)\otimes e(z)dz \tag{4.10}$$

where $f(t)$ is the impulse response of the loop filter, K is the VCO sensitivity gain, and $\otimes$ denotes convolution. Combining (4.8) and (4.10), the stochastic differential equation which describes the dynamic behaviour of the tracking loop is obtained:

$$\frac{d\varepsilon}{dt} = -K\left\{\sqrt{P_o}S(\varepsilon)+n_T(t)\right\}\otimes f(t). \tag{4.11}$$

In deriving (4.11), we assumed that the channel is slowly time varying such that the Doppler spread can be neglected. A first-order code tracking loop is considered in the following analysis and therefore $f(t)=\delta(t)$ where $\delta(t)$ is a unit impulse function given by $\delta(t)=\infty$ for $t=0$ and $\delta(t)=0$ otherwise.

4.3.2 Linear Analysis

The RMS tracking error for the DLL can be obtained by linear loop analysis and details of this analysis for conventional system can be found in [9, Chapter 4]. The results so obtained are generally considered as a good approximation of the actual performance in the region of SNR which we are interested in. The high SNR assumption will result in smaller tracking error for the DLL and this enables us to approximate the S-curve $S(\varepsilon)$ as linear in the vacinity of $\varepsilon \approx 0$. Thus, the S-curve $S(\varepsilon)$ is substituted by $S'(0)\varepsilon$ for all three chip waveforms and (4.11) can be rewritten for small tracking errors as

$$\frac{d\varepsilon}{dt} = -K\left\{\sqrt{P_o}S'(0)\varepsilon+n_T(t)\right\}. \tag{4.12}$$

The auto-correlation function of the white noise $n(t)$ is a delta function, denoted by $\delta(\tau)$, and so is the autocorrelation function of the total noise process $n_T(t)$ [9]. Note that the noise process $n_T(t)$ is white but not Gaussian, as shown by

$$R_{n_T}(\tau) = E\left\{n_T(t)n_T(t+\tau T_c)\right\} = \left\{R_{\hat{c}}(0)-R_{\hat{c}}(2\Delta)\right\}N_0\delta(\tau) \tag{4.13}$$

where $R_{\hat{c}}(\tau)$ is the normalized auto-correlation function of locally generated PN code $\hat{c}(t)$. In deriving (4.13), the expectations $E\left\{\hat{d}^2(t)\right\}=1$ is used. Denote the power spectral density of the total noise process $n_T(t)$ by $S_{n_T}(f)$ and taking the Fourier transform of $R_{n_T}(\tau)$ gives $S_{n_T}(f)$ which is flat with density of $S_{n_T}(0)$, equal to $\left\{R_{\hat{c}}(0)-R_{\hat{c}}(2\Delta)\right\}N_0$. The closed loop transfer function of the DLL with first order loop is

$$H(s) = \frac{\sqrt{P_o}S'(0)K}{s+\sqrt{P_o}S'(0)K}. \tag{4.14}$$

The closed loop transfer function and the closed loop bandwidth are varying with the early-late spacing for all three chip waveforms due to the change in $S'(0)$. The closed loop bandwidth of the DLL is given by

$$W_L = \int_0^\infty |H(j2\pi f)|^2 df = \frac{\sqrt{P_o}S'(0)K}{4}. \tag{4.15}$$

Thus, the rms tracking error (σ) of the DLL can be obtained using linear loop analysis for non-bandlimited received signals:

$$\sigma = \sqrt{\frac{2\{R_{\hat{c}}(0)-R_{\hat{c}}(2\Delta)\}W_L T_d}{S'(0)^2}\frac{N_o}{E_b}}. \tag{4.16}$$

where $E_b/N_o = P_o T_d/N_o$. We note that the linear analysis is valid only under normal operating conditions where the received SNR is high enough. Under these circumstances the MTLL is not of much interest since it is much larger than the transmission time.

4.3.3 Non-linear Analysis

Nonlinear analysis is found useful when the system operates at threshold or below. In these situations, the MTLL is the prime consideration. The MTLL is defined by the time taken for the timing error ε to reach one of its boundaries either $\varepsilon = \varepsilon_{\min}$ or $\varepsilon = \varepsilon_{\max}$ from the stable equilibrium point $\varepsilon = 0$. Since we consider arbitrary early-late spacing, the range of the S-curve becomes $\varepsilon_{\min}$ to $\varepsilon_{\max}$. The average behaviour of the system is characterised by the stationary probability density function (pdf) $p(\varepsilon)$ of the phase error ε. We follow the analysis given in [21-22] in which an aperiodic phase error characteristic function of the DLL determines $p(\varepsilon)$. In practical situations where the bandwidth of the noise process $n_T(t)$ is much greater than the closed loop bandwidth W_L, the function $G(\varepsilon)$ satisfies the following Fokker-Plank equation [21-23].

$$\frac{\partial}{\partial \varepsilon}\{K_1(\varepsilon)G(\varepsilon)\}-\frac{1}{2}\frac{\partial^2}{\partial \varepsilon^2}\{K_2(\varepsilon)G(\varepsilon)\} = \pi(\varepsilon) \tag{4.17}$$

where $\pi(\varepsilon)$ is the initial pdf of the phase error ε when the acquisition unit hands over to the tracking unit and the parameters $K_1(\varepsilon)$ and $K_2(\varepsilon)$ are defined by

$$K_1(\varepsilon) = -K\sqrt{P_o}S(\varepsilon), \tag{4.18}$$

$$K_2(\varepsilon) = K^2\int_{-\infty}^{+\infty} R_{n_T}(\delta)d\delta = K^2 S_{n_T}(0). \tag{4.19}$$

In (4.18) and (4.19), $S(\varepsilon)$ and $S_{n_T}(0)$ depend on the chip waveform and hence $K_1(\varepsilon)$ and $K_2(\varepsilon)$ depend on the choice of chip waveform. The in-lock region of the loop is denoted by $\Re \underline{\underline{\Delta}}(\varepsilon_{\min}, \varepsilon_{\max})$ in which the acquisition unit hands over to the tracking unit for fine synchronisation. The function $G(\varepsilon)$ in (4.17) can be shown to be an even function, $G(\varepsilon) = G(-\varepsilon)$ for $\varepsilon_{min} \le \varepsilon \le 0$, using $\varepsilon_{min} = -\varepsilon_{max}$, $K_1(\varepsilon)$ and $K_2(\varepsilon)$ are odd and even functions of ε [21,23]. When the SNR is low, the lock detector indicates the true lock condition when sufficient code correlation is present. This happens near the zero offset position and with very low probability elsewhere. Therefore, a suitable expression for $\pi(\varepsilon)$ is a delta function [21, pp. 33]. Furthermore, according to the MTLL definition, the first passage of ε starts from $\varepsilon = 0$ and therefore the solution of the differential equation in (4.17) with $\pi(\varepsilon) = \delta(\varepsilon)$ can be used to evaluate both rms tracking error and MTLL. Assuming $\pi(\varepsilon) = \delta(\varepsilon)$, the following expression for $G(\varepsilon)$ is obtained [14]:

$$G(\varepsilon) = \frac{e^{-U(\varepsilon)}}{K_2(\varepsilon)}\int_{\varepsilon}^{\varepsilon_{\max}} e^{U(z)}dz \quad 0 \le \varepsilon \le \varepsilon_{\max} \tag{4.20}$$

where

$$U(\varepsilon) = -2\int_{\varepsilon_{\min}}^{\varepsilon} \frac{K_1(z)}{K_2(z)}dz. \tag{4.21}$$

The probability density function $p(\varepsilon)$ is related to the MTLL, which is denoted by $\overline{T}$, and the function $G(\varepsilon)$ by:

$$p(\varepsilon) \;=\; \{1 - P_{LD}(\varepsilon)\}\frac{G(\varepsilon)}{\overline{T}}, \quad \varepsilon_{\min} \le \varepsilon \le \varepsilon_{\max}. \tag{4.22}$$

where $P_{LD}(\varepsilon)$ is the probability that the tracking unit declares out of lock when ε is in the region of $\Re$ and it is assumed that $P_{LD}(\varepsilon_{\min}) = P_{LD}(\varepsilon_{\max}) = 1$, i.e., the boundaries are absorbing. The following expressions for MTLL ($\overline{T}$) and rms tracking error (σ) are derived using (4.20) and (4.22):

$$\overline{T} \;=\; \int_{\varepsilon_{\min}}^{\varepsilon_{\max}} \{1 - P_{LD}(\varepsilon)\} G(\varepsilon) d\varepsilon \tag{4.23}$$

and

$$\sigma^2 \;=\; \int_{\varepsilon_{\min}}^{\varepsilon_{\max}} \varepsilon^2 p(\varepsilon) d\varepsilon. \tag{4.24}$$

By assuming a suitable function for $P_{LD}(\varepsilon)$, the solutions of (4.21), (4.20), (4.23), (4.22) and (4.24) are obtained in the respective order by using numerical integration.

4.4 BAND-LIMITED RECEIVED SIGNAL

In practice, all received signals are bandlimited. Generally, the transmission channel consists of a transmitter filter, a receiver filter, and a propagation medium. The equivalent transfer functions of these are denoted by $H_T(f)$, $H_R(f)$, and $H_M(f)$, respectively. As a result, the equivalent low-pass transfer function of the transmission channel, $H_C(f)$, becomes

$$H_C(f) = H_T(f) H_M(f) H_R(f) \tag{4.25}$$

Note that the transfer function of the propagation medium depends on the channel environment. We assume that the propagation medium is ideal and therefore $H_M(f) = 1$. This assumption can be justified since our intention is to study the effect of band-limiting on the received signal. The impulse response of the transmission channel is denoted by $h(t)$ and this is the inverse Fourier transform of $H_C(f)$. In practice, due to the influence of $H_C(f)$, the received chip waveform $\overline{p}(t)$ will not be limited in the range of $0 \le t < T_c$. As a result, the expression for the cross correlation $R(\tau)$ given in (4.4), which is still valid, is not limited to $|\tau| < 1$. The band-limited received signals are modelled by assuming ideal low pass transfer function for $H_C(f)$. Even though this is a rough approximation of $H_C(f)$, it serves the purpose of showing the effects of band-limiting on the received signals. For illustration purpose, assume

$$H_C(f) = \begin{cases} 1, & |f| \le \dfrac{2}{T_c} \\ 0, & \text{otherwise} \end{cases} \tag{4.26}$$

i.e., the bandwidth of the channel is assumed to be twice that of the main lobe spectrum of the transmitted PN sequence. Using the expression for $h(t)$ of (4.26), the received PN sequence $c(t) \otimes h(t)$, denoted by $\overline{c}(t)$, has the chip waveform $\overline{p}(t)$ where $\overline{p}(t) = p(t) \otimes h(t)$. Replacing $\overline{p}(t)$ for $p(t)$, $R(\tau)$ in (4.4) for band-limited received signals is obtained.

In the case of linear analysis, the band-limiting on the received signals affects the parameter $S'(0)$ in the rms tracking error expression given in (4.16). The parameter $S'(0)$ is computed using numerical integration. For the nonlinear analysis, the S-curve $S(\varepsilon)$ in (4.7) has to be obtained using $\overline{c}(t)$. Thus, replacing new $K_1(\varepsilon)$ in (4.18), the MTLL and the tracking error in (4.23) and (4.24) are computed using numerical integration. We note that the change in $S'(0)$ due to band-limitation will also cause variation in the loop transfer function $H(j2\pi f)$ in (4.14) and the closed loop bandwidth W_L in (4.15) for both linear and nonlinear analyses. In practice, $H(j2\pi f)$ and W_L can be adjusted by varying the sensitivity gain K of the voltage-controlled clock in the DLL.

4.5 EXAMPLES

When the received signal is not band-limited, the parameter $S'(0)$, which is given by $R'(-\Delta)-R'(+\Delta)$, can be obtained by differentiating the cross-correlation functions $R_s(\tau)$ in (4.5) and $R_t(\tau)$ in (4.6) with respect to the phase shift τ. For band-limited received signals, the values of $S'(0)$ are obtained by numerical integration for all chip waveforms. The auto-correlation values $R_{\hat{c}}(\tau)$ for rectangular and half-sine chip waveforms are $1-\tau$ and $\{\pi(1-\tau)\cos(\pi\tau)+\sin(\pi\tau)\}/2\pi$ respectively when $0\le\tau\le1$. The values of $R_{\hat{c}}(\tau)$ for triangular chip waveform is $1/3-2\tau^2+2\tau^3$ for $0\le\tau\le1/2$ and $2(1-\tau)^3/3$ for $1/2<\tau\le1$. The value of $R_{\hat{c}}(\tau)$ is zero for all chip waveforms when $\tau>1$. The value of $W_L T_d=0.1$ is used in the following numerical examples.

Following the procedures explained in Section 4.3.3, the MTLL and the rms tracking errors are obtained for all three chip waveforms using the non-linear theory with the assumption

$$P_{LD}(\varepsilon) = \begin{cases} 0, & |\varepsilon|<1 \\ 1, & \text{otherwise} \end{cases} \tag{4.27}$$

When the phase error ε of the incoming and locally generated PN sequence exceeds 1, the data demodulation cannot be performed since the cross-correlation value $R(\tau)$ in (4.4) vanishes. The lock detector therefore may declare the loop out of lock. Thus, the assumption made in (4.27) is clearly justifiable when the lock detector has no faulty triggering.

4.5.1 Early-Late Spacing

The effect of the early-late spacing (2Δ) on the performance of the DLL is first investigated. Due to the wide range of Δ being considered, non-linear analysis yields more accurate results and is therefore used. Fig. 4.4 plots the rms tracking error against the early-late spacing for E_b/N_o equals to -5 and 15 dB. Figure 4.5 shows the MTLL against the early-late spacing for E_b/N_o equals to -5 and 0 dB. These graphs indicate firstly within the range of 0.7 to 1.3 chip time, the half-sine and triangular chip waveforms reduce tracking error and increase MTLL. Clearly, this range of early-late spacing depends on the shapes of local and incoming chip waveforms. Secondly, band-limitation affects the tracking performance for all three waveforms. In particular, rectangular waveform has an increase in tracking error for small Δ and decrease in MTLL for large Δ since the band-limitation changes the shape of incoming chip waveform. Thirdly, the conventional rectangular despreading waveform has better tracking performance when 2Δ is less than 0.7, however band-limitation severely degrades its performances as Δ decreases. From the acquisition point of view, one chip early-late spacing is preferred. It is also clear from Figs. 4.4 and 4.5 that the performance of the DLL with one chip early-late spacing would be improved by using triangular

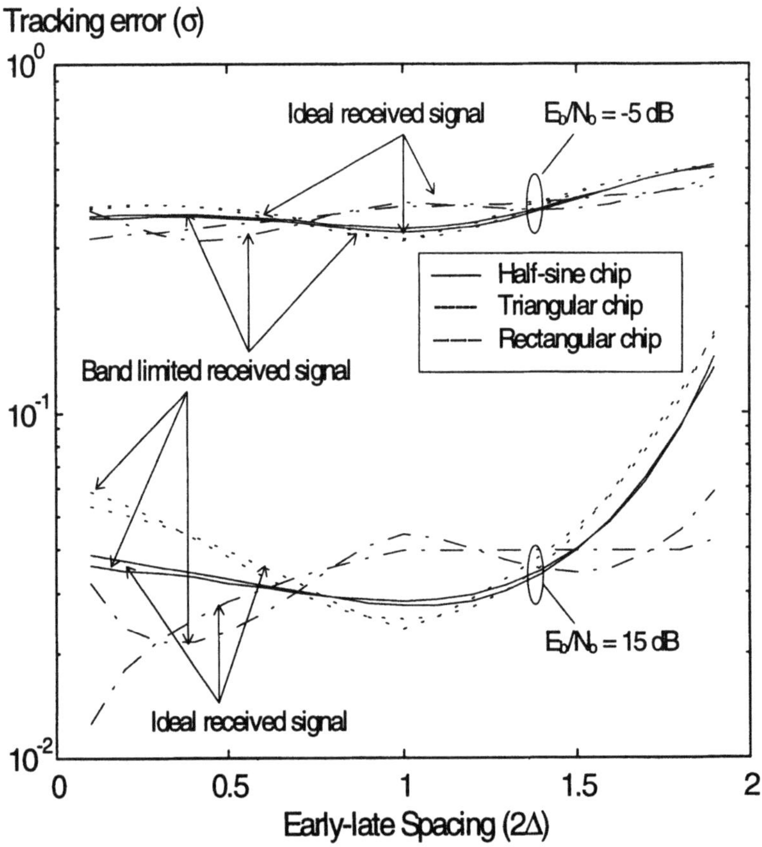

Fig. 4.4 The rms tracking error against the early-late spacing (2Δ).

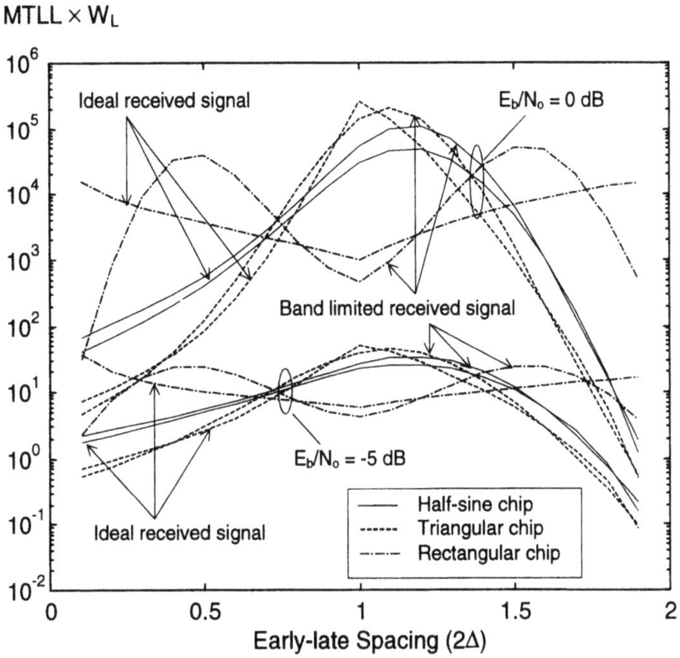

Fig. 4.5 The MTLL against the early-late spacing (2Δ).

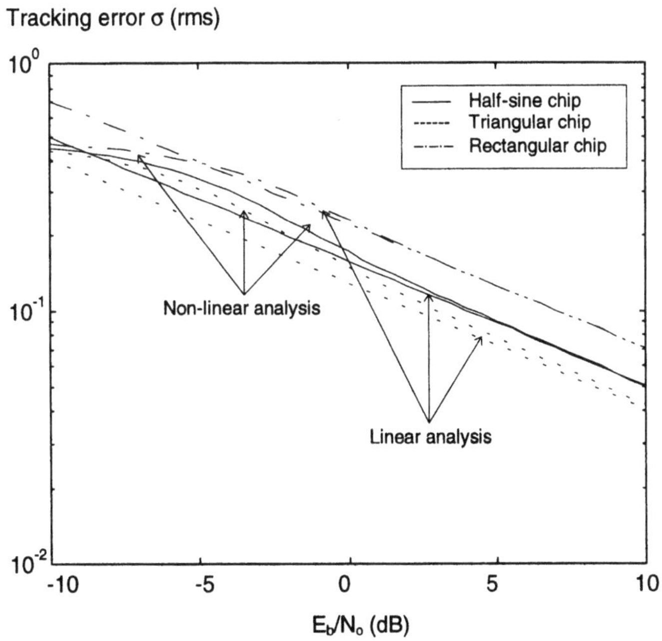

Fig. 4.6 The rms tracking error against E_b/N_o for ideal signals.

and half-sine chip waveforms. In the following section, we shall evaluate the performance of the DLL with commonly used $\Delta = 1/2$.

4.5.2 DLL with One Chip Time Early-Late Spacing

The rms tracking error of the DLL, obtained by both linear and non-linear analysis, against E_b/N_0 is plotted for the three cases in Fig. 4.6 for ideal received signal and in Fig. 4.7 for band-limited received signals. These results indicate good agreement between linear and non-linear analysis under the conditions considered. The triangular chip waveform has the smallest tracking error among the three chip waveforms. Both ideal and band-limited signals give the same conclusions. It is clear from Fig. 4.6 that for a specific tracking error, triangular and half sine chip waveforms reduce the required E_b/N_0 by approximately 4.77 dB and 3 dB, respectively. However, from Fig. 4.7, band-limitation of the received signal requires an increase of E_b/N_0 by approximately 1.2 dB for the rectangular and 0.9 dB for the triangular chip waveforms. The rms tracking error for the half-sine chip waveform does not seem to be affected much by band-limiting.

Fig. 4.8 shows the curves of MTLL for both ideal and band-limited received signals. For a specific E_b/N_0, the triangular chip waveform has the longest MTLL while the rectangular chip waveform has the lowest MTLL among the three chip waveforms. A band-limited received signal reduces the MTLL as expected. However, the effect of the half-sine chip waveform is minimal.

4.5.3 Simulations

The DLL scheme using the three different despreading waveforms was simulated in a computer for band-limited received signal using a maximal-length sequence with $N = 255$. The rms tracking error is obtained by using 10,000 sample points for each E_b/N_0 at -10 dB, -5 dB, 0 dB, 5 dB, and 10 dB. The results are marked in Fig. 4.7 and they show a close agreement with that obtained analytically. The simulated results for MTLL using the mean of 1000 times to loose lock for E_b/N_0 at 1 dB intervals are also marked in Fig. 4.8. These simulated results for MTLL are higher than that obtained by analytical means. However, the differences among the MTLL by using different chip waveforms are approximately the same as that obtained analytically.

4.6 EFFECT OF MULTIPATH SIGNALS

It is well known that the DLL will introduce a tracking offset when multipath signals are present [14-16]. The following investigates the effect of tracking offset when the received signal consists of two rays, one direct and one reflected ray. Band-limiting of the received signals is not considered

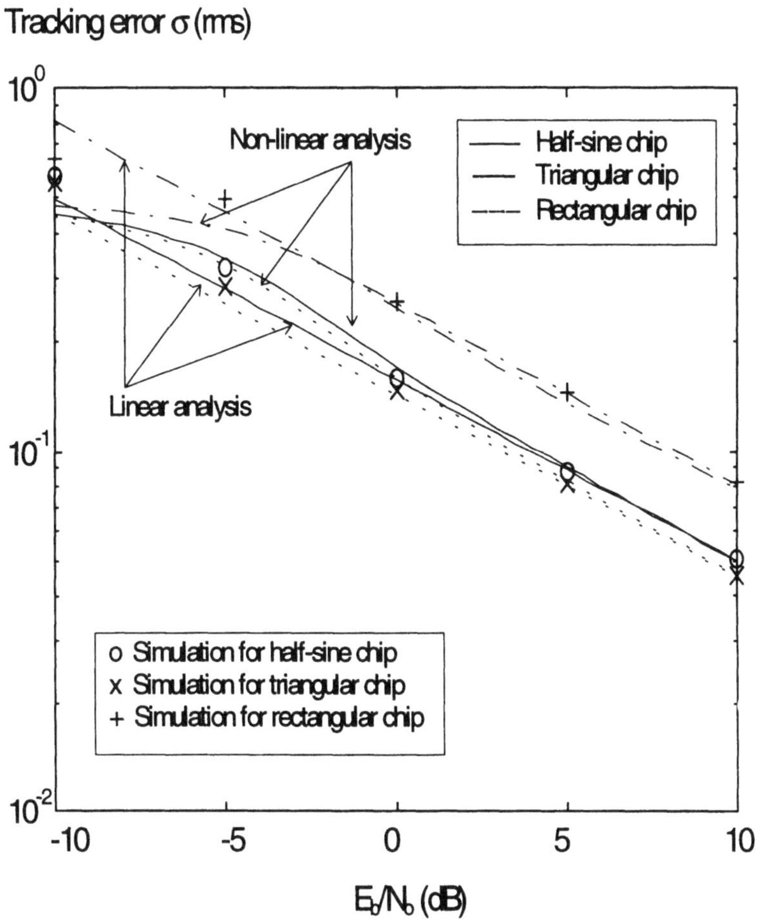

Fig. 4.7 The rms tracking error against E_b/N_o for band limited signals.

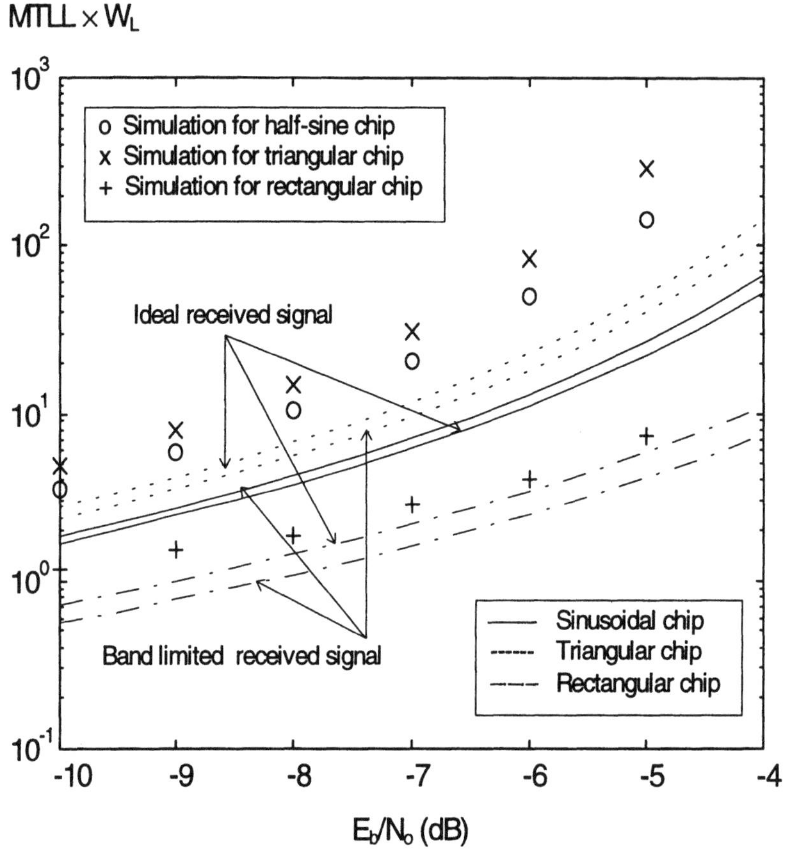

Fig. 4.8 The plots of MTLL against E_b/N_o.

in this analysis. The received signal in (4.1) for the two ray models can be expressed as

$$r(t) = \sqrt{P}\{d(t)c(t) + \alpha d(t-\beta T_c)c(t-\beta T_c)\cos(\zeta)\}\cos(\eta) + n(t) \qquad (4.28)$$

where α is the attenuation, β is the normalised (T_c) time delay, and ζ is the carrier phase difference of the reflected ray with respect to the direct ray. Note that the values of $\overline{d(t)\hat{d}(t)}$ and $\overline{d(t-\beta T_c)\hat{d}(t)}$ are approximately the same since the delay βT_c is negligible compared with the PN code length NT_c. Thus, the new S-curve of the DLL $S_m(\varepsilon)$ becomes

$$S_m(\varepsilon) = S(\varepsilon) + \alpha S(\varepsilon+\beta)\cos(\zeta) \tag{4.29}$$

The maximum tracking offset occurs when the reflected ray is in phase or out of phase with the specular signal. In this case, the carrier phase estimate will not have any significant error since the carrier tracking loop tracks the sum of the specular and diffuse path signals [25]. The tracking offset for any value of phase difference ζ and delay β is the solution of $S_m(\varepsilon)=0$. The absolute value of the maximum tracking offset for the worst case of phase difference ζ and delay β is denoted by ε_o and this happens when the second term in (4.29) has the maximum value for all possible ζ and β. Therefore ε_o is the solution of the equation

$$S(\varepsilon) - \alpha \times \max|S(\varepsilon)| = 0 \tag{4.30}$$

For the three waveforms considered, we obtain $\varepsilon_o = \alpha/2$ for the rectangular chip waveform, $\varepsilon_o = \sin^{-1}(\alpha)/\pi$ for the half-sine chip waveform and $\varepsilon_o = \left(1-\sqrt{1-\alpha}\right)/2$ for the triangular chip waveform respectively. The plots of ε_o versus the attenuation α are given in Fig. 4.9. The curves show that the triangular chip waveform gives the smallest tracking offset, with the sine chip waveform coming closely behind for all values of α except for the points $\alpha = 0$ and $\alpha = 1$ where the three cases coincide. The tracking offset depends on the shape of the characteristic curve and it is changed by the different chip waveforms for early and late sequences.

4.7 DISCUSSIONS

Improvement in the DLL performance by using half-sine and triangular chip waveforms can be explained with reference to Fig. 4.2. Under normal operating conditions, the performance of the DLL are determined by the DLL sensitivity (slope of the S-curve at the origin) and the loop noise power. The energies of the early and late PN chips for both triangular and half-sine waveforms are concentrated around the zero crossings of the incoming PN chip waveform because of the one chip time early-late spacing. This results

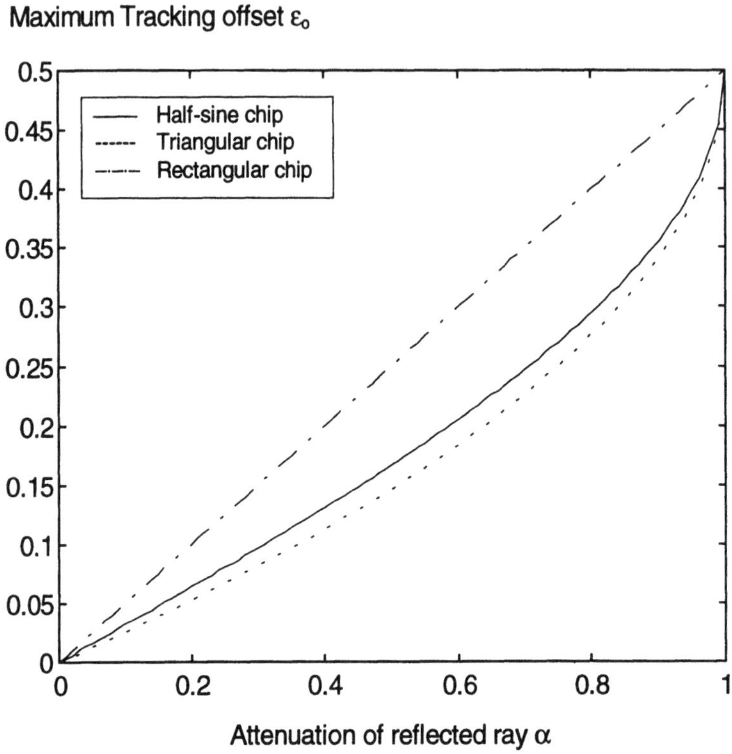

Fig. 4.9 The tracking offset ε_0 against attenuation (α).

in only minor effect on the DLL sensitivity compared to that of using rectangular waveform. On the other hand, the loop noise power is uniformly distributed within the chip time when rectangular waveform is used. The half-sine and triangular waveforms reduce the noise power on both sides of the zero crossings and hence reduce the effective loop noise power. Thus, the rms tracking error and the MTLL are respectively decreased and increased when the proposed chip waveforms are used. For band-limited signals, the energy distribution is reduced at the zero crossings of the incoming chip. This changes the DLL sensitivity and causes performance degradation.

The improved performance of the proposed DLL scheme may not have a significant improvement on BER of the system. However, the DS/SS applications which require more accurate timing, such as Global Positioning Systems, can be benefited from the proposed schemes. Also note that the digital implementation of the proposed schemes require at least three samples per chip.

For more performance analysis of the proposed DLL in CDMA environments, please refer to [31-32].

REFERENCES

[1] H. Meyr, "Nonlinear analysis of correlative tracking systems using renewal process theory," *IEEE Trans. Commun.*, vol. COM-23, Feb. 1975.

[2] H. Meyr, "Delay-lock tracking of stochastic signals," *IEEE Trans. Commun.*, vol.COM-24, Mar. 1976.

[3] W. C. Lindsey and H. Meyr, "Complete statistical description of the phase-error process generated by correlative tracking systems," *IEEE Trans. Inform. Theory*, vol. IT-23, Mar. 1977.

[4] A. Polydoros and C. L. Weber, "Analysis and optimization of correlative code-tracking loops in spread spectrum systems," Commun. Sci. Inst., Univ. Southern California, Los Angeles, Rep. CSI-82-11-05, Aug. 1982.

[5] A. Polydoros and C. L. Weber, "Analysis and Optimization of Correlative Code-Tracking Loops in Spread-Spectrum Systems," *IEEE Trans. Common.*, vol. COM-33, pp. 30-43, January 1985.

[6] H. P. Hartman, "Analysis of a Dithering Loop for PN Code Tracking," *IEEE Trans. Aerosp. Electron. Syst.*, Vol. AES-10, pp. 2-9, January 1974.

[7] P. M. Hopkins, "Double Dither Loop for Pseusonoise Code Tracking," *IEEE Trans. Aerosp. Electron. Syst.*, Vol. AES-13, pp. 644-650, November 1977.

[8] S. Haykin, *Communication Systems*, 3rd ed., New York, John Willy & Sons, Inc., 1994.

[9] R. L. Peterson, R. E. Ziemer and D. E. Borth, *Introduction to Spread Spectrum Communications*, Upper Saddle River, N.J, Prentice-Hall, 1995.

[10] R. A. Yost and R. W. Boyd, "A Modified PN Code Tracking Loop: Its performance Analysis and Comparative Evaluation," *IEEE Trans. Commun.*, vol. COM-30, pp. 1027-1036, May 1982.

[11] W. M. Bowles, "GPS Code Tracking and Acquisition Using Extended Range Detectors," Charles Stark Draper Lab., Inc., Apr.1979; see also *Proc. Nat. Telecommun. Conf.*, Houston, TX, Dec. 1980, pp. 24.1.1-24.1.5.

[12] Wakabayashi K., Nakagawa M. and Tsunogae T., "Tracking Performance of Improved Delay-Locked Loop," *Proc. Nat. Telecommun. Conf.*, Houston, TX, Dec. 1980.

[13] A. Wilde, "Modified Coherent PN-Code Tracking Dealy-Locked loop," *Proc. IEEE Intern. Symp. on Personal, Indoor, and Mobile Radio Commun.*, Den Haag, 1994.

[14] A. Wilde, "Extended Tracking Range Delay-Locked Loop," *Proc. Intern. Conf. on Commun.*, pp. 1051-1054, June, 1995.

[15] A. Wilde, "Delay-Locked Loop with generalized Detector Characteristic," *Proc. Intern. Symp. on Spread Spectrum Techn. and Applic.*, pp. 450-454, September, 1996.

[16] K. Jamaleddine, D. Vizireanu, F. Malassenet, and S. Halunga, "A New Delay Locked Loop Structure with Enhanced Tracking Behaviour," *Proc. of the 7th Intern. Symp. on personal, Indoor, and Mobile Commun.*, pp. 1116-1119, October, 1996.

[17] R. Anjaria and R. Wyrwas, "The Effect of Chip Waveform on the Performance of CDMA Systems in Multipath Fading Noisy Channels," *Proc. of the 42nd IEEE VTS Conf.*, pp. 672-675, May 1992.

[18] M. A. Landolsi and W. E. Stark, "DS-CDMA Chip waveform Design for Optimal Power-Bandwidth Performance," *Proc. IEEE Intern. Symp. on Personal, Indoor and Mobile Radio Commun.*, pp. 706-709, September, 1995.

[19] Y. Huang and T.S. Ng, "Performance of Coherent Receiver with Weighted Despreading Sequence for DS-CDMA," *Electron. Lett.*, Vol. 33, pp. 23-25, Jan. 1997.

[20] R. D. Gaudenzi and M. Luise, "Decision-Directed Coherent Delay-Lock Tracking Loop for DS-Spread-Spectrum Signals," *IEEE Trans. Commun.*, vol. COM-39, pp. 758-765, May 1991.

[21] A. Polydoros and C. L. Weber, "Analysis and Optimization of Correlative Code-Tracking Loops in Spread-Spectrum Systems," *IEEE Trans. Common.*, vol. COM-33, pp. 30-43, January 1985.

[22] W. H. Sheen and G. L. Stuber, "A New Tracking Loop for Direct Sequence Spread Spectrum Systems on Frequency-Selective Fading Channels," *IEEE Trans. Common*, vol. 43, pp. 3063-3072, December 1995.

[23] Jack K. Holmes and Leon Biederman, "Delay-Lock-Loop Mean Time to Lose Lock," *IEEE Trans. Commun.*, vol. COM-26, pp. 1549-1556, November 1978.

[24] W. H. Sheen and G. L. Stuber, "Effects of Multipath Fading on Delay Locked Loops for Spread Spectrum Systems," *IEEE Trans. Commun.*, vol. 42, pp. 1947-1956, March 1994.

[25] D. J. R. Van Nee, "Reducing Multipath Tracking Errors in Spread-Spectrum Ranging Systems," *Electron. Lett.*, Vol. 28, pp. 729-731, April 1992.

[26] D. J. R. van Nee, "Multipath and Interference in Spread-Spectrum Tracking Systems," *Proc. IEE colloq. on spread spectrum techn. for radio commun.*, pp. 13/1-13/5, London, 1992.
1996.

[27] A. Wilde and U. P. Bernhard, "Mean Time to Lose Lock for a Second-Order Extended Tracking Range Dealy-Locked loop," *Proc. IEEE Intern. Symp. on Personal, Indoor, and Mobile Radio Commun.*, Toronto, 1995.

[28] A. Wilde, "Mean Time to Lose Lock for a Coherent Extended Tracking Range Dealy-Locked loop," *Proc. IEE Radio Receivers and Associated Systems*, pp. 176-180, Sept. 1995.

[29] A. Wilde, "Reduced Complexity Delay-Locked Loop," *Electron. Lett.*, Vol. 31, pp. 1979-1980, Nov. 1995.

[30] A. L. Welti and U. P. Bernhard, "Novel PN-Code Tracking Loop with Periodic Discriminator Characteristic," *Electron. Lett.*, Vol. 29, pp. 583-584, April 1993.

[31] S. Thayaparan, T.S. Ng and J. Wang, "Half-Sine and Triangular Despreading Chip Waveforms for Coherent Delay-Locked Tracking in DS/SS Systems," *IEEE Trans. on Commun.*, vol. COM-48, Aug. 2000.

[32] S. Thayaparan, "Delay-Locked Loop Techniques in Direct Sequence Spread Spectrum Receivers," Ph.D. Thesis, University of Hong Kong, 1999.

Chapter 5

CDMA OVERLAY

Abstract This chapter is concerned with a direct sequence code division multiple access (DS-CDMA) system sharing common spectrum with a narrowband waveform for wireless LAN applications. A suppression filter at a receiver is employed to reduce the narrowband interference. The average up-link bit error rate (BER) performance is investigated how to be influenced by various parameters, such as the number of taps of the suppression filter, the number of multiple-access users, the ratio of narrowband interference bandwidth to the spread spectrum bandwidth, the interference power to signal power ratio, the ratio of the offset of the interference carrier frequency from the spread spectrum carrier frequency to the half spread spectrum signal bandwidth, and so on.

Keywords CDMA Overlay, Narrowband Interference Suppression, Notch Filtering. Wireless LAN

5.1 OVERVIEW

Wireless LAN has been proposed to operate in the ISM band (902-928 MHz, 2.4-2.4835 GHz and 5.725-5.85 GHz). Since the ISM has been partially occupied by narrowband signals, coexistence of narrowband systems and wireless LAN systems should be required. CDMA overlay may be a good choice for the wireless LAN applications.

A number of authors have discussed the interference rejection techniques in spread spectrum [1-7, 9-27, 30]. Among them, Milstein's tutorial paper [1] is the most informative. He discusses in depth two classes of interference rejection schemes: estimator/subtractor methods that perform time-domain notch filtering; transform-domain methods that operate to block or suppress narrowband energy in the frequency domain. Further, he states that both schemes can use tapped delay line implementations, and both can be made adaptive. Poor [30] and Poor /Rush [3] provide excellent overviews of NBI suppression in spread spectrum CDMA. They divide CDMA interference suppression into linear techniques, nonlinear techniques, and multi-user detection techniques. In multi-user CDMA environment, since not only the code sequences of different users are not perfectly orthogonal also due to the near-far problem, there would be serious multi-access interference (MAI).

5.1.1 Narrowband Interference Suppression Techniques

The spreading of the data signal's energy over a large bandwidth allows it to overlay on existing narrowband signals with only minimum of interference for either signal. The broadband noise-like character of spread spectrum signal assures that it will cause little damage to the narrowband signal. On the other hand, although the narrowband signal has high spectral density, this energy is concentrated near one frequency and is of very narrow bandwidth. The despreading operation of the spread spectrum receiver reverses the narrowband signal into wideband noise or low spectral density signal. Thus, spread spectrum communications is inherently resistant to the NBI. However, when the processing gain is insufficient due to bandwidth restriction to allow satisfactory communications to take place, NBI suppression technique may be a vital solution to improve the system performance. It has also been demonstrated that the performance of spread spectrum systems in the presence of narrowband signals can be enhanced significantly through a use of narrowband suppression techniques before despreading [1]. The NBI techniques can be roughly categorized by liner techniques and nonlinear techniques.

5.1.2 Linear Techniques

Here, we mainly review NBI suppression techniques based on linear signal processing criteria. This methodology is quite well developed, and it has been investigated extensively. It has focussed on two basic types of technique. The basic idea in both types is to notch out or flattens the spectrum of interference. The review also includes the adaptive version of this methodology.

First, Prediction/estimation type filters perform an operation to mitigate NBI in time-domain. The interfering signal, being narrowband, can be predicted accurately on time scale of the PN signal based on past values of received signal. The predicted value then will be subtracted from the actual received signal. A tapped delay line can implement either a one-sided prediction filter or a two-sided prediction filter. The prediction of one-sided prediction filter, also called Wiener filter, is based on past values of received signal. On the other hand, the two-sided prediction filter, also called interpolating linear filter, is based on both past and future values. Later technique was found to give even greater interference suppression. Iltis and Milstein [4] investigate the performance of DS spread spectrum systems in presence of either tone or narrowband Gaussian interference. Analytical expressions for bit error rate are derived for two types of DS spread spectrum systems.

If the interference is wide-sense stationary and the statistics are known, the Levinson-Durbin algorithm can be used to recursively solve for the optimal filter coefficients. For unknown statistics, several techniques exist to determine the optimal filter coefficient adaptively. However, least mean square (LMS) estimation technique is widely used to update the filter weights due to ease of analysis and implementation. Iltis and Milstein [11] presents an approximate statistical analysis of the Widrow LMS algorithm, applied to DS spread spectrum system in the presence of NBI. Exact expressions for the steady state tap weight covariance matrix and resulting average excess of mean square error are developed.

Several researchers have analyzed the impact of adaptive algorithms on performance. Bershad [6] investigates the effects of the LMS Adaptive line enhancer (ALE) weight misadjustment errors on the bit error rate for a DS-SS binary communication system in the presence of strong NBI. The converged ALE weights are modeled as the parallel connection of a straight wire, a deterministic finite impulse response (FIR) filter, and a random FIR filter. The statistics of the random filter are derived, assuming the output of the random filter to be primarily due to the jammer convolved with random filter weights, yielding a non-Gaussian output that causes significant error rate degradation in comparison to a Gaussian model. Wang and Milstein [10] evaluate analytically the average bit error rate and investigate the performance is influenced by the parameters such as number of taps of suppression filter, number of active users in cellular system, factor due to adjacent cell interference, and so on.

Performing notch filtering operation in a completely different manner from prediction/estimation type filters, transform domain processing system utilize, as a basic building block, a device that performs a real time Fourier transform [1]. The principal approach is to take Fourier transform of the received signal, to apply a mask in the frequency domain to notch out the NBI, and then to inversely transform the result back to the time domain [3]. The basic idea is to choose a transform such that the undesired signal or interference is nearly an impulse function in the transform domain while the desired signal is transformed to a waveform that is very flat or orthogonal with respect to the transformed interference. A single exciser, that sets the portions of transform which are jammed to zero, can remove the interferer without removing a significant amount of desired signal energy. Then, the inverse transform produces the nearly interference free desired signal [1].

Gevargiz, Das and Milstein [12] demonstrates the advantage of an intercept receiver that employs a transform domain processing filter and detects DS BPSK spread spectrum signals in the presence of NBI by employing adaptive narrowband interference rejection techniques. The receiver employs one of the two transform domain techniques. In the first technique, the NBI is detected and excised in the transform domain by using an adaptive notch filter. In the second technique, the interference is suppressed using soft limiting in the transform domain. Saulnier [13] investigates the use of transform domain adaptive filtering for the suppression of narrowband in a DS spread spectrum receiver. This suppressor combines some of features of time domain adaptive filter and the transform domain exciser. BER results are presented to illustrate the performance of for a single tone jammer and a jammer consisting of a second DS signal having a lower chip rate.

Medley, Saulnier and Das [14-15] extend transform domain processing to include wavelets as basis functions, in order to excise the jamming signals from spread spectrum systems. Implementation of the exciter using multirate digital filtering filters is also discussed. Recently, [16] presents a thorough analysis and evaluation of the lapped transform domain excision process as a means of mitigating NBI. [17] discusses the use of pre- and post-correlation transform domain LMS algorithms to suppress the interference while minimizing the mean square error between received signal and original data message. [16-17] also investigate the convergence performance of both conventional and normalized pre- and post-correlation transform domain LMS algorithms as it relates to jammer frequency and misadjustment noise.

Tazeby and Akansu [18-19] propose a smart adaptive time-frequency (ATF) exciser that evaluates both the time and frequency domain properties of time varying signals and decides the domain of excision where the interference is more localized. Adaptive subband transforms are utilized for frequency domain excision. A tutorial on wavelet and subband transforms, which includes the interference excision, is presented recently [20].

The prediction of NBI in the presence of non-gaussian noise is hardly possible by linear techniques and non-linear techniques may give better performance. Single user frequency bands can be modeled with some accuracy as having a gaussian environment, but presence of additional data signals causes non-gaussian behavior. Vijayan and Poor [21] proposed a nonlinear technique of predicting NBI and led to significant increases in the SNR improvement due to filtering. In [3], the narrowband signal is modeled as an autoregressive process. When the statistics of this autoregressive process are unknown to the receiver, the parameter can be estimated by an adaptive nonlinear filter that uses standard LMS adaptation algorithm to predict the interferer by incorporating a nonlinearity that takes the form of a soft decision feedback of an estimate of the spread spectrum signal. As discussed, the narrowband prediction is subtracted from the input, leaving the spread spectrum signal and AWGN.

5.1.3 Lattice Filter

Though lattice structure is linear technique, we discuss it separately in order to provide more details in this section. Lattice structures for adaptive estimation and filtering are the result of investigations of linear prediction speech modeling and parametric spectrum estimation [22]. Although these early efforts were focused on optimal modelling and estimation of stationary or quasi-stationary signals using the minimum mean square error criterion, it was soon recognized that the lattice form also provides a natural and computationally efficient structure for exact least square estimation of time series signals [23]. Since then, numerous research papers have been devoted to various lattice algorithms, their applications to adaptive filtering, estimation and parametric modelling and investigation of their properties. Among many attractive properties of lattice algorithms, the most distinctive is recursiveness of order.

The lattice form of an adaptive filter has several advantages over the direct form implementation. The decoupling of stages of the filter allows a simpler calculation of the filter parameters. This decoupling also improves convergence time. The convergence time of a direct form filter is proportional to the ratio of maximum to minimum eigenvalues of the data's autocorrelation matrix [24]. For data sequence where the ratio is large, a long conversion time would occur. The convergence time of lattice filter is independent of eigenvalue spread. Lattice filters also show a low sensitivity to parameter quantization and round off noise [22].

Guilford and Das [25] investigate the use of two stages of adaptive lattice filter in NBI rejection. Computer simulation results are presented. Saulnier, Yum and Das [26] present experimental results for both three and ten stages of adaptive lattice filter. In all these cases, the performance is improved significantly. He, Lei, Das and Saulnier [27] discuss the modified LMS algorithm for transversal filter structures and lattice filter structures and also present BER results by computer simulations. We will present some latest results of lattice filters in interference rejection for CDMA overlay in the next chapter.

5.1.4 Motivation of CDMA Overlay

[7] is the first paper to propose allowing a CDMA network to be overlaid on top of existing microwave narrowband users that occupy a part of the spread spectrum system bandwidth (BW). In fact, this is another advantage of CDMA over either TDMA or FDMA, because this will increase the overall spectrum capability. However, such an application must be considered carefully, because the CDMA (wideband) users and the narrowband users can interfere with one-another. This scenario is of particular interest in a CDMA cellular system, in which a given geographical region is divided into many small cells, where the number of cells depends on the size and geometry of the region. Each user has a unique pseudonoise (PN) sequence and communicates through central base-stations with other users in the same cell or different cells. The base-station is assumed to control the average transmission power within its particular cell, to overcome the near/far problem.

5.2 SYSTEM MODEL

Let K denote the number of active users. The transmitted signal of the kth mobile user $(1 \le k \le K)$ takes the form

$$S_k(t) = \sqrt{2P} b_k(t) a_k(t) \cos(2\pi f_o t + \theta_k) \qquad (5.1)$$

where P and f_0 denote the transmitted power and the carrier frequency, respectively, and are common to all users; θ_k is the phase angle introduced by the kth PSK modulator. The kth data signal, $b_k(t)$, is the kth source information sequence with a rate $1/T_b$. The λth square pulse of $b_k(t)$ has amplitude $b_k^{(\lambda)}$, taking values form {+1, -1} with equal probability. The spreading sequence waveform $a_k(t)$ has a rate $1/T_c$, and is modeled as a random binary sequence. The nth chip of $a_k(t)$ has amplitude $a_k^{(n)}$, and takes values from {+1, -1}. The autocorrelation function, $\rho_{T_c}(\tau)$, of $a_k(t)$ is given by

$$\rho_{T_c}(\tau) = \begin{cases} 1 - \dfrac{|\tau|}{T_c}, & |\tau| \le T_c, \\ 0, & |\tau| \ge T_c \end{cases} \tag{5.2}$$

We assume that there are N chips (processing gain) of the random sequence in each data bit ($T_b = NT_c$). Therefore, the spread spectrum system bandwidth (B_s) equals $2T_c^{-1}$.

It is assumed that the channel between the kth transmitter and the corresponding receiver at a central station is a frequency-nonselective Rayleigh fading channel, and is characterized by three random variables, β_k, τ_k and μ_k, which are, respectively, defined as the gain, delay and phase of the kth signal at the receiver. The gain β_k is an independent Rayleigh random variable with parameter $\rho = \rho_k = E[\beta_k^2]/2$ for all k, while the delay τ_k, also independent for each signal, has a uniform distribution in $[0,T_b]$. Further, we assume that the phase μ_k is an independent random variable, uniformly distributed in $[0, 2\pi]$. However, the results can be extended to multipath fading channels models.

The noise $n(t)$ at the input of the receiver is assumed to be additive white Gaussian noise (AWGN) with two-sided power spectral density of $N_o/2$. The interference $J(t)$ is assumed to be narrowband BPSK signal and is given by

$$J(t) = \sqrt{2J}\, j(t) \cos[2\pi (f_o + \Delta)t + \eta], \tag{5.3}$$

where Δ stands for the offset of the interference carrier frequency from the carrier frequency of CDMA signals. J and η denote the received interference power and phase, respectively. The information sequence $j(t)$ has a bit rate $1/T_j$, where T_j denotes the duration of one bit. The autocorrelation function $\rho_{T_j}(\tau)$ of j(t) is given by (5.2) with T_c replaced by T_j. Therefore, the interference bandwidth is $B_j = 2T_j^{-1}$ (we assume $B_j < B_s$). An important quantity is the ratio of the interference bandwidth to the spread spectrum system bandwidth (see Fig. 5.1):

$$p = \frac{B_j}{B_s} = \frac{T_c}{T_j} \tag{5.4}$$

Therefore, the received signal r(t) at a base station can be represented as

$$r(t) = \sqrt{2P}\sum_{k=1}^{K}\beta_k b_k(t-\tau_k)a_k(t-\tau_k)\cos(2\pi f_o t+\phi_k)+\sqrt{2J}\,j(t)\cos[2\pi(f_o+\Delta)t+\eta]+n(t), \tag{5.5}$$

where $\phi_k = \theta_k - \mu_k - 2\pi f_o \tau_k$. Note that the interference is assumed not to be fading, based on the fact that the transmission of the narrowband signal is approximately a direct path from a high microwave tower (or the top of a building) to the receiver of a high base station.

As shown in Fig.5.2, the receiver model of the DS system consists of a bandpass filter, a transversal filter (suppression filter), a DS despreader, a PSK demodulator and a hard decision device. The received signal *r(t)* enters a front-end ideal bandpass (BP) filter which removes out-of-band noise. The BP filter bandwidth (BW) equals the spread spectrum system BW B_s ($B_s = 2T_c^{-1}$). We assume that the CDMA signals can pass through the filter without distortion in order to simplify the performance analysis. Also, the received narrowband interference can pass through the BP filter without distortion, because the bandwidth B_j is assumed to be less than the BP filter bandwidth B_s.

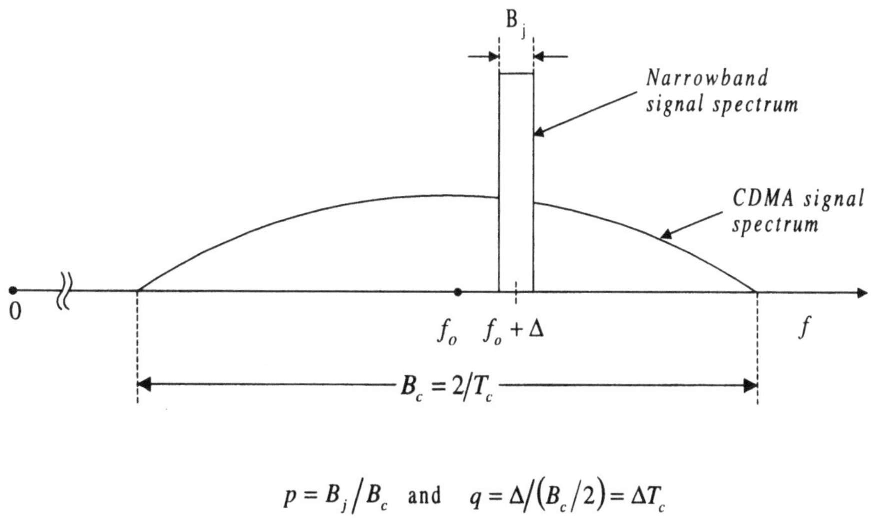

Fig. 5.1 CDMA and narrowband signal spectrums

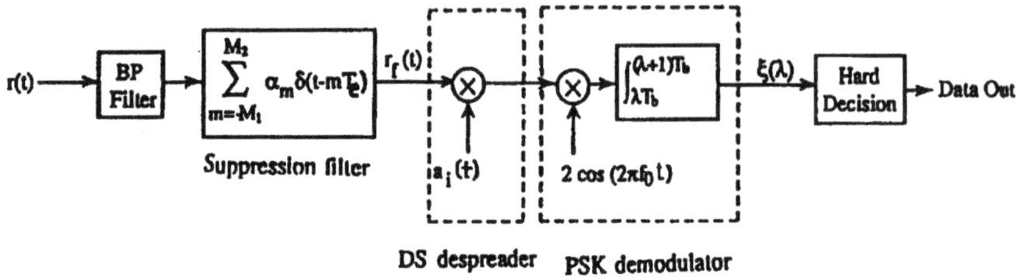

Fig.5.2 Receiver model of a DS system

The BP filter output signal enters a suppression filter whose impulse response is $\sum_{m=-M_1}^{M_2} \alpha_m \delta\ (t - mT_c)$, where $\alpha_o = 1, M_1 \geq 0, M_2 \geq 0$, and M_1 and M_2 stand for the numbers of taps on the left and right sides, respectively, of the center tap. For $M_1 \neq 0$ and $M_2 \neq 0$, the filter is a double-sided suppression filter, whereas for $M_1 = 0$ and $M_2 \neq 0$ or $M_1 \neq 0$ and $M_2 = 0$, the filter is a single-sided suppression filter. The suppression filter uses the past and/or future values of the BP filter output signal to predict its current value, and then subtracts the predicted value from its current (actual) value. The suppression filter output is given by

$$r_f(t) = \sum_{m=-M_1}^{M_2} \alpha_m \left\{ \sqrt{2P} \sum_{k=1}^{K} \beta_k b_k (t - \tau_k - mT_c) a_k (t - \tau_k - mT_c) \cos(2\pi f_o t + \phi_{k,m}) \right.$$

$$\left. + \sqrt{2J}\, j(t - mT_c) \cos[2\pi\ (f_o + \Delta\ )(t - mT_c) + \eta\ ] + \hat{n}(t - mT_c) \right\}, \tag{5.6}$$

where $\hat{n}(t)$ is the filtered AWGN.

The suppression filter signal $r_f(t)$ enters a DS despreader. We assume that user i is the reference user. The DS despreader output signal enters a conventional coherent PSK demodulator corresponding to user i. The output of the demodulator is a lowpass signal and gives rise to the decision variable $\xi(\lambda)$, where λ denotes the λth data bit of the reference user.

5.3 SYSTEM PERFORMANCE

The decision variable $\xi(\lambda)$ can be written as

$$\xi(\lambda) = \int_{\lambda T_b}^{(\lambda+1)T_b} r_f(t)\cdot a_i(t)\cdot 2\cos(2\pi f_o t)dt \tag{5.7}$$

where $2\cos(2\pi f_o t)$ is the recovered carrier and $a_i(t)$ is the spreading sequence of the reference user. It is assumed that $\phi_i = 0$ and $\tau_i = 0$ without loss of generality. Since the high frequency terms are removed by an integrator following the mixer of the PSK demodulator, the above expression reduces to

$$\xi(\lambda) = D(\lambda) + F(\lambda) + N(\lambda) + J(\lambda) + \sum_{\substack{k=1 \\ k\neq i}}^{K} I_{k,i}, \tag{5.8}$$

where the terms on the right hand side of (5.8) are described in detail below.

$D(\lambda)$ is the desired signal term of the reference user at the zero-th tap of the suppression filter, and is given by

$$D(\lambda) = \sqrt{2P}\beta_i T_b b_i^{(\lambda)}, \tag{5.9}$$

Given β_i , the useful signal power is equal to

$$S|_{\beta_i} = 2P\beta_i^2 T_b^2, \tag{5.10}$$

$F(\lambda)$ is an internal interference term, due to the reference user, which is caused by the taps of the suppression filter excluding the zeroth tap. It has been shown in [9-10] that $F(\lambda)$ can be neglected when $K>>1$.

$N(\lambda)$ is due to the noise $\hat{n}(t)$ of the front-end BP filter output. Its variance is given by

$$\sigma_2^2 = N_0 T_b \sum_{m=-M_1}^{M_2} \alpha_m^{\ 2} \tag{5.11}$$

$J(\lambda)$ is due to the narrowband interference and is given by

$$J(\lambda)=\sum_{m=-M_1}^{M_2}\alpha_m\int_{\lambda T_b}^{(\lambda+1)T_b}\sqrt{2J}\,j(t-mT_c)\cos[2\pi\Delta(t-mT_c)+\eta]a_i(t)dt$$

$$=\sum_{m=-M_1}^{M_2}\alpha_m\int_0^{T_b}\{\sqrt{2J}\,j(t+\lambda T_b-mT_c)\cos[2\pi\Delta(t+\lambda T_b-mT_c)+\eta]\}a_i(t)dt$$

(5.12)

The variance of J(λ) equals

$$\sigma_3^{\ 2}=T_b^{\ 2}(J/N)\sum_{m_1=-M_1}^{M_2}\sum_{m_2=-M_1}^{M_2}\alpha_{m_1}\alpha_{m_2}\,\sigma_j^{\ 2}(m_1,m_2), \tag{5.13}$$

where $\sigma_j^2(m_1,m_2)$ is given by

$$\sigma_j^{\ 2}(m_1,m_2)=\int_{-1}^{1}sign[1-p\,|\,xN-m_1+m_2\,|]\cos[2\pi q(xN-m_1+m_2)](1-|\,x\,|/N)(1-|\,x\,|)dx,$$

$$\approx\int_{-1}^{1}sign[1-p\,|\,xN-m_1+m_2\,|]\cos[2\pi q(xN-m_1+m_2)](1-|\,x\,|)dx,\qquad N>>1, \tag{5.14}$$

and where $N = T_b/T_c$, $p=T_c/T_j$ and $q = \Delta T_c$ denotes the ratio of the offset of the interference carrier frequency to the half spread spectrum bandwidth (see Figure 5.1). *sign*[x] is defined as *sign*[x]=x or 0 for x≥0 or x<0, respectively.

$I_{k,i}$ is the multiple access interference term due to the *k*th user ($k\neq i$) and is given by

$$I_{k,i}=\sqrt{2P}\beta_k\sum_{m=-M_1}^{M_2}\alpha_m\left\{b_k^{\ (\hat{\lambda}-1)}R_{k,i}[\tau(m)]+b_k^{\ (\hat{\lambda})}\hat{R}_{k,i}[\tau(m)]\right\}\cos(\phi_k), \tag{5.15}$$

where $b_k^{\ (\hat{\lambda}-1)}$ and $b_k^{\ (\hat{\lambda})}$ denote two consecutive bits of the *k*th user, and τ(m) is τ_k+mT_c, mod $T_b=NT_c$. It is obvious that τ(m) is uniformly distributed in [0, T_b]. $R_{k,i}(\tau)$ and $\hat{R}_{k,i}(\tau)$ are known as the continuous-time partial cross-correlation functions [28-29], defined as, respectively,

$$R_{k,i}(\tau)=\int_0^{\tau_k}a_k(t-\tau)a_i(t)dt=C_{k,i}(l-N)[(l+1)T_c-\tau]+C_{k,i}(l+1-N)(\tau-lT_c\}, \tag{5.16}$$

and

$$\hat{R}_{k,i}(\tau) = \int_{\tau}^{T_b} a_k(t-\tau)a_i(t)dt = C_{k,i}(l)[(l+1)T_c - \tau] + C_{k,i}(l+1)(\tau - lT_c\}, \qquad (5.17)$$

where l stands for the integer portion of τ/T_c. That is, l = int(τ/T_c), or $lT_c \leq \tau \leq (l+1)T_c$, and $C_{k,i}(l)$ is defined by [9]

$$C_{k,i}(l) = \begin{cases} \sum_{j=0}^{N-1-l} a_k^{(j)} a_i^{(j+l)}, & 0 \leq l \leq N-1, \\ \sum_{j=0}^{N-1+l} a_k^{(j-l)} a_i^{(j)}, & 1-N \leq l < 0 \end{cases} \qquad (5.18)$$

It is shown in [29] that

$$E[C_{k,i}(l)C_{k,i}(m)] = \begin{cases} N - |l|, & l = |m| \\ 0, & l \neq |m| \end{cases} \qquad (5.19)$$

The variance of $I_{k,i}$ is given by

$$\sigma_4^2 = 2P\rho \sum_{m_1=-M_1}^{M_2} \sum_{m_2=-M_1}^{M_2} \alpha_{m_1}\alpha_{m_2} \{E[R_{k,i}[\tau(m_1)]R_{k,i}[\tau(m_2)] + E[\hat{R}_{k,i}[\tau(m_1)]\hat{R}_{k,i}[\tau(m_2)]]\} \qquad (5.20)$$

Consider first $E[R_{k,i}[\tau(m_1)]R_{k,i}[\tau(m_2)]]$.

$$E[R_{k,i}[\tau(m_1)]R_{k,i}[\tau(m_2)]] = E\{[C_{k,i}(l_1 - N)[(l_1+1)T_c - \tau(m_1)] + C_{k,i}(l_1+1-N)[\tau(m_1) - l_1T_c]]$$

$$\cdot [C_{k,i}(l_2 - N)[(l_2+1)T_c - \tau(m_2)] + C_{k,i}(l_2+1-N)[\tau(m_2) - l_2T_c]]\}, \qquad (5.21)$$

where l_1 and l_2 equal int[$\tau(m_1)/T_c$] and int[$\tau(m_2)/T_c$], respectively, and are uniformly distributed in [0, N-1], because $\tau(m_1)$ and $\tau(m_2)$ have uniform distributions in [0, T_b] = [0, NT_c]. For $\tau(m_1) = \tau_k+m_1T_c$ and $\tau(m_2) = \tau_k+m_2T_c$, both mod $T_b = NT_c$, we have $\tau(m_1) = \tau(m_2)+(m_2- m_1)T_c$, mod T_b. That is, $l_1 = l_2+m_2- m_1$, mod N. Also, $[(l_1+1)T_c-\tau(m_1)]$ and $[(l_2+1)T_c-\tau(m_2)]$ are uniformly distributed in [0, T_c], and $[(l_1+1)T_c-\tau(m_1)] = [(l_2+1)T_c-\tau(m_2)] = \delta$. Equivalently, $\tau(m_1)-l_1T_c = \tau(m_2)-l_2T_c = T_c-\delta$, which is independent of both m_1 and m_2. Thus, when $m_1 = m_2$ (or $l_1 = l_2$), using (5.19), the above expression reduces to

$$E[R_{k,i}[\tau(m_1)]R_{k,i}[\tau(m_2)]]\Big|_{m_1=m_2} = NT_c^{\,2}/3 = T_b^{\,2}/(3N). \qquad (5.22)$$

When $m_2 = m_1+1$ (or $l_2 = l_1+1$), (5.21) reduces to

$$E[R_{k,i}[\tau(m_1)]R_{k,i}[\tau(m_2)]]\Big|_{m_2=m_1+1} \approx T_b^{\,2}/(12N). \qquad (5.23)$$

When $m_2 = m_1-1$ or $l_2 = l_1-1$, the same results as for $m_2 = m_1+1$ apply, and

$$E[R_{k,i}[\tau(m_1)]R_{k,i}[\tau(m_2)]]\Big|_{m_2=m_1-1} = T_b^{\,2}/(12N) \qquad (5.24)$$

When $|\, m_2 - m_1| \geq 2$,

$$E[R_{k,i}[\tau(m_1)]R_{k,i}[\tau(m_2)]]\Big|_{|\,m_2-m_1|\geq 2} = 0 \qquad (5.25)$$

Also, identical results can be obtained for $E[\hat{R}_{k,i}[\tau(m_1)]\hat{R}_{k,i}[\tau(m_2)]]$. Therefore, σ_4^2 is given by

$$\sigma_4^{\,2} = 2P\rho T_b^{\,2}\left\{\frac{2}{3N}\sum_{m=-M_1}^{M_2}\alpha_m^{\,2} + \frac{1}{3N}\sum_{m=-M_1}^{M_2}\alpha_m\alpha_{m+1}\right\}, \qquad (5.26)$$

where α_{M_2+1} is defined to be α_{-M_1}.

Collecting the above results and neglecting F(λ), the decision variable ξ(λ), given by (5.8), consists of a useful term D(λ) and a total interference term (Gaussian noise N(λ) + narrowband interference J(λ) + multiple access interference $\sum_{\substack{k=1\\k\neq i}}^{K} I_{k,i}$). The total interference has the variance,

$$\sigma^2 = N_0T_b\sum_{m=-M_1}^{M_2}\alpha_m^{\,2} + T_b^{\,2}(J/N)\cdot\sum_{m_1=-M_1}^{M_2}\sum_{m_2=-M_1}^{M_2}\alpha_{m_1}\alpha_{m_2}\sigma_j^{\,2}(m_1,m_2)$$
$$+(K-1)2P\rho T_b^{\,2}\left[\frac{2}{3N}\sum_{m=-M_1}^{M_2}\alpha_m^{\,2} + \frac{1}{3N}\sum_{m=-M_1}^{M_2}\alpha_m\alpha_{m+1}\right]. \qquad (5.27)$$

Given β_i, the signal-to-noise ratio now becomes

$$\hat{r}_b\left(\beta_i^2\right) = s|_{\beta i}/2\sigma^2 \qquad (5.28)$$

The BER is given by [8]

$$\hat{p}_e = \frac{1}{2}\left[1 - \sqrt{\frac{r_b}{1+r_b}}\right] \approx \frac{1}{4r_b}, \tag{5.29}$$

where r_b denotes the average signal-to-noise ratio(SNR), with "noise" taken to be the sum of the contributions due to the narrowband interference, the multiple-access interference and the thermal noise, and is given by

$$r_b = \frac{1}{2}\frac{E[S \mid \beta_i]}{\sigma^2} = \frac{2P\rho T_b^2}{\sigma^2}$$

$$= \{\left[\frac{\overline{E}_b}{N_o}\right]^{-1} \sum_{m=-M_1}^{M_2} \alpha_m^2 + \frac{1}{N}\left[\frac{J}{S}\right] \sum_{m=-M_1}^{M_2} \sum_{m=-M_1}^{M_2} \alpha_{m_1}\alpha_{m_2}\sigma_j^2(m_1, m_2)$$

$$+ \frac{(K-1)}{3N}\left[\sum_{m=-M_1}^{M_2} 2\alpha_m^2 + \sum_{m=-M_1}^{M_2} \alpha_m \alpha_{m+1}\right]\}^{-1} \tag{5.30}$$

where $\overline{E}_b = 2P\rho T_b$ is the average energy-per-bit, and $J/S = J/(2P\rho)$ denotes the interference power to signal power ratio. The accuracy of the above approximation will be demonstrated in the next section.

The coefficients of the suppression filter can be determined by solving the following Wiener-Hopf equations [1]:

$$\sum_{\substack{m=-M_1 \\ m \neq 0}}^{M_2} \alpha_m R[(n-m)T_c] + R(nT_c) = 0, n = -M_1, \ldots, -1, 1, \ldots, M_2; n \neq 0 \tag{5.31}$$

In (5.31), $R(\tau)$ is the auto-correlation function of the input signal of the suppression filter and can be expressed as

$$R(\tau) = \frac{1}{2}\rho(\tau)\cos(2\pi f_o \tau) \tag{5.32}$$

where $\rho(\tau)$ is the low-pass version of $R(\tau)$. Therefore, for $f_0 T_c$ an integer, the Wiener Hopf equations become

$$\sum_{\substack{m=-M_1 \\ m\neq 0}}^{M_2} \alpha_m \rho[(n-m)T_c] + \rho(nT_c) = 0, n = -M_1,\ldots,-1,1,\ldots M_2; n \neq 0. \tag{5.33}$$

$\rho(lT_c)$ can be written as

$$\rho(lT_c) = \rho_s(lT_c) + \rho_{\hat{n}}(lT_c) + \rho_j(lT_c), \tag{5.34}$$

where $\rho_s(lT_c), \rho_{\hat{n}}(lT_c)$ and $\rho_j(lT_c)$ are, respectively, the low-pass versions of the DS signal, noise and interference functions of the input signal to the suppression filter. The DS term is given by

$$\rho_s(lT_c) = P \cdot \sum_{k=1}^{K} E[\beta_k^2] E[b_k(t-\tau_k) b_k(t+lT_c-\tau_k)] E[a_k(t-\tau_k) a_k(t+lT_c-\tau_k)] \tag{5.35}$$

Since $E[a_{k1}(t-\tau_{k1}) a_{k2}(t+lT_c-\tau_{k2})] = 0$ for $k_1 \neq k_2$, $\rho_{Tc}(\tau)$ is given by (5.2), and further $E[a_k(t-\tau_k) a_k(t+lT_c-\tau_k)]$ is given by

$$E[a_k(t-\tau_k) a_k(t+lT_c-\tau_k)] = \begin{cases} 1, & l = 0 \\ 0, & l \neq 0 \end{cases}$$

one obtains

$$\rho_s(lT_c) = \begin{cases} 2KP\rho, & l = 0 \\ 0, & l \neq 0 \end{cases} \tag{5.36}$$

The filtered-noise term $\hat{n}(t)$ has a flat power spectral density of $N_0/2$ with the bandwidth of 2/Tc. Since the auto-correlation function of a signal is the Fourier transform of its power spectral density function, the low-pass version of the auto-correlation of the noise term is given by

$$\rho_{\hat{n}}(lT_c) = \frac{N_0}{\pi} \frac{\sin(\pi B_c lT_c)}{lT_c} = \frac{N_0}{\pi} \cdot \frac{\sin(2\pi l)}{lT_c} = \begin{cases} 2N_0/T_c, & l = 0 \\ 0, & l \neq 0 \end{cases}. \tag{5.37}$$

Since the baseband signal of the narrowband interference is a binary sequence, its auto-correlation is a triangular function. Therefore, the lowpass version of the auto-correlation of the narrowband interference is given by

$$\rho_j(lT_c)=\begin{cases} J(1-|l|p)\cos(2\pi lq), & |l|\le \mathrm{int}[1/p] \\ 0, & |l|>\mathrm{int}[1/p] \end{cases}, \tag{5.38}$$

where *int*[x] is defined as the integer part of x. Therefore, from the above, one obtains

$$\rho(lT_c)=2P\rho\hat{\rho}(lT_c), \tag{5.39}$$

where $\hat{\rho}(lT_c)$ is defined as

$$\hat{\rho}(lT_c)\equiv\begin{cases} K+2N\left[\overline{E}_b/N_0\right]^{-1}+J/S, & l=0 \\ (J/S)(1-|l|p)\cos(2\pi lq), & |l|\le \mathrm{int}[1/p] \\ 0, & |l|>\mathrm{int}[1/p] \end{cases} \tag{5.40}$$

where $S=2P\rho$ and $\overline{E}=2P\rho T_b$. By replacing ρ(τ) with $\hat{\rho}(\tau)$ in (5.33), we can obtain the coefficients α_m, and then we can obtain the average BER performance.

5.4 NUMERICAL RESULTS

In this section, using the above analytical results, we provide some representative numerical curves illustrating the BER performance of the DS-CDMA system in the presence of narrowband interference. Unless noted otherwise, it is assumed that the ratio of the offset of the interference carrier frequency to the half spread spectrum bandwidth q=0; that is the carrier frequencies of the CDMA signals and BPSK narrowband signal are identical (i.e., $\Delta=0$).

Fig.5.3 shows the BER performance of the system with a double-sided suppression filter as a function of the signal-to-noise ratio $\overline{E}_b/N_0$. Also, the BER of the system both with a single-sided suppression filter and without a suppression filter are plotted in order to allow for comparison. The number of active users K=10, the number of chips per bit N=255 and the interference power-to-signal power ratio J/S=20dB. It is assumed that the two-sided filter has three taps and is symmetric (i.e.,$M_1=M_2=M=1$), and the single-sided filter has the same number of taps (i.e., $M_1=2M=2$ and $M_2=0$, or $M_1=0$ and $M_2=2M=2$). Finally, the narrowband interference occupies 10 percent (p=0.1) of the spread spectrum bandwidth. As expected, the BER performance can be improved by using a suppression filter. In particular, tremendous improvement can be achieved by using a double-sided filter. The double sided filter is superior to the single-sided filter for the same number of the total taps because the former uses the autocorrelation function of the narrowband interference more efficiently than does the latter.

Fig.5.4 illustrates the asymptotic ($\overline{E_b} / N_0 \rightarrow \infty$) BER of the system with double-sided filters, single-sided filters and without suppression filters as a function of (J/S). As expected, the BERs of the systems with or without the suppression filters increase as J/S increases. When J/S is large (i.e., J/S $\geq$ 10dB) the system without a suppression filter degrades significantly, whereas the system is much more tolerant of the interference when the suppression filter is present.

In Fig. 5.5, the asymptotic BER of the system with and without the suppression filter is shown as a function of the ratio (p) of the interference bandwidth to the spread spectrum bandwidth. The system parameters are again K=10, N=255, and J/S=20dB, and the number of total taps is again 3. It is seen from the figure that the CDMA system with a double-sided filter can suppress the narrowband interference very well when p is small, but is ineffective in suppressing the interference when p is large. On the other hand, the BER performance of the system without a suppression filter is almost insensitive to the value of p, although it becomes slightly better with increasing p for fixed J/S. For p=1, the BERs are the same for the systems with and without suppression filters. This is because the autocorrelation function of interference, $\rho_j(lT_c)$, is 0 for $l\neq 0$ (or $\alpha_m=0$ for $m\neq 0$).

Fig.5.6 illustrates the asymptotic BER for a double-sided suppression filter as a function of the number of taps on each side ($M_1=M_2=M$). It is seen from the figure that the BER decreases first when M=1 and then keeps a value with increasing M. Indeed, for $M_1=M_2=0$, (without the suppression filter) the narrow-band interference J/S=20 dB is the dominating disturbance; the bit errors are caused by the narrow-band interference. For $M_1 = M_2 \geq 1$, the

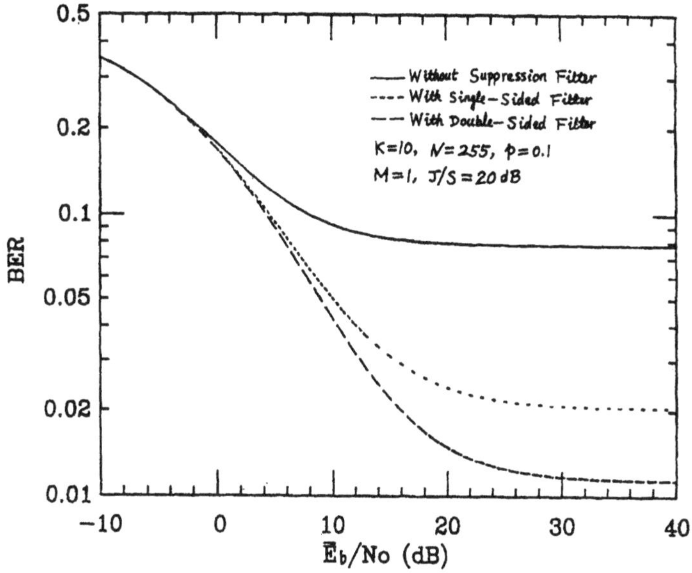

Fig.5.3 BER performance of DS-CDMA systems with and without suppression filters.

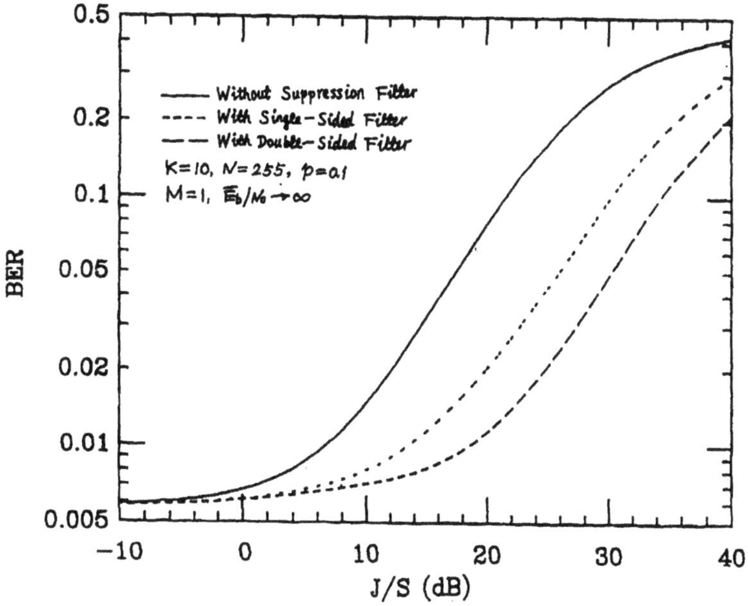

Fig. 5.4 The asymptotic $(\overline{E}_b / N_o \rightarrow \infty)$ BER performance of DS-CDMA systems with and without suppression filters as a function of the interference power-to-signal power ratio (J/S).

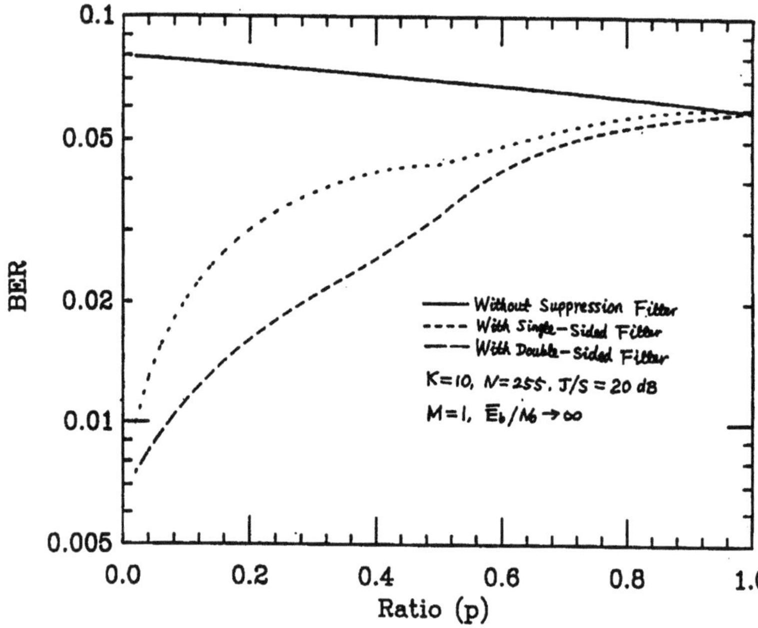

Fig.5.5 The asymptotic $(\overline{E}_b / N_o \rightarrow \infty)$ BER of DS-CDMA systems with and without suppression filters as a function of the ratio(p) of the interference BW to the spread spectrum BW.

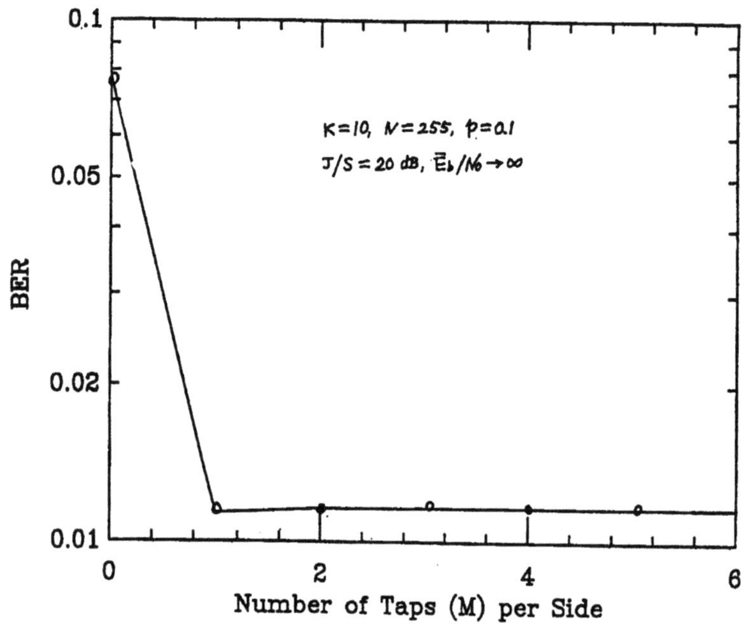

Fig.5.6 The asymptotic BER of DS-CDMA systems with double-sided suppression filter as a function of number of taps on each side (M_1=M_2=M).

Table I ($K = 10, N = 255, p = 0.1, E_b/N_0 \to \infty$, J/S = 20 dB)

Number of taps each side (M)	Coefficients (α_m)							A	B
	α_{-3}	α_{-2}	α_{-1}	α_0	α_1	α_2	α_3		
0	0	0	0	1	0	0	0	0.38	0.023
1	0	0	-0.474	1	-0.474	0	0	0.021	0.026
2	0	-0.119	-0.373	1	-0.373	-0.119	0	0.025	0.023
3	-0.032	-0.1	-0.367	1	-0.367	-0.1	-0.032	0.026	0.023

$$A = \frac{1}{N}\left(\frac{J}{S}\right)\sum_{m_1=-M}^{M}\sum_{m_2=-M}^{M}\alpha_{m_1}\alpha_{m_2}\sigma_2^2(m_1,m_2)$$

$$B = \frac{K-1}{3N}\sum_{m=-M}^{M}\alpha_m[2\alpha_m+\alpha_{m+1}]$$

narrowband interference is reduced by the suppression filter, and increasing M beyond 1 does not noticeably reduce the narrowband interference, because $\hat{\rho}(T_c)$ and $\hat{\rho}(-T_c)$ are almost identical to $\hat{\rho}(0)$ in (5.40), assuming J/S>>K. Thus, the suppression filter with $M_1=M_2=1$ is sufficient to reject the interference in this case (small p and q=0), and more taps than one on each side are unnecessary. Table I shows the tap coefficients of the double-sided suppression filters for different numbers of taps each side ($M_1= M_2=0,1,2,3$). As shown in Table I, for M=0, A>>B, where A represents the effect of the narrowband interference and B represents the effect of the multiple access interference; in this case, the narrowband interference dominates. However, for $M \geq 1$, neither A nor B varies very much. Therefore, we conclude that when the carrier frequencies of the narrowband interference and CDMA signals are the same, the suppression filter with three total taps is sufficient to reduce the narrowband interference, and increasing the number of total taps beyond three is not necessary.

Fig.5.7, the asymptotic BER of the DS-CDMA system with double-sided filters is plotted as a function of active users (K) for different values of J/S. The system parameters are N=255, $M_1= M_2=1$, and p=0.1. Also, the BER of the system without the suppression filter is shown for comparison. It is seen that, for a given BER, the system with the double-sided filter can support many more users than can the system without the suppression filter, especially when the J/S is large. When the interference is not too large (i.e., J/S ≤ 10dB) the system employing the suppression filter can support almost the same number of users as can the system in the absence of the interference.

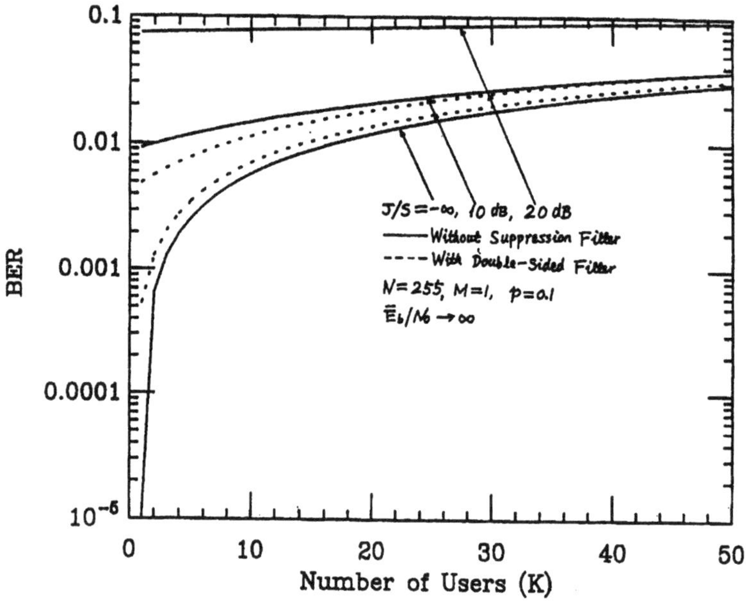

Fig.5.7 The asymptotic BER of DS-CDMA systems with and without double-sided filters as a function of the number of active users (K).

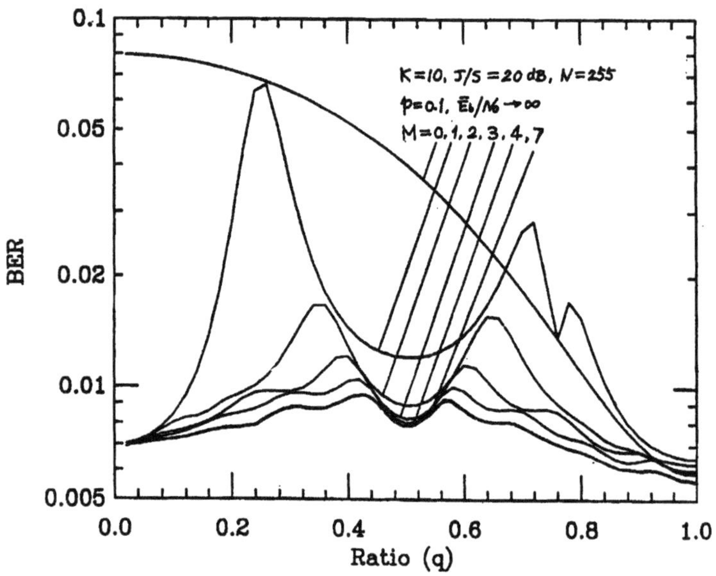

Fig. 5.8 The asymptotic BER of the system with double-sided filters as a function of the ratio (q) of the interference carrier frequency to the half spread spectrum BW.

Fig. 5.8 illustrates the BER performance of the systems with double-sided suppression filters as a function of the ratio (q) of the offset of the interference carrier frequency to the half spread spectrum bandwidth for different number of the taps on each side. It is seen that when q is very small, the BERs of the system with double-sided filters for $M \geq 1$ are almost identical, as was the case in Fig.6. However, the BER performance improves as M increases when q increases, and the BER for large M is also more robust to a change in q. Further, when q is very large (i.e., $q \geq 90\%$), the suppression filter is unnecessary. Also, for $q \approx 0.25$, the BERs of the systems with the double-sided suppression filter ($M_1= M_2=1$) and without the suppression filter are almost identical, because the low-pass version of the autocorrelation of the interference is zero, (i.e., $\rho_j=(lT_c)=0$ for $l \neq 0$), so that all off-center tap coefficients are zero.

5.5 CONCLUSIONS

The BER performance of a DS-CDMA system overlaying a narrowband BPSK waveform, and employing a suppression filter in the receiver, has been evaluated. From the numerical results the following conclusions can be drawn.

- A large improvement in BER can be achieved by using double-sided suppression filters. The double-sided suppression filter is superior to the single-sided suppression filter for the same number of total taps, because the double-sided suppression filter is more effective in using the autocorrelation function of the narrowband interference in order to predict and subtract out that interference.
- When the ratio (q) of the interference carrier frequency to half of the spread spectrum bandwidth is very small, that is, when the interference carrier frequency is very close to the spread spectrum carrier frequency, the suppression filter with three total taps is sufficient to reduce the narrowband interference and increasing the number of total taps beyond three is not necessary. However, spread spectrum systems with a large number of taps (i.e., $M_1= M_2 > 1$) is preferable when the ratio is large. Further, when the ratio q is greater than 90% , the suppression filter is not useful.

REFERENCES

[1] L. B. Milstein, "Interference rejection techniques in spread spectrum communications," Proceedings of IEEE, vol. 76, pp. 657-671, June 1988.

[2] E. Masry and L. B. Milstein, "Performance of DS spread-spectrum receiver employing interference-suppression filters under a worst-case jamming condition," IEEE Trans. Commun., vol. COM-34, pp. 13-21, January 1986.

[3] L. A. Rusch and H. V. Poor, "Narrowband interference suppression in CDMA spread spectrum communications," IEEE Trans. Commun., vol. COM-42, pp. 1969 - 1979, April 1994.

[4] R. A. Iltis and L. B. Milstein, "Performance analysis of narrow-band interference rejection techniques in DS spread spectrum systems," IEEE Trans. Commun., vol. COM-32, pp. 1169-1177, November 1984.

[5] J. W. Ketchum and J. G. Proakis, "Adaptive algorithms for estimating and suppressing narrow-band interference in PN spread spectrum systems," IEEE Trans. Commun., vol. COM-30, pp. 913-923, May 1982.

[6] N. J. Bershad, "Error probabilities for DS spread-spectrum systems using an ALE for narrowband interference rejection," IEEE Trans. Commun., vol. COM-36, pp. 588-595, May 1988.

[7] R. L. Pickholtz., L. B. Milstein and D. L. Schilling, "Spread spectrum for mobile communications," IEEE Trans. Vehicular Technology, May 1991.

[8] J. G. Proakis, Digital Communications. New York: McGraw-Hill, 1983.

[9] J. Wang, "On the use of a suppression filter for CDMA overlay situations," *IEEE Transactions on Vehicular Technology, vol. VT-48, No. 2, March 1999.*

[10] J. Wang and L. B. Milstein, "CDMA overlay situations for microcellular mobile communications," *IEEE Transactions on Communications*, vol. COM-43, pp. 603-614, Feb. 1995.

[11] Iltis, R. and Milstein, L.B., "An Approximate Statistical Analysis of the Widrow LMS Algorithm with Application to Narrowband Interference Rejection," *IEEE Trans. Commun.*, vol. COM-33, pp. 121-130, Feb. 1985.

[12] Gevargiz, J., Das, P.K. and Milstein, L.B., "Adaptive Narrowband Interference Rejection in a DS Spread spectrum Intercept receiver Using Transform Domain Signal Processing Techniques," *IEEE Trans. Commun.*, vol. COM-37, pp. 1359-1366, December 1989.

[13] Saulnier, G.J., "Suppression of Narrowband Jammers in a Spread Spectrum Receiver Using Transform-Domain Adaptive Filtering," *IEEE J. Select. Areas Commun.*, vol. JSAC-10, pp. 742-749, April 1992.

[14] Medley, M.J., Saulnier, G.J. and Das, P.K., "Applications of the wavelet transform in Spread spectrum communications systems,"

Proc. of SPIE, Inter. Society for Optical Engineering, vol. 2242, pp. 54-68, 1994.

[15] Medley, M.J., Saulnier, G.J. and Das, P.K., "Radiometric detection of Direct sequence spread spectrum signals with interference excision using the wavelet transform," *IEEE Inter. Conf. Commun.*, vol. 3, pp. 1648-1652, May 1994.

[16] Medley, M.J. and Saulnier, G.J., "Transform domain LMS convergence in tone interference," *IEEE MILCOM 97'*, vol. 2, pp. 872-876, Nov 1997.

[17] Medley, M.J., Saulnier, G.J. and Das, P.K., "Narrowband interference excision in spread spectrum systems using the lapped transform," *IEEE Trans. Commun.*, vol. COM-45, pp. 1444-1366, November 1997.

[18] Tazebay, M.V. and Akansu, A.N., "A Smart time-frequency exciser for spread spectrum communications," *Proc. IEEE Conf. Acoustic, Speech, Signal Processing*, pp. 1209-1212, May 1995.

[19] Tazebay, M.V. and Akansu, A.N., "Adaptive subband transforms in Time-frequency excisers for DSSS communications systems," *IEEE Trans. Signal Processing,* Vol. 43, pp. 2776-2782, November 1995.

[20] Akansu, A.N., Tazebay, M.V., Medley, M.J. and Das, P.K., "Wavelet and subband transforms: Funtamentals and communication applications," *IEEE Communications Magazine*, Vol. 35, pp. 104-115, December 1997.

[21] Vijayan, R. and Poor, H.V., "Nonlinear techniques for Interference Suppression in Spread Spectrum Systems," *IEEE Trans. Commun.*, vol. COM-38, pp. 1060-1065, July 1990.

[22] Friedlander, B., "Lattice filters for adaptive processing," *IEEE Proc.*, vol. 70, no. 8, pp. 829-867, August 1982,.

[23] Morf, M. and Lee, D.T. and Friedlander, B., "Recursive least squares ladder estimation algorithms," *IEEE Trans. Acoustic, Speech, Signal Processing*, vol. ASSP-34, pp. 1589-1602, 1986.

[24] Griffiths, L.J., "A continuously adaptive filter implemented as a lattice structure, " *Proc. IEEE Conf. Acoustic, Speech, Signal Processing*, pp. 683-685, May 1977.

[25] Guilford, J. and Das, P.K, "The use of the adaptive lattice filter for Narrowband Jammer rejection in DS Spread Spectrum Systems," in Proc. *IEEE International Conf. on Commun.*, pp. 822-826, June 22-26, 1985.

[26] Saulnier, G.J., Yum, K. and Das, P.K., "Narrowband Jammer Suppression using an Adaptive Lattice Filter," *IEEE International Conference on Commun.*, pp. 2113-2117, 1987.

[27] He, Y., Lei, S. F., Das, P.K. and Saulnier, G.J., "Suppression of narrowband jammers in a spread spectrum receiver using modified

adaptive filtering technique," IEEE *Global Telecommun. Conf and Exbib.,* pp. 540-545, Nov 28-Dec 1 1988.

[28] M. B. Pursley, "Spread spectrum multiple access communications," in Multi-User Communications, G. Longo, Ed. Vienna and New York: Springer-Verlag, pp139-199, 1981.

[29] J. Wang, M. Moeneclay and L. B. Milstein, "DS-CDMA with predetection diversity for indoor radio communications," IEEE Trans. Commun., pp. 1929-1938, April 1994.

[30] H. V. Poor, "Active interference suppression in CDMA overlay systems," IEEE Journal on Selected Areas in Communications, Jan. 2001.

Chapter 6

ADAPTIVE FILTERING IN CDMA OVERLAY

Abstract This chapter presents the behavior of reflection coefficients of a stochastic gradient lattice (SGL) filter applied to a CDMA overlay system. Analytic expressions for coefficients for a two stage filter are derived with the presence of narrowband interference and additive white gaussian noise (AWGN). It is shown that the coefficients of the lattice filter exhibit separate tracking and convergent properties and that compared to a LMS filter, the lattice filter provides fast rate of convergence, while having good capability of narrowband interference suppression.

Keywords Narrowband interference suppression, lattice filters, CDMA

6.1 INTRODUCTION

Direct sequence code division multiple access (DS-CDMA) communications is a popular approach in cellular mobile communications, due to its efficient utilization of channel bandwidth, the relative insensitivity to multipath interference and the potential for improved privacy. In addition to providing multiple-access capability and multipath rejection, spread spectrum communications also offers the possibility of further increasing the overall spectrum efficiency by overlaying a CDMA network on the existing narrowband users [1-5]. Such a procedure must be done very carefully so as not to cause intolerable interference for either the existing narrowband users or the CDMA users.

A number of studies have been performed using LMS filters to reject narrowband interference in DS Spread spectrum (or CDMA) systems [1-8]. The LMS algorithm performs well except for its slow rate of convergence. A lattice algorithm has been proposed as an alternative and efficient solution since it provides improved rate of convergence when applied to a linear chirp FM signal [9]. This chapter studies the performance of a lattice filter applied to rejecting narrowband interference in a CDMA overlay situation. The reflection coefficients of a SGL filter are updated by a gradient-based algorithm. At each time step, new reflection coefficients are calculated based on the previous values of reflection coefficients and the current values of input signals. The behavior of reflection coefficients of the SGL filter will be described in the presence of narrowband interference and channel noise in a Rayleigh fading channel.

6.2 FILTER COEFFICIENTS

As shown in Fig. 6.1, the receiver can be categorized into following parts: a bandpass filter, a lattice filter, a de-spreader and a hard decision device. It is assumed that the channel between the CDMA mobile user and its base station is a flat Rayleigh fading channel. However, the analytical results can be extended to frequency selective channels. The received signal $r(t)$ at the base station consists of the sum of K independently-fading CDMA signals, a narrowband interfering signal, and band-limited AWGN:

$$r(t)=\sqrt{2P}\sum_{k=1}^{K}\beta_k b_k(t-\tau_k)a_k(t-\tau_k)\cos(2\pi f_o t+\phi_k)+j(t)+n(t), \tag{6.1}$$

where P represents signal power, f_o denotes the CDMA carrier frequency, $b_k(t)$ is the kth user binary information sequence with bit duration T_b, $a_k(t)$ is a random spreading sequence of the kth user with chip duration T_c and processing gain N $(N = T_b/T_c)$. The random gain β_k and phase ϕ_k of the kth user have a Rayleigh distribution with $E(\beta_k^2) = 2\rho$ for all k $(k = 1,\cdots,K)$, and a uniform distribution in $[0,2\pi)$, respectively. The path delay, τ_k, is uniformly distributed in $[0,T_b)$. The CDMA signal bandwidth is $B_c = 2/T_c$. For simple analysis, $j(t)$ is assumed to be a Gaussian narrowband signal (or interference), instead of BPSK signal, as assumed in Chapter 5, given by

$$j(t) = j_c(t)\cos[2\pi(f_o + \Delta)t] + j_s(t)\sin[2\pi(f_o + \Delta)t], \tag{6.2}$$

where $j_c(t)$ and $j_s(t)$ stand for the lowpass quadrature terms of the narrowband interference with bandwidth $B_j/2$, where B_j is the bandwidth of $j(t)$. Δ denotes the frequency offset of the interference from the CDMA carrier frequency. Further, it is assumed that the parameters p $(p = B_j/B_c)$ and q $(q = \Delta T_c)$ are the ratio of the interference bandwidth to the spread bandwidth and the ratio of the offset of the interference carrier frequency to half of the spread bandwidth, respectively (see Fig 5.1). $n(t)$ is band-limited AWGN with two-sided power spectral density $N_o/2$ and bandwidth B_c.

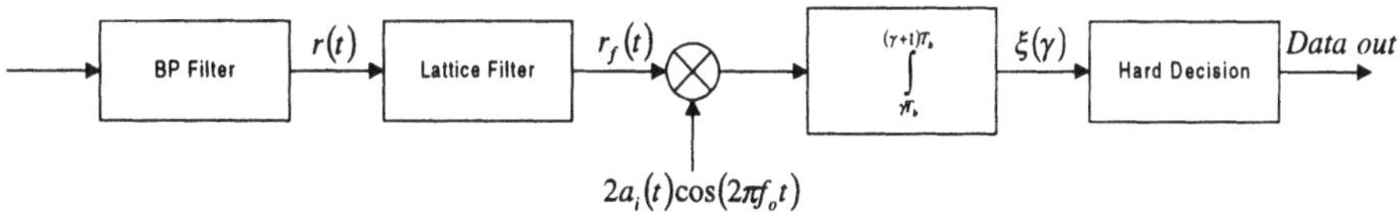

Fig. 6.1 CDMA overlay receiver model.

The detail of the lattice filter is shown in Fig. 6.2 and its input signal is given by

$$e_f(t + sT_c, 0) = e_b(t + sT_c, 0) = r(t + sT_c) = r(sT_c), \tag{6.3}$$

where T_c stands for the delay (or chip duration) of delay element and "s" is an integer. For simple notations, "t" is neglected. The lattice filter can be described by

$$e_f(sT_c,n)=e_f(sT_c,n-1)-R_n(sT_c)e_b((s-1)T_c,n-1) \tag{6.4}$$

and

$$e_b(sT_c,n)=e_b((s-1)T_c,n-1)-R_n(sT_c)e_f(sT_c,n-1), \tag{6.5}$$

where $e_f(sT_c,n)$ and $e_b(sT_c,n)$ are forward and backward prediction errors, n denotes the nth stage of the lattice filter, $R_n(sT_c)$ is the reflection coefficient of the nth stage of the filter. Note that the lattice filter is assumed symmetric (i.e. the forward and backward coefficients of the nth stage are the same). Finally, assuming that the number of stages is M, the output of the filter is

$$r_f(sT_c)=e_f(sT_c,M)=e_b(sT_c,M). \tag{6.6}$$

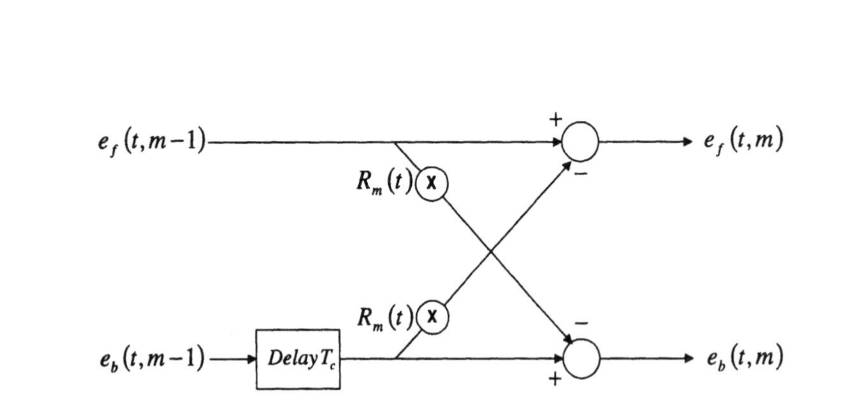

Fig. 6.2 Lattice structure

6.2.1 First stage reflection coefficient

From (6.1) and (6.3), a general expression for the input of the filter can be written as

$$e_f((s-x)T_c,0)=\sqrt{2P}\sum_{k=1}^{K}\beta_k b_k((s-x)T_c-\tau_k)a_k((s-x)T_c-\tau_k)$$

$$\cdot\cos[2\pi f_o(s-x)T_c+\phi_k]+j((s-x)T_c)+n((s-x)T_c). \quad (6.7)$$

Similarly, $e_b((s-y)T_c,0)$ also can be obtained by replacing x with y on the right hand side of (6.7). Note that both x and y are integers. The cross-correlation of the forward and backward prediction errors can be written as

$$E[e_f((s-x)T_c,0)e_b((s-y)T_c,0)]=Z_s(x,y)+Z_j(x,y)+Z_n(x,y). \quad (6.8)$$

Since $E\{a_{k_1}((s-x)T_c-\tau_{k_1})a_{k_2}((s-y)T_c-\tau_{k_2})\}=0$ for $k_1\neq k_2$, $Z_s(x,y)$ is given by

$$Z_s(x,y)=2PK\rho\varepsilon(x,y)\cos[2\pi f_o(x-y)T_c]\delta(x,y)$$

$$=2PK\rho\delta(x,y), \quad (6.9)$$

where $E[\beta_k^2]=2\rho$, $\delta(x,y)=\begin{cases}1, & x=y\\ 0, & otherwise\end{cases}$ and $\varepsilon(x,y)$ is the auto-correlation function of the baseband data, $b_k(t)$, with $\varepsilon(x,y)$=1 for x=y. Assuming that f_oT_c is an integer, the term $Z_j(x,y)$ is given by

$$Z_j(x,y)=E[j((s-x)T_c)j((s-y)T_c)],$$

$$=J\frac{\sin(\pi(x-y)T_cB_j)}{\pi(x-y)T_cB_j}\cos[2\pi\Delta(x-y)T_c]$$

$$=J\sin c[2\pi(x-y)p]\cos[2\pi(x-y)q], \quad (6.10)$$

where J stands for the average power of the narrowband signal. The function $\sin c[x]$ is defined as $\sin c[x] = \sin(x)/x$. Finally, the term $Z_n(x, y)$ is given by

$$Z_n(x, y) = E[n((s-x)T_c)n((s-y)T_c)]$$

$$= \frac{N_o}{2} \cdot 2B_c\delta(x, y) = N_oB_c\delta(x, y). \tag{6.11}$$

From (6.9-6.11), $E[e_f((s-x)T_c,0)e_b((s-y)T_c,0)]$ is an even function against $x - y$. Therefore, defining

$$c(|x-y|T_c) = E[e_f((s-x)T_c,0)e_b((s-y)T_c,0)], \tag{6.12}$$

one obtains from (6.8-6.11),

$$c(0) = 2PK\rho + J + N_oB_c, \tag{6.13}$$

$$c(T_c) = J \sin c(2\pi p)\cos(2\pi q), \tag{6.14}$$

$$c(2T_c) = J \sin c(4\pi p)\cos(4\pi q). \tag{6.15}$$

According to [9] and Fig. 6.2, the general recursive equation for reflection coefficient of stage n is given by

$$R_n((j+1)T_a) = [1 - \alpha e_b^2(jT_a - T_c, n-1)]R_n(jT_a) + \alpha e_f(jT_a, n-1)e_b(jT_a - T_c, n-1) \tag{6.16}$$

where α is a convergence factor (or adaptation step size), "j" represents the jth adaptation (iteration) of the coefficient and T_a stands for the adaptation period. The minimum value of T_a is the value which guarantees that both the current (the jth adaptation) and previous (the $(j-1)$th adaptation) input signals of the filter are statistically independent. Normally, $T_a \geq 2/B_j$, where $2/B_j$ is the approximate correlation time of the narrowband signal. Most often, via a central limit theorem, it is argued that the steady-state coefficients of the filter are jointly Gaussian [6,10] for small adaptation step size. In (6.16), the nth stage derivative of reflection coefficients in terms of forward and backward errors is shown in the following

$$R_n(jT_a) = \left[1 - \alpha e_b^2((j-1)T_a - T_c, n-1)\right] R_n((j-1)T_a) + \alpha e_f((j-1)T_a, n-1) \cdot e_b((j-1)T_a - T_c, n-1)$$

$$\vdots$$

$$R_n(2T_a) = \left[1 - \alpha e_b^2(T_a - T_c, n-1)\right] R_n(T_a) + \alpha e_f(T_a, n-1) e_b(T_a - T_c, n-1)$$

$$R_n(T_a) = \left[1 - \alpha e_b^2(-T_c, n-1)\right] R_n(0) + \alpha e_f(0, n-1) e_b(-T_c, n-1)$$

Assuming $R_n(T_a) = 0$ (initialization), one obtains

$$R_n((j+1)T_a) = \alpha e_f(jT_a, n-1) e_b(jT_a - T_c, n-1) + \alpha \sum_{x=1}^{j} e_f((j-x)T_a, n-1) \cdot e_b((j-x)T_a - T_c, n-1) \prod_{r=1}^{x} \left[1 - \alpha e_b^2((j-r)T_a - T_c, n-1)\right]. \quad (6.17)$$

Substituting n with 1 in (6.17), the first stage reflection coefficient $R_1((j+1)T_a)$ can be obtained

$$R_1((j+1)T_a) = \alpha e_f(jT_a, 0) e_b(jT_a - T_c, 0) + \alpha \sum_{x=1}^{j} e_f((j-x)T_a, 0) e_b((j-x)T_a - T_c, 0) \cdot \prod_{r=1}^{x} \left[1 - \alpha e_b^2((j-r)T_a - T_c, 0)\right]. \quad (6.18)$$

Since the jth and (j-1)th adaptation input signals of the filter are independent, the expectation of $R_1((j+1)T_a)$ can be written as

$$E[R_1((j+1)T_a)] = \alpha E\left[e_f(jT_a, 0) e_b(jT_a - T_c, 0)\right] + \alpha \sum_{x=1}^{j} E\{e_f((j-x)T_a, 0) e_b((j-x)T_a - T_c, 0) \cdot \left[1 - \alpha e_b^2((j-x)T_a - T_c, 0)\right]\} \prod_{r=1}^{x-1} \{1 - \alpha E\left[e_b^2((j-r)T_a - T_c, 0)\right]\}, \quad (6.19)$$

where $E\left[e_b^2((j-r)T_a-T_c,0)\right]=c(0)$, given by (6.13), represents the average power of the input of the lattice filter. It is assumed that

$$\lambda=\alpha c(0) \tag{6.20}$$

and

$$g_1=1-\lambda, \tag{6.21}$$

where λ is the scaled adaptation step size and practically, $0<\lambda<<1$ (or $0<g_1<1$). Thus,

$$E[R_1((j+1)T_a)]=\alpha E\left[e_f(jT_a,0)e_b(jT_a-T_c,0)\right]+\alpha\sum_{x=1}^{j}E\left[e_f((j-x)T_a,0)e_b((j-x)T_a-T_c,0)\right]g_1^{x-1}$$

$$-\alpha^2\sum_{x=1}^{j}E\left[e_f((j-x)T_a,0)e_b^3((j-x)T_a-T_c,0)\right]g_1^{x-1}, \tag{6.22}$$

where from (6.12) $E\left[e_f(jT_a,0)e_b(jT_a-T_c,0)\right]=c(T_c)$. Based on the central-limit theorem, when the number of active users, K, is large $(K>>1)$, the sum of all CDMA signals can be approximated by a Gaussian random variable. Since the narrowband signal is also Gaussian, the input signals $e_f((j-x)T_a,0)$ and $e_b((j-x)T_a-T_c,0)$ are Gaussian. There is a well-known result, $E[X_1X_2X_3X_4]=E[X_1X_2]E[X_3X_4]+E[X_1X_3]E[X_2X_4]+E[X_1X_4]E[X_2X_3]$, when all X_1, X_2, X_3, X_4 are Gaussian. Therefore, one obtains

$$E\left[e_f((j-x)T_a,0)e_b^3((j-x)T_a-T_c,0)\right]=3E\left[e_b^2((j-x)T_a-T_c,0)\right]$$

$$\cdot E\left[e_f((j-x)T_a,0)e_b((j-x)T_a-T_c,0)\right]$$

$$=3c(0)c(T_c). \tag{6.23}$$

Therefore, (6.22) becomes

$$E[R_1((j+1)T_a)]=\alpha c(T_c)+\left[\alpha c(T_c)-3\alpha^2c(0)c(T_c)\right]\cdot\sum_{x=1}^{j}g_1^{x-1}$$

$$= \alpha c(T_c) + \left[\alpha c(T_c) - 3\alpha^2 c(0)c(T_c)\right]\left[\frac{1-g_1^j}{1-g_1}\right]. \tag{6.24}$$

It can be seen from (6.24) that the separate steady state and transient components of the coefficient are shown. Since $\lim_{j\to\infty} g_1^j = 0$, from (6.13-6.14) and (6.20-6.21) the steady state mean of the first stage coefficient is given by

$$E[R_1] = \lim_{j\to\infty} E[R_1((j+1)T_a)]$$

$$= \frac{c(T_c) - 2\alpha c(0)c(T_c)}{c(0)},$$

$$= (1-2\lambda)\cdot\frac{(J/S)\cdot \sin c(2\pi p)\cos(2\pi q)}{K + J/S + 2N(E_b/N_o)^{-1}}, \tag{6.25}$$

where $J/S = J/(2P\rho)$ is the ratio of narrowband interference to signal power, $N = T_b/T_c$ is processing gain and $E_b = 2P\rho T_b$.

6.2.2 Second stage reflection coefficient

Since finding closed-form solutions to the coefficients is impossible for the case of the second stage, the second-stage recursion equation is presented. Substituting n with 2 in (6.16), the second stage reflection coefficient is given by

$$R_2((j+1)T_a) = \left[1 - \alpha e_b^2(jT_a - T_c,1)\right]R_2(jT_a) + \alpha e_f(jT_a,1)e_b(jT_a - T_c,1) \tag{6.26}$$

It is seen that $R_2((j+1)T_a)$ is a function of the forward and backward errors of the first stage. Since it is only a function of the past inputs of the second stage, $R_2(jT_a)$ is independent of current inputs, $e_f(jT_a,1)$ or $e_b(jT_a - T_c,1)$. Therefore,

$$E[R_2((j+1)T_a)] = \left\{1 - \alpha E\left[e_b^2(jT_a - T_c,1)\right]\right\}E[R_2(jT_a)] + \alpha E\left[e_f(jT_a,1)e_b(jT_a - T_c,1)\right] \tag{6.27}$$

In order to obtain the mean of the coefficient, the following equations will be used (see Fig. 6.2)

$$e_f(jT_a,1) = e_f(jT_a,0) - R_1(jT_a)e_b(jT_a - T_c,0) \tag{6.28}$$

and

$$e_b(jT_a,1) = e_b(jT_a - T_c,0) - R_1(jT_a)e_f(jT_a,0). \tag{6.29}$$

Both $E[e_f(jT_a,1)e_b(jT_a - T_c,1)]$ and $E[e_b^2(jT_a - T_c,1)]$ in (6.27) are given by [11]

$$\alpha E[e_f(jT_a,1)e_b(jT_a - T_c,1)] = \lambda\left[\frac{c(2T_c)}{c(0)} + A + B(j)\right] \tag{6.30}$$

$$\alpha E[e_b^2(jT_a - T_c,1)] = \lambda[1 + A + B(j)], \tag{6.31}$$

where $c(\cdot)$ is given by (6.12), A and $B(j)$ represent the steady-state and convergent components, respectively, and are given by

$$A = -2\lambda\,(1-\lambda)[c(T_c)/c(0)]^2 + \lambda^2 - 2(1-\lambda)(1-3\lambda)[c(T_c)/c(0)]^2$$

$$+\frac{\lambda}{2-3\lambda}\left\{1 + 2[c(T_c)/c(0)]^2 - 2\lambda E[Z_1^2 Z_2^4] + \lambda^2 E[Z_1^2 Z_2^6]\right\}$$

$$+2[c(T_c)/c(0)]\cdot\frac{(1-3\lambda)}{(2-3\lambda)}\left\{[c(T_c)/c(0)](1-6\lambda) + \lambda^2 E[Z_1 Z_2^5]\right\} \tag{6.32}$$

and

$$B(j) = 2(1-\lambda)(1-3\lambda)[c(T_c)/c(0)]^2\, g_1^{j-1} - \lambda\,\frac{g_2^{j-1}}{2-3\lambda}$$

$$\cdot\left\{1 + 2[c(T_c)/c(0)]^2 - 2\lambda E[Z_1^2 Z_2^4]\right.$$

$$\left.+\lambda^2 E[Z_1^2 Z_2^6]\right\} - 2[c(T_c)/c(0)]\left[g_1^{j-1} - \frac{g_2^{j-1}}{(2-3\lambda)}\right]$$

$$\cdot\left\{ [c(T_c)/c(0)](1-6\lambda)+\lambda^2 E[Z_1 Z_2^5] \right\}, \tag{6.33}$$

where g_2 is defined as

$$g_2 = 1 - 2\alpha c(0) + 3\alpha^2 c^2(0) = 1 - 2\lambda + 3\lambda^2, \tag{6.34}$$

and 0< g_2 <1 when $\lambda < 0.8$. Z_1 and Z_2 are zero-mean Gaussian variables with $E[Z_1 Z_2] = c(T_c)/c(0)$ and $E[Z_1^2] = E[Z_2^2] = 1$. Starting with the initial value of *j=1* and setting $E[R_2(T_a)] = 0$, substitute the mean of $E[R_2(jT_a)]$ at each iteration in (6.27) in order to obtain the mean of the second stage coefficient, $E[R_2((j+1)T_a)]$. Finally, by letting j approach infinity in (6.27), one obtains

$$\lim_{j\to\infty} E[R_2((j+1)T_a)] = \lim_{j\to\infty}\{[\, 1-\alpha E(e_b^2(jT_a - T_c,1)) \,]E[R_2(jT_a)]$$

$$+\alpha E[e_f(jT_a,1)e_b(jT_a - T_c,1)]\}. \tag{6.35}$$

Since $E[R_2] = \lim_{j\to\infty} E[R_2((j+1)T_a)] = \lim_{j\to\infty} E[R_2(jT_a)]$ and $\lim_{j\to\infty} B(j) = 0$, the steady state mean of the second stage coefficient is given by

$$E[R_2] = \frac{\lim_{j\to\infty} \alpha E[e_f(jT_a,1)e_b(jT_a - T_c,1)]}{\lim_{j\to\infty} \alpha E[e_b^2(jT_a - T_c,1)]} = \frac{c(2T_c)/c(0) + A}{1 + A}. \tag{6.36}$$

6.3 SIGNAL TO NOISE RATIO

The output of the lattice filter is given by

$$r_f(t) = \sum_{m=0}^{M} w_m r(t - mT_c) = e_f(t,M), \tag{6.37}$$

where M denotes the number of stages of the lattice filter, w_m are weights which can be obtained from reflection coefficients. The transformation between the weights and the set of reflection coefficients is highly nonlinear. However, both implementations are mathematically equivalent. But there are some practical differences such as convergence and tracking performances. The relationship between coefficients and weights can easily be derived.

For a two-stage filter (M=2), from Fig. 6.2, the first stage forward and backward errors are given by

$$e_f(t,1) = r(t) - R_1(t) \cdot r(t - T_c) \tag{6.38}$$

and

$$e_b(t,1) = r(t - T_c) - R_1(t) \cdot r(t). \tag{6.39}$$

The second stage forward error is given by

$$e_f(t,2) = e_f(t,1) - R_2(t) \cdot e_b(t - T_c, 1). \tag{6.40}$$

Substitute (6.38) and (6.39) into (6.40) and one obtains

$$r_f(t) = e_f(t,2) = r(t) - [R_1(t) - R_1(t)R_2(t)] \cdot r(t - T_c) - R_2(t) \cdot r(t - 2T_c)$$

$$= \sum_{m=0}^{2} w_m r(t - mT_c). \tag{6.41}$$

Then, the corresponding weights are

$$\begin{cases} w_0 = 1 \\ w_1 = -R_1(t)(1 - R_2(t)) \\ w_2 = -R_2(t) \end{cases} \tag{6.42}$$

Assuming that the ith user is the reference user and $\tau_i = \phi_i = 0$, the despreader output is given by

$$\xi(\gamma) = \int_{\gamma T_b}^{(\gamma+1)T_b} r_f(t) 2a_i(t) \cos(2\pi f_o t) dt, \tag{6.43}$$

where $2\cos(2\pi f_o t)$ is the local recovered carrier and $a_i(t)$ is the spreading sequence of the reference user. Since the high frequency terms are removed by the integrator, the above expression reduces to

$$\xi(\gamma) \approx \sqrt{2P}\beta_i T_b b_i^{(\gamma)} + J(\gamma) + N(\gamma) + \sum_{\substack{k=1 \\ k \neq i}}^{K} I_k \,. \tag{6.44}$$

where $b_i^{(\gamma)}$ stands for the γth bit of the ith user. The first term on the right hand side of (6.44) is the desired signal and the useful signal power is $4P\rho T_b^2$. $J(\gamma)$ is due to the Gaussian narrowband signal, given by

$$J(\gamma) = \sum_{m=0}^{M} w_m \int_{\gamma T_b}^{(\gamma+1)T_b} \{ j_c(t - mT_c)\cos[2\pi(f_o + \Delta)(t - mT_c)] + j_s(t - mT_c)$$

$$\cdot \sin[2\pi(f_o + \Delta)(t - mT_c)]\} \cdot 2a_i(t)\cos(2\pi f_o t)dt \,.$$

Since the double frequency components are removed by the integrator, $J(\gamma)$ is approximated by

$$J(\gamma) \approx \sum_{m=0}^{M} w_m \sum_{n=0}^{N-1} a_i^{(n)} \int_{nT_c}^{(n+1)T_c} \{ j_c(t + \gamma T_b - mT_c)\cos[2\pi\Delta(t + \gamma T_b - mT_c)]$$

$$+ j_s(t + \gamma T_b - mT_c)\sin[2\pi\Delta(t + \gamma T_b - mT_c)] \}dt$$

$$= \sum_{m=0}^{M} w_m \sum_{n=0}^{N-1} a_i^{(n)} \int_{nT_c}^{(n+1)T_c} \hat{j}(t + \gamma T_b - mT_c)dt \,, \tag{6.45}$$

where $a_i^{(n)}$ represents the *n*th chip of sequence $a_i(t)$ and $\hat{j}(t)$ is the low-frequency version of $j(t)$, given by (6.2), and is also Gaussian, defined by

$$\hat{j}(t) = j_c(t)\cos(2\pi\Delta t) + j_s(t)\sin(2\pi\Delta t) \,. \tag{6.46}$$

The auto-correlation function of $\hat{j}(t)$ is the same as (6.10) and the variance of $J(\gamma)$ is given by [14]

$$\sigma_J^2 = \frac{J \cdot T_b^2}{N} \sum_{m_1=0}^{M} \sum_{m_2=0}^{M} E\left[w_{m_1} w_{m_2}\right] Q(m_1, m_2) \tag{6.47}$$

where $Q(m_1, m_2)$ is defined as

$$Q(m_1, m_2) = \int_{-1}^{1} (1 - |\tau|) \sin c\left[2\pi p(\tau - m_1 + m_2)\right] \cos\left[2\pi q(\tau - m_1 + m_2)\right] d\tau . \tag{6.48}$$

In (6.44), $N(\gamma)$ is due to the thermal noise with variance

$$\sigma_N^2 = N_o T_b \sum_{m=0}^{M} E\left[w_m^2\right]. \tag{6.49}$$

Finally, analogous to [2], the variance of the multiple access interference, $\left(\sum_{k=1}^{K} I_k \right)$, is given by

$$\sigma_I^2 = \frac{4P(K-1)\rho T_b^2}{3N} \left[\sum_{m=0}^{M} E\left[w_m^2\right] + \frac{1}{2} \cdot \sum_{m=0}^{M} E\left[w_m w_{m+1}\right] \right]. \tag{6.50}$$

The output signal to noise ratio is given by

$$SNR = \frac{4P\rho T_b^2}{\sigma_J^2 + \sigma_N^2 + \sigma_I^2}$$

$$= \left\{ \frac{1}{2} \cdot \left(\frac{E_b}{N_o} \right)^{-1} \sum_{m=0}^{M} E\left[w_m^2\right] + \frac{(J/S)}{2N} \sum_{m_1=0}^{M} \sum_{m_2=0}^{M} E\left[w_{m_1} w_{m_2}\right] Q(m_1, m_2) \right.$$

$$\left. + \frac{(K-1)}{3N} \left[\sum_{m=0}^{M} E\left[w_m^2\right] + \frac{1}{2} \cdot \sum_{m=0}^{M} E\left[w_m w_{m+1}\right] \right] \right\}^{-1}. \tag{6.51}$$

6.4 LMS FILTERING

By replacing a lattice filter with a LMS filter, the CDMA receiver with an adaptive LMS filter is constituted (Fig. 6.3). As in [1], the adaptive LMS filter is modelled as consisting of a Wiener filter and a misadjustment filter

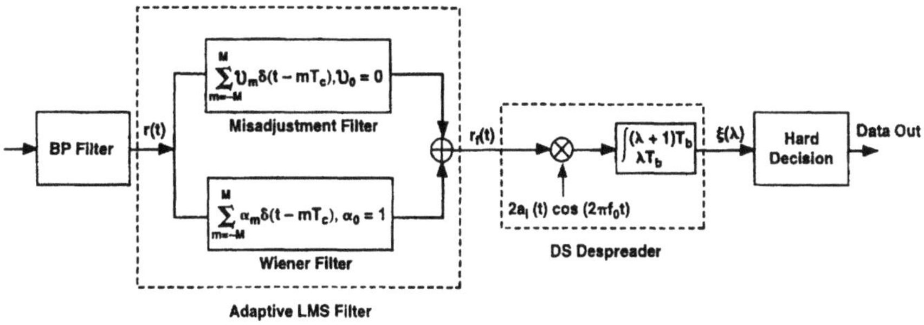

Fig. 6.3 Receiver model of CDMA overlay with LMS filters

operating in parallel. The output of the LMS filter is given by

$$r_f(t) = \sum_{m=-M}^{M} (\alpha_m + v_m) \cdot r(t - mT_c) = \sum_{m=-M}^{M} w_m r(t - mT_c), \qquad (6.52)$$

where

$$w_m = \alpha_m + v_m, \qquad (6.53)$$

α_m denotes the *m*th tap coefficient (deterministic) of the Wiener filter, v_m denotes the *m*th steady-state tap coefficient (random) of the misadjustment filter. Note that v_0 is always zero, because the center tap of the Wiener filter is fixed at 1 (i.e., $\alpha_0 = 1$). It is assumed that $\alpha_m = \alpha_{-m}$ and $v_m = v_{-m}$ for double-sided filters. Since each stage of a lattice filter has two taps (reflection coefficients), the number of taps on each side of the double-sided LMS filter is set the same as the number of stages of the lattice filter for fair comparison (the same number of total taps for both filters). Therefore, the sum in (6.52) is from $-M$ to M for LMS, whereas the sum in (6.37) is from 0 to M for lattice.

From the LMS adaptation algorithm, the tap coefficient vector of the misadjustment filter can be presented as

$$V(j+1) = [I - \alpha X(j)X^T(j)]V(j) + \alpha e^*(j)X(j),$$

$$= \left[I - \frac{\lambda}{c(0)} X(j)X^T(j)\right]V(j) + \frac{\lambda}{c(0)} e^*(j)X(j) \tag{6.54}$$

where $V(j)$ is the column tap weight vector on the *j*th adaptation, *I* stands for an identity matrix, $X(j)$ is the column sample vector of the input signal on the *j*th adaptation, $\lambda = \alpha c(0)$ is the scaled adaptation step size, which is assumed the same for both lattice and LMS filters, and $e^*(j)$ is the prediction error at the Wiener filter output on the *j*th adaptation, and is given by

$$e^*(j) = \sum_{m=-M}^{M} \alpha_m r(t - mT_c) \tag{6.55}$$

Analogous to (6.8-6.15), the steady-state variance of $e^*(j)$ can easily be derived as

$$E[(e^*)^2] = (2KP\rho + N_o B_c) \sum_{m=-M}^{M} \alpha_m^2$$

$$+ J \sum_{m_1=-M}^{M} \sum_{m_2=-M}^{M} \alpha_{m_1} \alpha_{m_2} \operatorname{sin c}(2\pi(m_1 - m_2)p) \cos(2\pi(m_1 - m_2)q) \tag{6.56}$$

When the input signal is Gaussian, the steady-state tap coefficient covariance matrix has been shown in equ. (A13) of [1] as

$$[VV^T] = \lim_{j \to \infty} E[V(j+1)V^T(j+1)] \approx \frac{\lambda}{2c(0)} E[(e^*)^2] I \tag{6.57}$$

With the Gaussian assumption of the input signal, the tap coefficient covariance matrix is a diagonal matrix, which completely defines the statistics of the misadjustment filter for the Gaussian input signal. That is, in the steady state, the variance of different tap coefficients are equal and different tap coefficients are uncorrelated and independent. The performance (SNR) of LMS filtering can be obtained from (6.51) by substituting w_m with $\alpha_m + v_m$ and 0 in the sum with –*M*, respectively, i.e.,

$$\mathrm{SNR}=\left\{\frac{1}{2}\cdot\left(\frac{E_b}{N_o}\right)^{-1}\left(\sum_{m=-M}^{M}E\left[\alpha_m^2\right]+M\frac{\lambda}{c(0)}E[(e^*)^2]\right)\right.$$

$$+\frac{(J/S)}{2N}\left(\sum_{m_1=-M}^{M}\sum_{m_2=-M}^{M}E\left[\alpha_{m_1}\alpha_{m_2}\right]Q(m_1,m_2)+M\frac{\lambda}{c(0)}E[(e^*)^2]Q(0,0)\right)$$

$$\left.+\frac{(K-1)}{3N}\left[\sum_{m=-M}^{M}E\left[\alpha_m^2\right]+M\frac{\lambda}{c(0)}E[(e^*)^2]+\frac{1}{2}\cdot\sum_{m=-M}^{M}E\left[\alpha_m\alpha_{m+1}\right]\right]\right\}^{-1} \quad (6.58)$$

6.5 NUMERICAL AND SIMULATION RESULTS

It is assumed that the ratio of interference bandwidth to that of spread spectrum is $5\%\,(p=0.05)$ and the ratio of the offset of the interference carrier frequency to half that of spread spectrum is 0 $(q=0)$. That is, the narrowband interference is centered at the CDMA spectrum. The processing gain is 750 and the number of stages of the filter is $2\,(M=2)$. The number of simultaneous users is 30 (K=30). The normalised adaptation step size is 0.05 $(\lambda=5\%)$. The ratio of narrowband interference to signal is 25 dB (*J/S*=25 dB).

Figs. 6.4(a) and 6.4(b) illustrate the mean of the first stage and the second stage reflection coefficients of the lattice filter, respectively, as a function of the number of iterations. Also, simulation results with single runs are shown. It is seen from the solid curves of the two figures that it takes about 100 iterations for the coefficients to complete the convergent state. Since the ratio of interference bandwidth to that of spread spectrum is 5% $(p=0.05)$, the bandwidth of the narrowband interference is $0.05\times 2/T_c$. Therefore, the correlation time of the narrowband interference is $20T_c$. Assuming the minimum adaptation period $T_a=20T_c$, 100 iterations means $100T_a=2000T_c\approx 8T_b$. That is, it needs around 8 bits for completion of convergence. It is also seen from the dotted curves of both figures that the simulated coefficients fluctuate around the analytic means. Note that

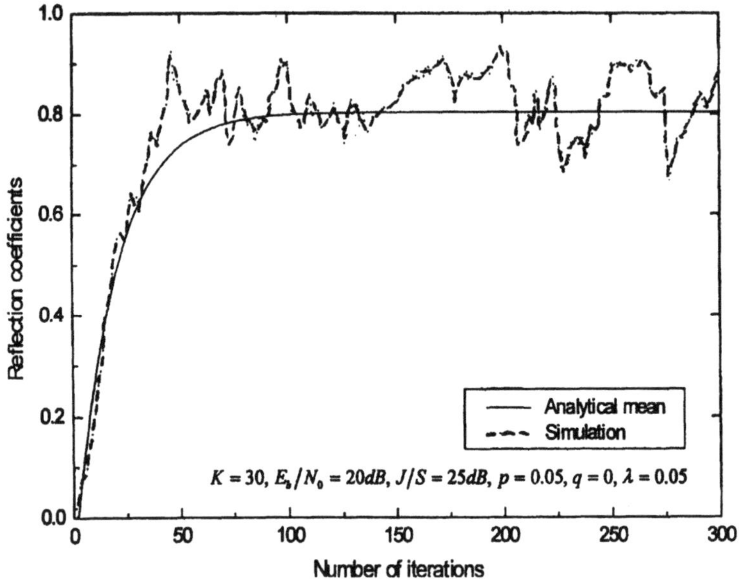

(a) First stage

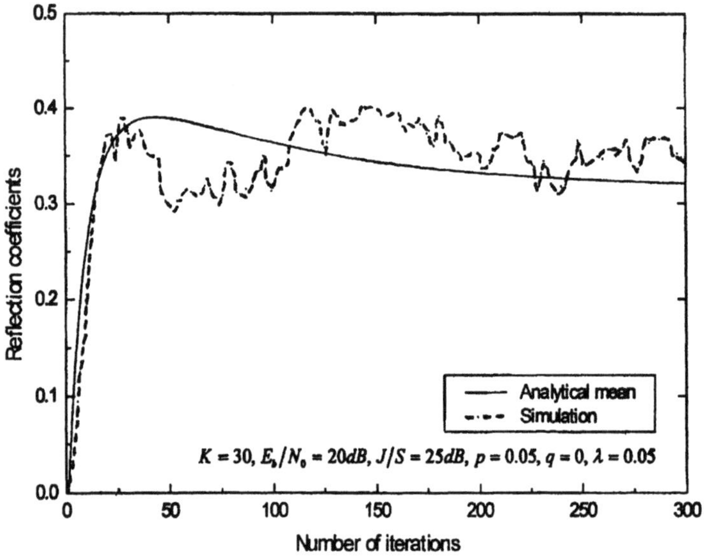

(b) Second stage

Fig. 6.4 Analytic mean and simulation of the reflection coefficients of lattice filters (a) first stage; (b) second stage.

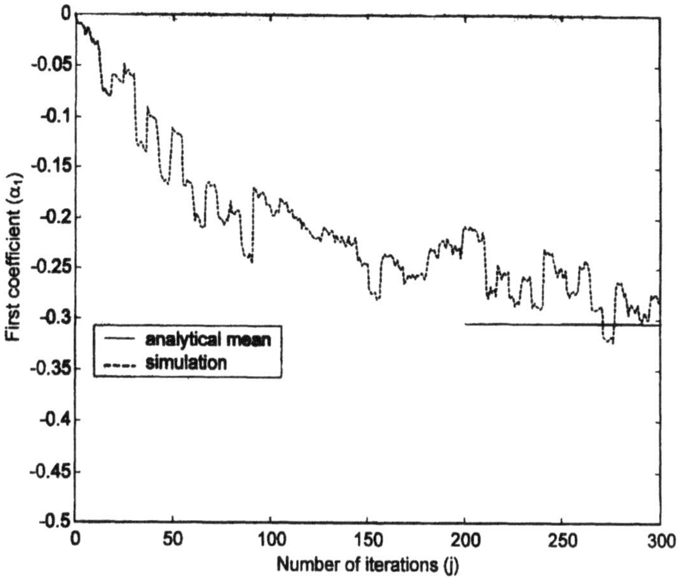

(a) first coefficient (α_1)

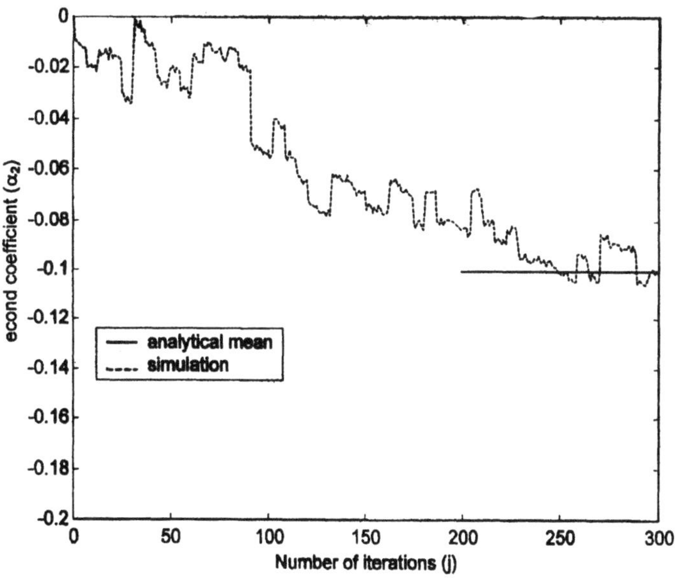

(b) second coefficient (α_2)

Fig. 6.5 Analytic mean and simulation of the coefficients of LMS filters. (a) first coefficient; (b) second coefficients

recursive equation (6.17) is used for the simulation with n=1 and n=2, respectively, $\alpha = \lambda / c(0)$ and $R_1(0) = R_2(0) = 0$.

For comparison, Figs. 6.5(a) and 6.5(b) show the coefficients of the LMS filter with the same parameters as in Figs 6.4(a-b). It is seen from Figs. 6.5(a-b) that it takes more than 200 iterations for the coefficients to converge. That is, the convergence of LMS filters takes longer (about twice) than that of lattice filters. The reason for the lattice to converge fast is that each stage of the lattice converges individually (see (6.16)), independent of the remaining stages (i.e., the various stages of the lattice are decoupled from one another). However, the adaptation of the LMS is related with all taps (or coefficients) together (see (6.54)).

Fig. 6.6 shows the output SNRs of CDMA overlay systems with both lattice and LMS filters. It can be seen that the lattice filter converges faster (about twice) than LMS filter, which is consistent with Figs. 6.5(a-b). However, the SNR performance for both filters is very close in the stable state (i.e., the iteration number j >200). By use of the filters, the SNR performance in the stable state can be improved by as much as 10 dB, compared to that without filters (i.e., j=0).

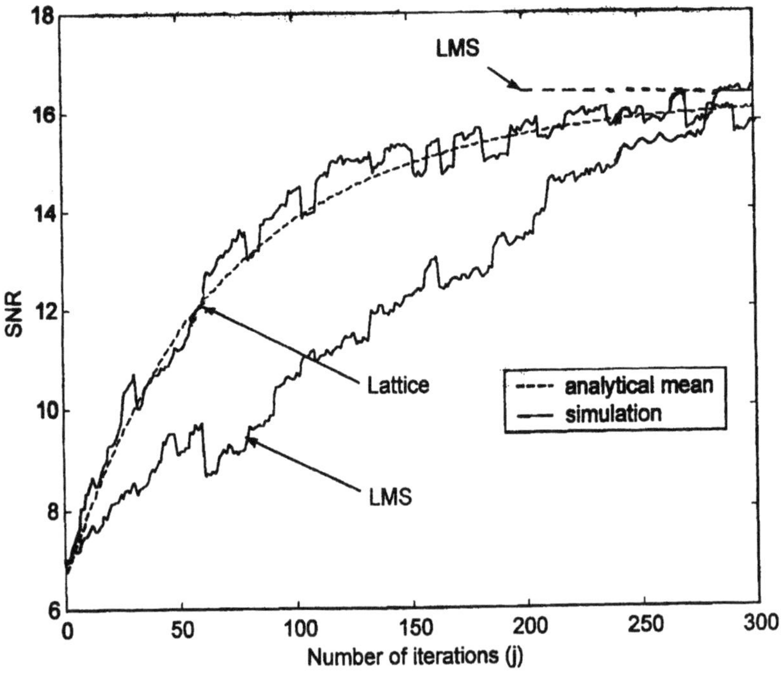

Fig. 6.6 SNR of the CDMA overlay system against the number of iterations

Since the optimal performance of lattice [9] in stable state should be the same as that of LMS, the stable state performance of LMS is shown from [1]. The channel model is multipath Rician fading. Unless otherwise noted, the numerical results for the BERs of the adaptive CDMA with LMS are presented for the following common parameters. It is assumed that the ratio of the interference bandwidth to the spread spectrum bandwidth is 10% $(p=0.1)$ and the ratio of the offset of the interference carrier frequency to half of the spread spectrum bandwidth is either 0 or 20% $(q=0,0.2)$. The processing gain and the number of taps on each side of a suppression filter are set at $N=255$ and $M=2$, respectively. The number of paths, the propagation exponent and the ratio of the specular component power to the fading component power are assumed to be $L=3$, $\gamma=3$ and $H=7\ dB$, respectively. Note that $\gamma=3$ means that the adjacent cell interference $\zeta(r)=.97$. That is, the interference from all adjacent cells is 97% of the interference of the cell-of-interest. Notice that cellular model is used here (see [1]). Finally, the adaptation step size is selected as

$$\mu=\frac{1}{10\lambda_{\max}}\leq\frac{1}{20M\rho_r(0)}, \tag{6.59}$$

where $\lambda_{\max}$ and $\rho_r(0)$ are the maximum eigenvalue of the covariance matrix of the input signal and the power of the input signal, respectively.

Fig. 6.7 illustrates the BERs of the adaptive CDMA overlay system as a function of E_b/N_0. Fig. 6.7(a) and Fig. 6.7(b) correspond to q=0 and q=0.2, respectively. It is seen that, as expected, the adaptive LMS filter is very effective in rejecting the narrowband interference. Note that for q=0, the adaptive filter is not as effective as it is for q=0.2, compared to the Wiener filter.

Fig. 6.8 shows the asymptotic $(E_b/N_0\rightarrow\infty)$ BERs of the CDMA system as a function of $(J/S)/[(1+\zeta(\gamma))KL]$, which is roughly the ratio of J/S to all multiple access interference (see [1]). When J/S is sufficiently small (i.e., $(J/S)/[(1+\zeta(\gamma))KL]$<<0dB), neither a Wiener filter nor an adaptive filter is necessary, because the multiple access interference dominates, whereas when J/S is sufficiently large, an adaptive filter can provide a large improvement in performance.

In Fig. 6.9, the asymptotic BER of the CDMA overlay system is plotted as a function of the number of the active users K per cell, for various values of J/S. It is clearly seen that the system using a suppression filter can support many more users than can the system without a filter.

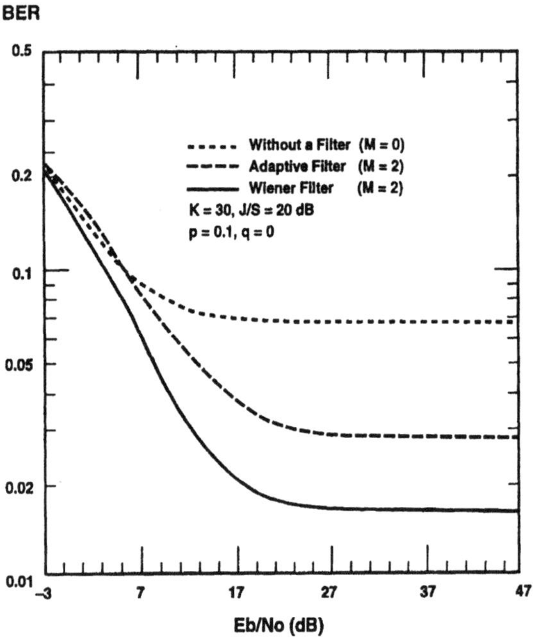

(a) q = 0

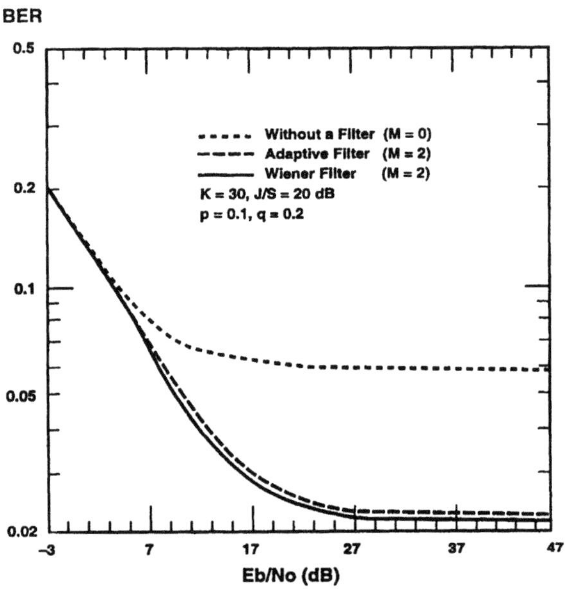

(b) q = 0.2

Fig. 6.7 BER comparison of the CDMA system with a Wiener two-sided filter both for cases that wideband signals are and are not disturbed by the filter.

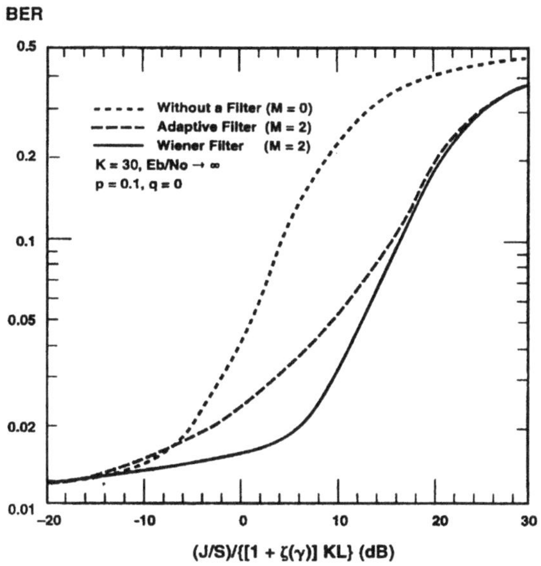

(a) q = 0

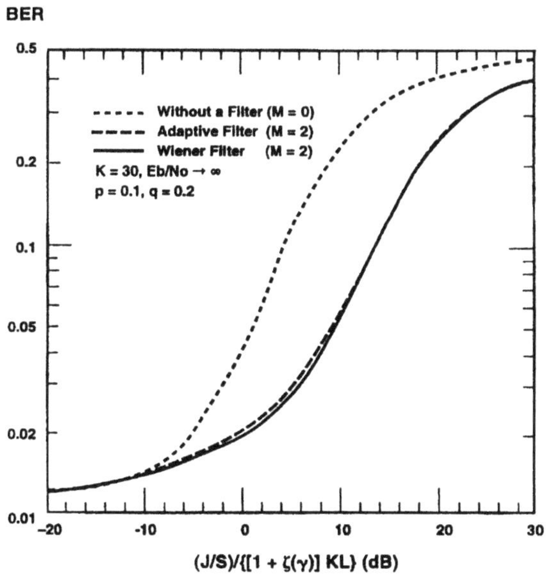

(b) q = 0.2

Fig. 6.8 BER performance of the CDMA overlay system vs. E_b/N_0, (a) q=0, (b) q=0.2.

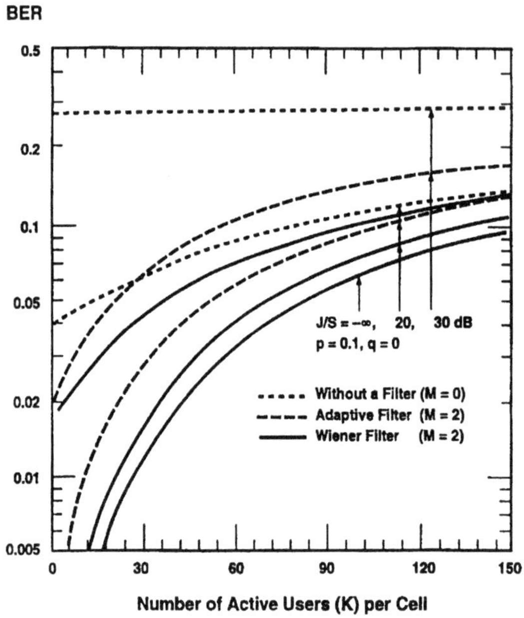

(a) q = 0

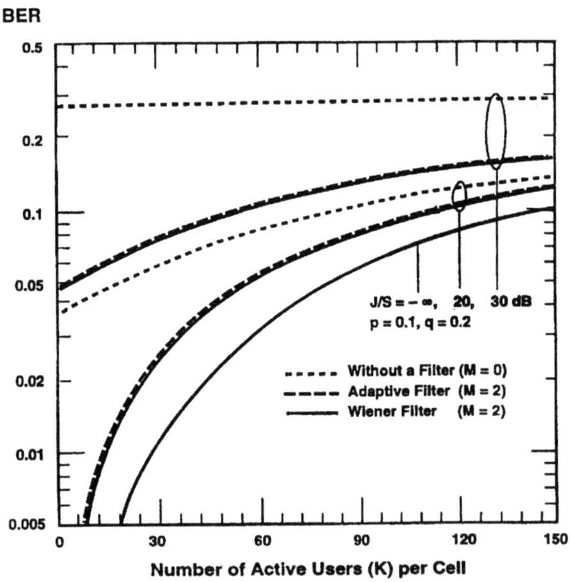

(b) q = 0.2

Fig. 6.9 The asymptotic BER of CDMA overlay system as a function of the number of active users (K) each cell, (a) q=0, (b) q=0.2.

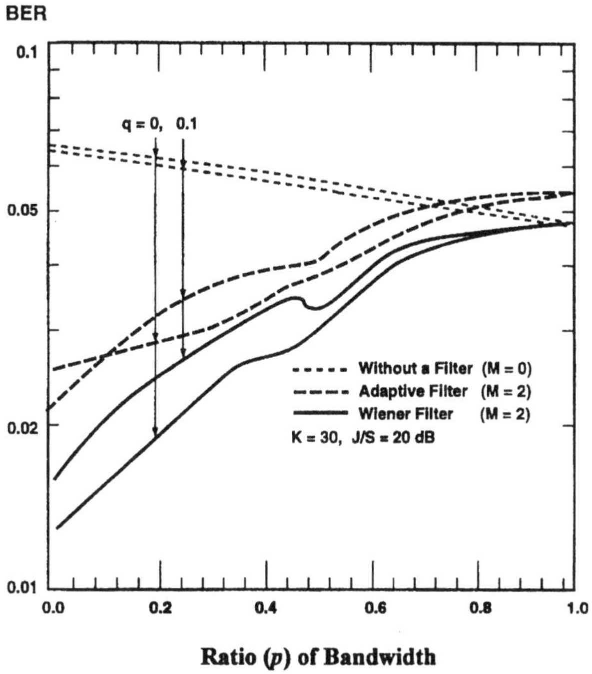

(a) q = 0, 0.1

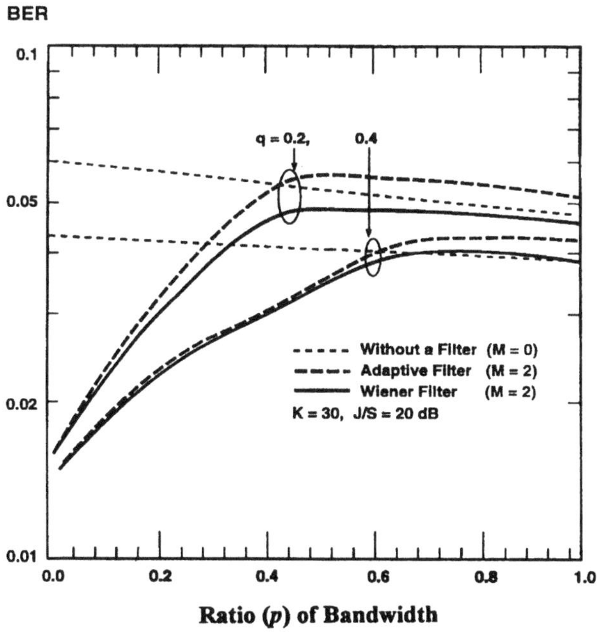

(b) q = 0.2, 0.4

Fig. 6.10 The asymptotic BER of CDMA overlay system as a function of the ratio (p) of the interference BW to spread spectrum BW, (a) q=0, 0.1, (b) q=0.2, 0.4.

In Fig. 6.10, the asymptotic BER of the CDMA overlay system is shown as a function of the ratio (p) of the narrowband bandwidth to the spread spectrum bandwidth for q=0, 0.1, 0.2, and 0.4. It is seen that when p is small, an adaptive filter is very effective.

Fig. 6.11 illustrates the BER performance of the overlay system as a funciton of the ratio (q) of the offset of theinterference carrier frequency to the half spread spectrum bandwidth for different number of taps on each side. It is seen that when q is small, the adaptive filter performs worse than the Wiener filter. However, when q Is large, the BERs of the systems with adaptive filter and Wiener filter are almost identical. That is, the noise caused by the misadjustment filter is negligible.

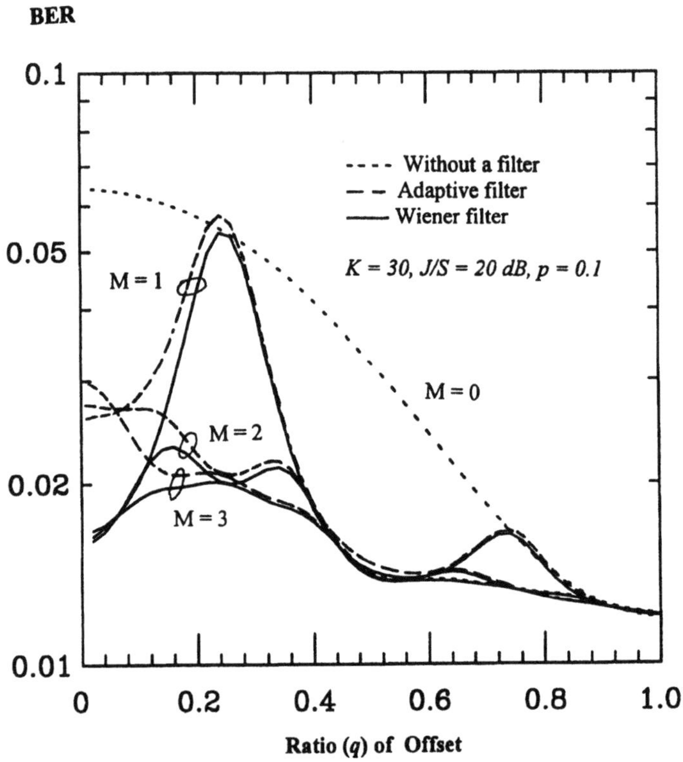

Fig. 6.11 BER's as a function of the ratio (q) of the interference carrier frequency to the half spread spectrum bandwidth.

In conclusion, both lattice and LMS provide good capability of narrowband interference suppression in CDMA overlay. However, because each stage of the lattice converges individually, independent of the remaining stages, the lattice filter provides fast rate of convergence, while having good capability of narrowband interference suppression.

REFERENCES

[1] J. Wang and L. B. Milstein, "Adaptive LMS filters for cellular CDMA Overlay situations", *IEEE Journal on Selected Areas in Communications*, vol. JSAC-14, pp. 1548-1559, October 1996.

[2] J. Wang and L. B. Milstein, "CDMA Overlay situations for Microcellular Mobile Communications," *IEEE Trans. Commun.*, vol. 43, pp. 603-614, Feb./Mar./April 1995.

[3] P. Wei, J. R. Zeidler and W. H. Ku, "Adaptive Interference Suppression for CDMA Overlay Systems", *IEEE Journal on Selected Areas in Communications*, vol. 12, pp. 1510-1523, December 1994.

[4] B. J. Rainbolt and S. L. Miller, "CDMA transmitter filtering for cellular overlay systems," To appear in *IEEE Trans. on Communications*.

[5] H.V. Poor and L.A. Rush "Narrowband Interference Suppression in Spread Spectrum Communications", *IEEE Personal Communications,* pp. 14-27, Third Quarter, 1994.

[6] N. J. Bershad, "Error probabilities of DS Spread Spectrum Systems using an ALE for narrowband interference rejection", *IEEE Trans. on Communications*, vol. 36, No. 5, pp. 588-595, May 1988.

[7] R. Iltis and L.B. Milstein, "An approximate statistical analysis of the Widrow LMS algorithm with applications to narrowband interference rejection," *IEEE Trans. Commun.*, vol. COM-33, pp. 121-130, Feb. 1985.

[8] G. J. Saulnier, "Suppression of Narrowband Jammers in a Spread Spectrum Receiver Using Transform-Domain Adaptive Filtering," *IEEE J. Select. Areas Commun.*, vol. JSAC-10, pp. 742-749. April 1992.

[9] K. C. Chew, T. Soni, J. R. Zeidler and W. H. Ku, "Tracking model of adaptive lattice filter for a linear chirp FM signal in noise", *IEEE Trans. Signal Processing*, vol. 42, pp. 1939-1951, August 1994.

[10] B. Widrow and S. D. Stearns, Adaptive Signal Processing, Englewood Cliffs, NJ: Prentice Hall, 1985

[11] V. Prahatheesan, "A lattice filter for CDMA overlay," M. Phil. Thesis, The University of Hong Kong, 1998.

[12] M. Lops, G. Ricci and A. M. Tulino, "Narrowband interference suppression in multiuser CDMA systems," IEEE Trans. Commun., vol. COM-46, pp. 1163-1175, Sept. 1998.

[13] G. Gelli, L. Paura and A. M. Tulino, "Cyclostationarity-based filtering for narrowband interference suppression in direct sequence

spread spectrum systems," IEEE Journal on Selected Areas in Commun., vol. 16, pp. 1747-1755, Dec. 1998.

[14] J. Wang and V. Prahatheesan, "Adaptive lattice filters for CDMA overlay," *IEEE Transactions on Communications,* vol. COM-48, pp.820-829, May 2000.

Chapter 7

Open-Loop Power Control in CDMA Overlay

Abstract Adaptive power control has widely been used in DS/CDMA systems to overcome the so-called "near-far" problem. This chapter studies the adaptive open-loop power control of a cellular CDMA system, which is overlaid in downlink by a narrowband signal. The effects of downlink power allocation schemes to power control error in the presence of narrowband interference are analyzed and numerically computed. In order to get a minimum power control error in the CDMA overlay situations, an optimum downlink power allocation scheme is used, which works well for wide range of signal to narrowband interference ratio.

Keywords CDMA, Open-Loop Power Control, Downlink Power Allocation, Narrowband Interference Suppression.

7.1 OVERVIEW

Because all users use the same frequency channel, CDMA systems have intrinsic disadvantages. Among them, the two most prominent are *self-jamming* and the *near-far* effect [32].

Self-jamming is also called multiple-access interference (MAI) caused by non-orthogonal code sequences. In cellular mobile systems, mobile users transmit independently from each other and their signals arrive asynchronously at the base station. Since their relative time delays are randomly distributed, the cross-correlation between the received signals coming from different users is nonzero. To achieve a low level of interference, the assigned PN sequences need to have low cross-correlation values for all relative time delays. A low cross-correlation between sequences is obtained by designing a set of orthogonal sequences. However, there is no set of code sequences that is completely orthogonal when used in an asynchronous system. The non-orthogonal components of signals of other users appear in the demodulated desired signal as interference.

The near-far effect is another major disadvantage of CDMA. It happens when a weak signal received at the base station from a far mobile user is overpowered by a strong signal coming from a nearby interference. An interfering signal with a power of n times higher than that of the desired signal will approximately have the same effect on system capacity as n interferes of the same power as the desired signal. To combat the near-far problem, power control must be used in most DS-CDMA systems [1-3]. Power control is carried out by each base station, which periodically instructs the mobile users to adjust their transmitter powers so that all signals are received at the base station at roughly the same level. There have been numerous papers published addressing power control problem in CDMA systems.

7.1.1 Power Control

On the uplink of a DS-CDMA system, each signal is received by the base station with a different path loss due to (1) different distance between each user and the base station; (2) log-normal shadowing and (3) multipath fading. That is, if the all mobile users transmit at the same power, their signals will arrive at the base station with vastly different power levels.

If a particular signal received at the base station is too weak, it will be masked by other stronger signals. On the other hand, if it has too high power, it will cause interference to other user signals in the same cell and degrade their performance. The capacity of a conventional CDMA system is maximized if each user transmitted power is controlled so that at the base station the same power level is received from each mobile. Power control is used on the uplink to combat the near-far problem and thus minimize the effect of interference on system capacity. This is the primary benefit of power control. In addition to mitigating the near-far problem, in the process of power control, because users always try to keep the transmitted power at a minimum as long as certain BER level is maintained, power control has another benefit to minimize consumption of the transmitted power of the users.

Power control can be either centralized [4-5] or distributed [6-11]. Centralized power control reportedly is capable of achieving optimal performance, although it is costly and for a network of large-scale centralized power control becomes infeasible. Although distributed power control does not provide the best service, it is much less costly to operate. According to the function of power control mechanism, power control can be done in the manner of open or closed-loop.

7.1.2 Open-Loop Power Control (OLPC)

The open-loop power control [12] has two main functions: it adjusts the initial access channel transmission powers of mobile users and tries to equalize the powers of all mobile users received by the base station to avoid the near-far effect. Particularly, it compensates for large abrupt variations in the path loss attenuation where closed-loop power control can't catch up with it. In the open-loop power control algorithm, a mobile user determines an estimate of the path loss on downlink by measuring the received signal strength at the user. Then the mobile user controls its uplink transmitted power according to its downlink power estimation. The smaller the downlink received power, the larger the propagation loss. Thus a larger transmission power is generated, and vice-versa. The reason that the downlink power estimation can be used to adjust the uplink transmission power is that in a terrestrial CDMA communication system with frequency division duplex (FDD), the shadowing on uplink and downlink can be assumed be correlated.

One advantage of open-loop power control is that it has a larger dynamic range (about 80dB) than closed-loop power control (about 30dB). Another advantage of open-loop power control is that it has smaller loop delay than the closed loop power control. Apart from terrestrial CDMA systems, another application of open-loop power control is in the satellite communication [12], where the round-trip delay is larger than the channel correlation time and closed-loop power control will be ineffective. In a real terrestrial DS-CDMA system a combination of open and closed-loop power control can be used to get a better performance.

7.1.3 Closed-Loop Power Control (CLPC)

Because there is a frequency separation between uplink and downlink of a FDD DS-CDMA system, the multipath fading between the two links are uncorrelated. To account for the independence of the Rayleigh fading on the uplink and downlink, the base station should also control the mobile users' transmission powers. In closed-loop power control, the base station uses the averaged power of the signal received from a mobile user to issue a command to inform the mobile user how to adjust its transmitted power. A typical closed-loop power control model is shown in Fig. 7.1.

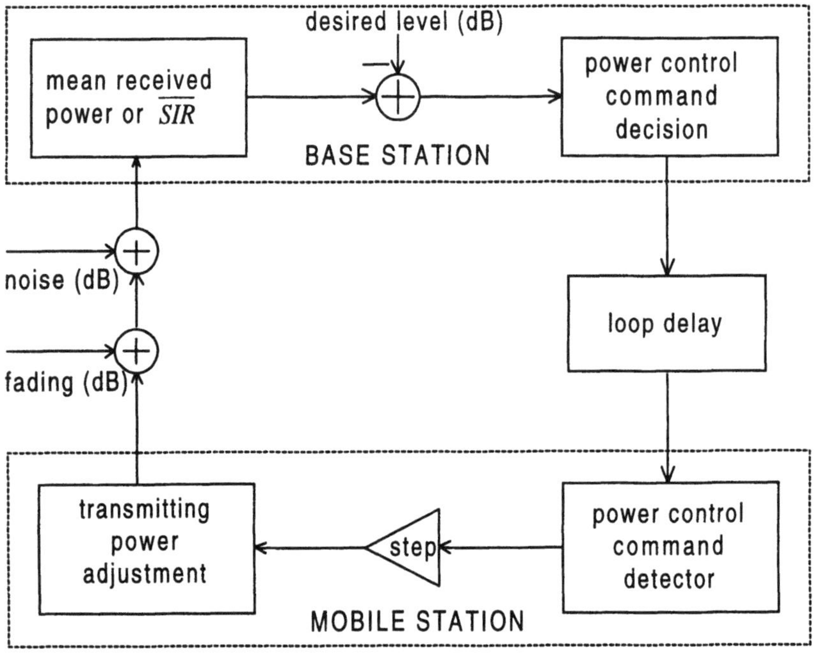

Fig. 7.1 Block disgram of the closed-loop power control [35].

In the design of a power control scheme, power control criteria, command update rate, step size and dynamic range are need to be considered. According to the measured signal to determine whether to raise or lower the transmitted power from a mobile user, closed-loop power control is divided into two categories: strength-based [7,13-15], and signal to interference ratio based (SIR based) [9,14]. The strength-based schemes is based on the original idea of power control — to keep the powers of all signals received at the base station as nearly the same as possible so as to mitigate the near-far effect. The SIR-based mechanism is based on the objective of directly controlling the communication quality or the system performance according to the SIR received at the base station.

SIR is clearly a more accurate indication of signal quality and provides better performance. However, one serious problem associated with SIR-based power control is the potential of positive feedback, which can endanger the stability of the system [9,14]. Positive feedback arises in the situation where one mobile user receives instructions from the base station to raise its transmitted power in order to maintain desirable SIR to the base station. But the increase of its power also results in increased interference to other mobile users, so that other mobile users are then forced to also raise power, and so on. In [16], the authors propose a strength and SIR combined adaptive power control to eliminate the potential positive feedback from SIR-based power control and also maintain good performance.

Command update rate is an another important parameter to determine the accuracy of power control. Faster update rate will make power control more effective. However, a faster update rate will need larger bandwidth on the return channel to carry it. The required command update rate is also relevant to the fading rate. A slower fading rate will need a relative slower update rate for a target performance. It is known that closed-loop power control is effective against multipath fading if the power command update rate is significantly higher than ten times the maximum fading rate [7].

The transmitted power update step size can be either fixed (fixed step size algorithm) [7] or made adaptive to the channel variations. Fixed step size algorithm is easy to be implemented and does not need much bandwidth on the downlink. Adaptive power control algorithm has better performance but need more downlink bandwidth. Several adaptive power control algorithms have been proposed. Schilling gives a delta modulation (DM) algorithm in [17]. A pulse-code-modulation (PCM) is studied in [14], and it has shown that PCM PC could achieve a low outage probability and higher link performance than DM and fixed step PC algorithm. In [18], a combination of fixed-step and adaptive step algorithms using RAKE receivers is proposed. An adaptive fuzzy proportional integral power control has been proposed in [19-20]. The fuzzy logic control overcomes the effect of integral power control. The integral power control scheme of [21] can cause a large overshoot. A trellis-based power control has also shown in [22]. All these adaptive algorithms have shown to achieve a better performance compared to the fixed step power control.

7.1.4 Downlink Power Control

The primary use of power control is on the uplink to combat the near-far effect. Power control can be used on the downlink too. On the downlink, a common pilot signal can be used for channel estimation and time synchronization. Furthermore, since all the signals of one base station can be received at the mobile with equal power, the downlink does not suffer from the near-far problem. So both MAI and near-far problem need not be considered in downlink. However, power control can be applied to improve the performance of mobile users near the cell boundary where the signal is weak and the interference from other base station signals is strong. Note that the downlink power control is of an entirely different nature with respect to that of the uplink; it requires only a limited dynamic range, and need not be fast.

In a downlink power control algorithm, a base station controls its transmitted power to the mobile users to let it have minimum interference to other cell's users and let the intracell users have the same minimum required carrier to interference ratio (CIR) to achieve high capacity. In [3], a simpler power control model has been suggested, based on the *n*th-power of the radial distance from the home base station to the user. However, in its analysis, the adjacent cell interference is neglected for users near the base station. In [23], an optimum model for the CIR for the users within a cell, taking into account the interference from the adjacent cells is proposed, which has a better performance than [3]. In [24], Kim provided a downlink power control algorithm and gave attention to the procedure of adjusting the allocated power levels and explicit the convergence of the CDMA downlink power control methods.

7.1.5 Power Control Error

One objective of power control is to equalize the received powers in the base station to eliminate the near-far effect. However, in practice it cannot eliminate the near-far effect completely due to imperfections in the power control loop. Thus power control error occurs and causes the capacity degradation.

The power control error mainly results from the power estimation inaccuracy, power control loop delay, transmitter's dynamic range. In a real-time power measurement, the square law estimation of the received signal power will cause some error. It has been shown in [25] that it may be possible for a mobile to do real-time power measurement in which the power control error is log-normally distributed with standard deviation of 3dB or less. The loop delay will cause that the power-controlled signal can't catch up fading signal. Increasing the command update rate and adopting the adaptive step size are methods to reduce the effect of loop delay. However, they are limited by the channel bandwidth. Dynamic range will limit the maximum and minimum transmitted power of transmitter and results in the power control error. [26] presents the analysis of power control error in the dynamic range.

The effect of imperfect power control has been studied widely. In [2,21], the performance in terms of system capacity degrades as the power control error increases. Empirical evidence suggests that power control error is log-normally distributed and that the standard deviation for CDMA employing both open-loop and closed-loop power control is between 1.5 and 2.5 dB [27].

Wideband CDMA can reduce the impact of imperfect power control. This is because better diversity results in less variation in the received SIR. In [28], power control feedback delay was shown to be less critical for wideband systems than narrowband systems.

7.1.6 Motivation

It has been shown that DS CDMA systems can share common spectrum with the currently operating cellular or fixed microwave systems [1] in order to achieve efficient bandwidth utilization. The signals of coexisting users would appear as narrowband interference in the spectrum of a wideband signal. As known, spread spectrum modulation has an inherent ability to reject narrowband interference. The effectiveness of narrowband interference rejection depends on the processing gain. When the processing gain is limited due to bandwidth constraints and the narrowband interference is very large, interference suppression can be substantially enhanced by filtering the received signal prior to the despreading.

Power control and narrowband interference rejection have been widely studied as mentioned. However, in previous studies, when authors analyzes the CDMA overlay system, it is always assumed that the narrowband interference is overlaid on the uplink. In fact, due to terrestrial DS-CDMA system use FDD operation, the narrowband interference may also appear on the downlink (Fig. 7.2). It will increase the open-loop power control error because open-loop power control loop uses the downlink estimated power. In closed-loop power control analysis, when authors analyze the overlay system, it is always assumed the power control is perfect or the power estimation at the power control loop is perfect. In fact, firstly, the power control cannot be perfect and the power estimation at the power control loop also can't be perfect in the overlay situation, especially when the narrowband interference is very large; secondly, the imperfect power control will increase the residue of the narrowband interference suppression filter. The relationship between the power control and narrowband interference has not been sufficiently analyzed. In this and next chapters, the effect of narrowband interference on both open-loop and closed-loop power control errors are investigated.

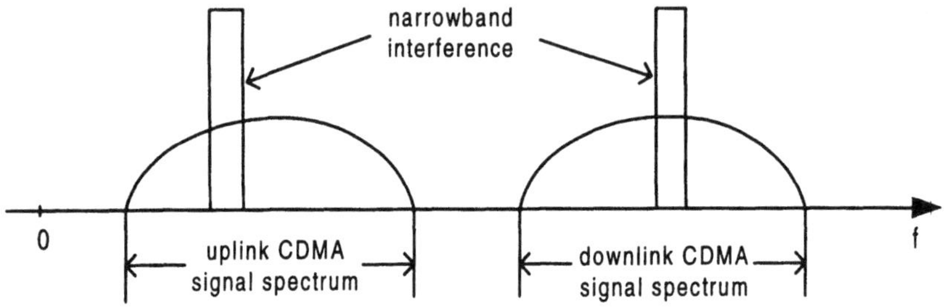

Fig. 7.2 Spectrums of CDMA overlay situation

7.2 DOWNLINK SYSTEM DESCRIPTION

This chapter studies the effect of power control error on performance of CDMA overlay system, which is overlaid on downlink by a narrowband signal (see Fig. 7.3). Neglecting the white noise on the downlink, the waveforms received by each mobile is composed of the desired signal, a number of interfering signals from the base station of the cell of interest and the adjacent-cell base stations, and the narrowband interference.

Assuming that all spreading codes of each cell are orthogonal and that the downlink transmissions of a cell are synchronized, the intracell interference can be ignored because of the same transmission delay. However, the interfering signals from adjacent-cell base stations cannot be assumed to be synchronized with the desired signal because of different transmission delays. This interference becomes significant when a mobile moves towards the boundary of its cell.

In order to study the downlink adjacent-cell interference, two layers of cells are considered (see Fig. 7.4). Each user is assumed to be located independently of all the other users and uniformly distributed over the area of the cell. The location of a reference user in the first cell (the cell of interest) is (r_{11}, θ_{11}) (see Fig. 7.4), where r_{11} and θ_{11} stand for the distance and angle of the mobile from its base station. The distance between the base station of the ith adjacent cell and the reference mobile of the first cell is given by

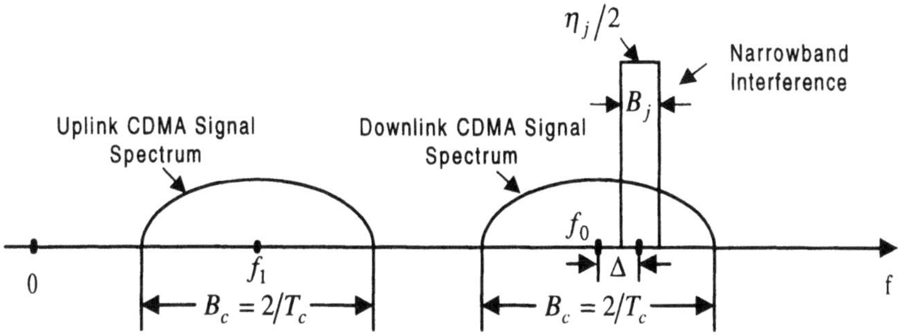

$$p = B_j / B_c = B_j T_c / 2 \quad \text{and} \quad q = 2\Delta / B_c = \Delta T_c$$

Fig.7.3. Power spectrum of the uplink and downlink CDMA signals and narrowband interference

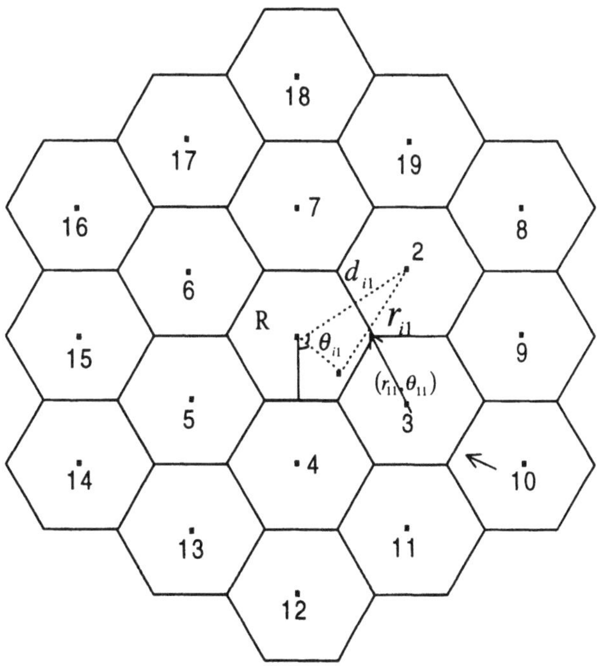

Fig.7.4. Downlink adjacent cell interference geometry [34].

$$r_{i1} = \sqrt{d_{i1}^2 + r_{11}^2 - 2d_{i1}r_{11}\cos\theta_{i1}}, \qquad i = 2, 3, ..., 19 \qquad (7.1)$$

where d_{i1} is the distance between the first base station and the ith base station, and

$$d_{i1} = \begin{cases} \sqrt{3}R, & i = 2, 3, ..., 7 \\ 2\sqrt{3}R, & i = 8, 10, 12, 14, 16, 18 \\ 3R, & i = 9, 11, 13, 15, 17, 19 \end{cases} \qquad (7.2)$$

where R stands for the radius of the hexagonal cell. In (7.1), θ_{i1} is the angle between d_{i1} and r_{11}. The relationship between θ_{i1} and θ_{11} for various values of i is given by

$$\cos(\theta_{i1}) = \begin{cases} \cos[\theta_{11} + (i+2)\pi/3], & 2 \le i \le 7 \\ \cos(\theta_{11} + i\pi/6), & 8 \le i \le 19 \end{cases}. \qquad (7.3)$$

The channel is modeled as a frequency nonselective channel with Rayleigh statistics and lognormal distributed shadowing. The mobile receiver is a conventional coherent receiver with a Wiener suppression filter at its front end [29]. Assuming that there are K active users for each cell, the received signal at mobile user 1 (reference user) can be written as

$$r(t|_{r_{11},\theta_{11}}) = \sum_{i=1}^{19}\sum_{k=1}^{K}\sqrt{2P_R\xi_{ik}r_{i1}^{-\gamma}}\,a_i(t)e^{\zeta_i(t)}c_{ik}(t-\tau_i)b_{ik}(t-\tau_i)\cos(2\pi f_0 t - \phi_i) + j(t) \qquad (7.4)$$

where P_R is the transmitted power of the base station when the reference user is located at a cell vertex $(r_{11} = R)$, and ξ_{ik} $(0 < \xi_{ik} \le 1)$ is the downlink link power adjustment factor (or power control function) and $\xi_{11} = 1$ when $r_{11} = R$. $r_{i1}^{-\gamma}$ represents the propagation path loss from the ith base station to the reference user and γ may take values between two and four. $a_i(t)$ is the Rayleigh distributed random variable with unitary power, i.e. $E[a_i^2(t)] = 1$. $e^{\zeta_i(t)}$ is the log-normal shadowing with random variable $\zeta_i(t)$, taking zero mean and a typical variance of 8dB. The single subscript of $a_i(t)$ and $\zeta_i(t)$ means that all the signals transmitted by each base station propagate on the same path, and undergo the same fading and shadowing. $c_{ik}(t)$ and $b_{ik}(t)$ represent the spreading sequence with chip duration T_c and the binary data sequence with duration T of user k of cell i, respectively. τ_i and ϕ_i are the corresponding time delay and phase, respectively. Note that τ_i and ϕ_i are the same for all users of cell i, because of the synchronous downlink transmission. f_0 is the CDMA carrier frequency on downlink. $j(t)$ represents the narrowband interference, which is modeled as a zero-mean Gaussian random process with a double-sided power spectral density of $\eta_j/2$ over the bandwidth of B_j (see Fig. 7.3), and is given by

$$j(t) = j_c(t)\cos[2\pi(f_0+\Delta)t] - j_s(t)\sin[2\pi(f_0+\Delta)t]. \quad (7.5)$$

The power of $j(t)$ is given by

$$P_j = E\left[j^2(t)\right] = E\left[j_c^2(t)\right] = E\left[j_s^2(t)\right] = \eta_j B_j. \quad (7.6)$$

In microcellular models, the cell size is very small, compared to conventional cellular models. Therefore, the microwave narrowband interference power can be assumed to be the same within a cell. Since $j(t)$ is Gaussian with bandwidth B_j and flat spectrum $\eta_j/2$, the auto-correlation function of $j(t)$ is given by

$$\begin{aligned} R_j(\tau) &= E[j(t)j(t+\tau)] \\ &= P_j \operatorname{sinc}(\pi B_j \tau)\cos[2\pi(f_0+\Delta)\tau] \\ &= P_j \operatorname{sinc}(2\pi p\tau/T_c)\cos(2\pi f_0\tau + 2\pi q\tau/T_c) \end{aligned} \quad (7.7)$$

where $\operatorname{sinc}(x) = \sin(x)/x$, $p = B_j T_c/2$ is defined as the ratio of the interference bandwidth to the spread spectrum bandwidth, $q = \Delta T_c$ is defined as the ratio of the offset (Δ) of the interference carrier frequency to the half $(1/T_c)$ of the spread spectrum bandwidth.

The suppression filter is assumed to be a double-sided Wiener filter with M taps on each side. The number of total taps is $2M+1$. Its impulse response is $\sum_{m=-M}^{M} \alpha_m \delta(t - mT_c)$, where $\alpha_0 = 1$ and $\alpha_m = \alpha_{-m}$. Therefore, the signal at the output of the suppression filter is given by

$$r_f\left(t\big|_{r_{11},\theta_{11}}\right)=\sum_{m=-M}^{M}\alpha_m r\left(t-mT_c\big|_{r_{11},\theta_{11}}\right). \tag{7.8}$$

Assuming $\tau_1=\phi_1=0$ for the first cell, the output of despreader/demodulator at time lT, is given by

$$Z\left(lT\big|_{r_{11},\theta_{11}}\right)=\frac{1}{T}\int_{(l-1)T}^{lT} r_f\left(t\big|_{r_{11},\theta_{11}}\right)\cdot 2c_{11}(t)\cos(2\pi f_0 t)dt \tag{7.9}$$

where $2\cos(2\pi f_0 t)$ and $c_{11}(t)$ are the recovered carrier and the spreading sequence of the reference user. Since high frequency terms are removed by the lowpass filter following the mixer of the demodulator, the above expression reduces to

$$Z\left(lT\big|_{r_{11},\theta_{11}}\right)\approx\sqrt{2P_R\xi_{11}r_{11}^{-\gamma}}\,a_1(lT)e^{\zeta_1(lT)}b_{11}(lT)+J(lT)+I(lT). \tag{7.10}$$

Note that, in the following, "lT" is replaced by "l" for simple notation. The first term on the right hand side of (7.10) is the desired component, corresponding to the zero-th tap of the suppression filter. However, the contribution of the desired signal through the non-zero-th taps is very minor, compared to the first term, and thus is neglected like [29]. The average desired signal power is

$$S=2P_R\xi_{11}r_{11}^{-\gamma}e^{2\sigma_{\zeta_1}^2}. \tag{7.11}$$

The second term on the right hand side of (7.10) is the narrowband interference component and is given by

$$J(l)=\frac{1}{T}\sum_{m=-M}^{M}\alpha_m\int_{(l-1)T}^{lT} j(t-mT_c)2c_{11}(t)\cos(2\pi f_0 t)dt.$$

$$=\frac{1}{T}\sum_{m=-M}^{M}\alpha_m\int_{(l-1)T}^{lT}\{j_c(t-mT_c)\cos[2\pi(f_0+\Delta)(t-mT_c)]$$

$$-j_s(t-mT_c)\sin[2\pi(f_0+\Delta)(t-mT_c)]\}\cdot 2c_{11}(t)\cos(2\pi f_0 t)dt. \tag{7.12}$$

The double frequency component in (7.12) can be removed by the integrator. Assuming $f_0 T_c$ is an integer, $J(l)$ becomes

$$J(l)=\frac{1}{T}\sum_{m=-M}^{M}\alpha_m\int_{(l-1)T}^{lT}\{j_c(t-mT_c)\cos[2\pi\Delta(t-mT_c)]$$

$$-\,j_s(t-mT_c)\sin[2\pi\Delta(t-mT_c)]\}c_{11}(t)dt$$

$$=\frac{1}{T}\sum_{m=-M}^{M}\alpha_m\int_{(l-1)T}^{lT}\hat{j}(t-mT_c)c_{11}(t)dt \qquad (7.13)$$

where $\hat{j}(t)$ is the low frequency version of the narrowband interference, $j(t)$, and is defined as

$$\hat{j}(t)=j_c(t)\cos(2\pi\Delta t)-j_s(t)\sin(2\pi\Delta t). \qquad (7.14)$$

The autocorrelation of $\hat{j}(t)$ is given by as (7.7), i.e. $R_{\hat{j}}(\tau)=R_j(\tau)$, with f_0 replaced by 0. Therefore,

$$J(l)=\frac{1}{T}\sum_{m=-M}^{M}\alpha_m\int_0^T\hat{j}(t-(l-1)T-mT_c)c_{11}(t-(l-1)T)dt$$

$$=\frac{1}{T}\sum_{m=-M}^{M}\alpha_m\sum_{n=1}^{N}c_{11}^{(n)}\int_{(n-1)T_c}^{nT_c}\hat{j}[t-(l-1)T-mT_c]dt$$

$$=\frac{1}{T}\sum_{m=-M}^{M}\alpha_m\sum_{n=1}^{N}c_{11}^{(n)}\int_0^{T_c}\hat{j}[t-(l-1)T-mT_c-(n-1)T_c]dt \qquad (7.15)$$

where $c_{11}^{(n)}$ represents the nth chip of random sequence $c_{11}(t)$ and N is the processing gain. The variance of $J(l)$ is given by

$$\sigma_J^2=E[J^2(l)]=\frac{1}{T^2}\sum_{m_1=-M}^{M}\alpha_{m_1}\sum_{m_2=-M}^{M}\alpha_{m_2}\sum_{n_1=1}^{N}\sum_{n_2=1}^{N}E\left[c_{11}^{(n_1)}c_{11}^{(n_2)}\right].$$

$$\int_0^{T_c}\int_0^{T_c} E\{\hat{j}[t_1-(l-1)T-m_1T_c-(n_1-1)T_c]\hat{j}[t_2-(l-1)T-m_2T_c-(n_2-1)T_c]\}dt_1dt_2 \tag{7.16}$$

Since $E\left[c_{11}^{(n_1)}c_{11}^{(n_2)}\right]=1$ and 0 for $n_1=n_2$ and $n_1\neq n_2$, respectively, σ_J^2 reduces to

$$\sigma_J^2=\frac{1}{T^2}\sum_{m_1=-M}^{M}\alpha_{m_1}\sum_{m_2=-M}^{M}\alpha_{m_2}\sum_{n=1}^{N}\int_{-T_c}^{T_c}R_j[\tau-(m_1-m_2)T_c](T_c-|\tau|)d\tau$$

$$=\frac{NT_c^2}{T^2}\sum_{m_1=-M}^{M}\alpha_{m_1}\sum_{m_2=-M}^{M}\alpha_{m_2}\int_{-1}^{1}R_j\{[\tau-(m_1-m_2)]T_c\}(1-|\tau|)d\tau$$

$$=\frac{P_j}{N}\sum_{m_1=-M}^{M}\alpha_{m_1}\sum_{m_2=-M}^{M}\alpha_{m_2}[Q(m_1,m_2)+Q(-m_1,-m_2)]$$

$$=\frac{P_j}{N}\hat{\sigma}_J^2 \tag{7.17}$$

where

$$\hat{\sigma}_J^2=\sum_{m_1=-M}^{M}\alpha_{m_1}\sum_{m_2=-M}^{M}\alpha_{m_2}[Q(m_1,m_2)+Q(-m_1,-m_2)] \tag{7.18}$$

and

$$Q(m_1,m_2)=\int_0^1(1-|\tau|)\text{sinc}\{2\pi[\tau-(m_1-m_2)]p\}\cos\{2\pi[\tau-(m_1-m_2)]q\}d\tau \tag{7.19}$$

The third term on the right hand side of (7.10) is the adjacent-cell interference, and is expressed as

$$I(l)=\sum_{i=2}^{19}a_i(l)e^{\zeta_i(l)}\sum_{k=1}^{K}\sqrt{2P_R\xi_{ik}r_{i1}^{-\gamma}}\,I_{i,1}(l)\cos\phi_i \tag{7.20}$$

where

$$I_{i,1}(l)=\frac{1}{T}\int_{(l-1)T}^{lT}b_{ik}(t-\tau_i)c_{ik}(t-\tau_i)c_{11}(t)dt. \tag{7.21}$$

When K is large, $I_{i,1}(l)$ can be approximated by a Gaussian random variable with the variance of $var[I_{i,1}(l)]=2/(3N)$ [30]. Then, the variance of $I(l)$ is given by

$$\sigma_I^2 = \frac{2K}{3N} e^{2\sigma_{\zeta i}^2} \sum_{i=2}^{19} P_R E[\xi_{ik}] r_{i1}^{-\gamma} \tag{7.22}$$

where $E[\xi_{ik}]$ stands for the mean of downlink power adjustment factor, ξ_{ik}, and is discussed in the following.

- ***Without downlink power control***

If there is no downlink power control $(\text{i.e.}, \xi_{ik} = 1)$, the transmit power at a base station is the same for all users of the ith cell. Thus

$$E[\xi_{ik}] = 1. \tag{7.23}$$

- ***With optimum downlink power control*** [34]

The criterion for optimum downlink power control is that the output signal to interference ratio (SIR) of a mobile receiver is the same wherever the mobile user is. In the presence of narrowband interference, the total downlink interference should include the adjacent-cell interference and narrowband interference. Therefore, the SIR at the output of the despreader of the mobile receiver is

$$SIR(r_{11}, \theta_{11}) = \frac{S}{\sigma_I^2 + \sigma_J^2}$$

$$= \frac{2P_R \xi_{11} r_{11}^{-\gamma} e^{2\sigma_{\zeta 1}^2}}{\frac{2KP_R}{3N} E[\xi_{ik}] e^{2\sigma_{\zeta 1}^2} \sum_{i=2}^{19} r_{i1}^{-\gamma} + \sigma_J^2}$$

$$= \frac{P_s \xi_{11} (r_{11}/R)^{-\gamma}}{\frac{KP_s}{3N} E[\xi_{ik}] \sum_{i=2}^{19} (r_{i1}/R)^{-\gamma} + \frac{P_j}{N} \hat{\sigma}_J^2} \tag{7.24}$$

where r_{i1} is given by (7.1), P_s is defined as the average received power at the cell vertices,

$$P_s = E\left[\left(\sqrt{2P_R R^{-\gamma}}\, a(t) e^{\zeta(t)}\right)^2\right]$$

$$= 2P_R R^{-\gamma} e^{2\sigma_\zeta^2} . \tag{7.25}$$

When the mobile is located at one of the cell vertices (i.e., $r_{11} = R$, $\theta_{11} = \pi/6$ and $\xi_{11} = 1$), (7.24) reduces to

$$SIR(R,\pi/6) = \frac{P_s}{\frac{KP_s}{3N} E[\xi_{ik}] \sum_{i=2}^{19} (r_{i1}/R)^{-\gamma} \Big|_{(r_{11}=R,\theta_{11}=\pi/6)} + \frac{P_j}{N}\hat{\sigma}_J^2} . \tag{7.26}$$

For the same SIR, (7.24) and (7.26) should be equal, i.e.,

$$SIR(r_{11},\theta_{11}) = SIR(R,\pi/6). \tag{7.27}$$

From (7.24), (7.26) and (7.27), one obtains

$$\frac{P_s \xi_{11} (r_{11}/R)^{-\gamma}}{\frac{KP_s}{3N} E[\xi_{ik}] \sum_{i=2}^{19} (r_{i1}/R)^{-\gamma} + \frac{P_j}{N}\hat{\sigma}_J^2}$$

$$= \frac{P_s}{\frac{KP_s}{3N} E[\xi_{ik}] \sum_{i=2}^{19} (r_{i1}/R)^{-\gamma} \Big|_{(r_{11}=R,\theta_{11}=\pi/6)} + \frac{P_j}{N}\hat{\sigma}_J^2} \tag{7.28}$$

From (7.28), the power adjustment factor ξ_{11} is derived

$$\xi_{11} = \frac{(r_{11}/R)^{\gamma} \left[\frac{KP_s}{3N} E[\xi_{ik}] \sum_{i=2}^{19} (r_{i1}/R)^{-\gamma} + \frac{P_j}{N}\hat{\sigma}_J^2\right]}{\frac{KP_s}{3N} E[\xi_{ik}] \sum_{i=2}^{19} (r_{i1}/R)^{-\gamma} \Big|_{(r_{11}=R,\theta_{11}=\pi/6)} + \frac{P_j}{N}\hat{\sigma}_J^2}$$

$$= \frac{\frac{K}{3} E\left[\xi_{ik}\right] \sum_{i=2}^{19} \left(r_{i1}/r_{11}\right)^{-\gamma} + \frac{P_j}{P_s} \hat{\sigma}_J^2 \left(r_{11}/R\right)^{\gamma}}{\frac{K}{3} E\left[\xi_{ik}\right] \sum_{i=2}^{19} \left(r_{i1}/R\right)^{-\gamma} \Big|_{(r_{11}=R, \theta_{11}=\pi/6)} + \frac{P_j}{P_s} \hat{\sigma}_J^2}. \tag{7.29}$$

In a special case, when P_j/P_s is very small (i.e., the narrowband interference is absent, $P_j/P_s = 0$), (7.29) reduces to

$$\xi_{11} \approx \frac{\sum_{i=2}^{19} \left(r_{i1}/r_{11}\right)^{-\gamma}}{\sum_{i=2}^{19} \left(r_{i1}/R\right)^{-\gamma} \Big|_{(r_{11}=R, \theta_{11}=\pi/6)}}, \tag{7.30}$$

which becomes the case discussed in [23].

Assuming that the position of each user is uniformly distributed in a cell, i.e.,

$$f(r_{11}, \theta_{11}) = \frac{1}{\pi R^2}, \qquad 0 \le r_{11} \le R, \quad 0 \le \theta_{11} \le 2\pi, \tag{7.31}$$

the expectation of ξ_{ik} is obtained by double integration of (7.29),

$$E\left[\xi_{ik}\right] = E\left[\xi_{11}\right] = \frac{1}{\pi R^2} \int_0^{2\pi} \int_0^R \xi_{11} r_{11} dr_{11} d\theta_{11}. \tag{7.32}$$

By numerical calculation, $E\left[\xi_{ik}\right]$ is shown in Table I for various values of propagation exponent and P_j/P_s.

7.3 POWER CONTROL ERROR

In the open-loop power control algorithm, the received signal power at the mobile is estimated by averaging the output square of a correlator [12]. The estimate at time LT is given by averaging the square of (7.10),

$$\hat{S}^2(L) = \frac{1}{L} \sum_{l=1}^{L} Z^2\left(l\big|_{r_{11}, \theta_{11}}\right)$$

Table I. $E[\xi_{ik}]$ with K=20, N=1000, p=0.2, q=0.2, M=4.

	$P_j/P_s = -\infty dB$	$P_j/P_s = 10dB$	$P_j/P_s = 15dB$	$P_j/P_s = 20dB$
$\gamma = 2$	0.4689	0.4801	0.4858	0.4912
$\gamma = 3$	0.3675	0.3836	0.3897	0.3936
$\gamma = 4$	03189	03279	03310	03321

$$= \frac{1}{L}\sum_{l=1}^{L} 2P_R\xi_{11}r_{11}^{-\gamma}a_1^{2}(l)e^{2\xi_1(l)} + \frac{2}{L}\sum_{l=1}^{L}\sqrt{2P_R\xi_{11}r_{11}^{-\gamma}}\,a_1(l)e^{\xi_1(l)}[I(l)+J(l)]$$

$$+\frac{2}{L}\sum_{l=1}^{L} I(l)J(l) + \frac{1}{L}\sum_{l=1}^{L} I^2(l) + \frac{1}{L}\sum_{l=1}^{L} J^2(l). \tag{7.33}$$

The mobile uses this estimate to adjust its transmit power, which is inversely proportional to the estimated power, e.g. $P_t = 1/\left(c \cdot \hat{S}^2(L)\right)$, where c is a weighted constant. Therefore, neglecting the loop delay on uplink, the desired signal component of the base station after despreading at the time LT, is given by

$$Z_1(L) = \sqrt{\frac{2r_{11}^{-\gamma}}{C \cdot \hat{S}^2(L)}}\,a_1'(L)e^{\zeta_1(L)}b_{11}'(L)$$

$$= \sqrt{2P_{err}}\,a_1'(L)b_{11}'(L) \tag{7.34}$$

where $a_1'(L)$, $e^{\zeta_1(L)}$ and $b_{11}'(L)$ stand for the Rayleigh fading, lognormal distributed shadowing and data bit, respectively, for the uplink transmission. Notice that the shadowing terms on both up and down links are the same, but the fading terms on the two links are different. P_{err} is the received power of the base station, defined as

$$P_{err} = \frac{r_{11}^{-\gamma} e^{2\zeta_1(L)}}{c \cdot \hat{S}^2(L)}. \tag{7.35}$$

Assuming the square-root of P_{err} is lognormally distributed, e.g.

$$\sqrt{P_{err}} = e^x, \tag{7.36}$$

one obtains

$$\begin{aligned}
\frac{1}{P_{err}} &= e^{-2x} \\
&= \frac{c}{L}\sum_{l=1}^{L} 2P_R\xi_{11}a_1^2(l)e^{2\xi_1(l)-2\zeta_1(L)} + \frac{2c}{Lr_{11}^{-\gamma/2}}\sum_{l=1}^{L}\sqrt{2P_R\xi_{11}}a_1(l)e^{\xi_1(l)-2\zeta_1(L)}\left[I(l)+J(l)\right] \\
&\quad + \frac{2c}{Lr_{11}^{-\gamma}}\sum_{l=1}^{L} e^{-2\zeta_1(L)}I(l)J(l) + \frac{c}{Lr_{11}^{-\gamma}}\sum_{l=1}^{L} e^{-2\zeta_1(L)}\left[I^2(l)+J^2(l)\right]
\end{aligned} \tag{7.37}$$

and

$$\begin{aligned}
\frac{1}{P_{err}^2} &= e^{-4x} \\
&= \frac{4c^2P_R^2\xi_{11}^2}{L^2}\sum_{l_1=1}^{L}\sum_{l_2=1}^{L} a_1^2(l_1)a_1^2(l_2)e^{2\zeta_1(l_1)+2\zeta_1(l_2)-4\zeta_1(L)} \\
&+ \frac{8c^2P_R\xi_{11}}{L^2r_{11}^{-\gamma}}\sum_{l_1=1}^{L}\sum_{l_2=1}^{L} a_1(l_1)a_1(l_2)e^{\zeta_1(l_1)+\zeta_1(l_2)-4\zeta_1(L)}\left[I(l_1)I(l_2)+J(l_1)J(l_2)\right] \\
&+ \frac{4c^2}{L^2r_{11}^{-2\gamma}}\sum_{l_1=1}^{L}\sum_{l_2=1}^{L} e^{-4\zeta_1(L)}I(l_1)I(l_2)J(l_1)J(l_2) \\
&+ \frac{c^2}{L^2r_{11}^{-2\gamma}}\sum_{l_1=1}^{L}\sum_{l_2=1}^{L} e^{-4\zeta_1(L)}\left[I^2(l_1)I^2(l_2)+2I^2(l_1)J^2(l_2)+J^2(l_1)J^2(l_2)\right] \\
&+ \frac{4c^2P_R\xi_{11}}{L^2r_{11}^{-\gamma}}\sum_{l_1=1}^{L}\sum_{l_2=1}^{L} a_1^2(l_1)e^{2\zeta_1(l_1)-4\zeta_1(L)}\left[I^2(l_2)+J^2(l_2)\right].
\end{aligned} \tag{7.38}$$

Since x is Gaussian distributed, $E\left[e^{x}\right]=\exp\left(\bar{x}+\sigma_x^2/2\right)$, where $\bar{x}$ and σ_x stand for the mean and standard deviation of power control error. In the following, it is shown how to obtain $\bar{x}$ and σ_x.

From (7.37) and (7.38), we have

$$E\left[e^{-2x}\right]=e^{-2\bar{x}+2\sigma_x^2}=cU\,, \tag{7.39}$$

and

$$E\left[e^{-4x}\right]=e^{-4\bar{x}+8\sigma_x^2}=c^2V \tag{7.40}$$

where U and V will be derived later. From (7.39) and (7.40), $\bar{x}$ and σ_x can be derived, i.e.

$$\bar{x}=\frac{1}{2}\ln\left(\sqrt{V}/cU^2\right), \tag{7.41}$$

and

$$\sigma_x=\frac{1}{2}\sqrt{\ln\left(V/U^2\right)}. \tag{7.42}$$

By choosing c appropriately $\left(\text{i.e. } c=\sqrt{V}/U^2\right)$, the mean of the power control error can be set to zero ($\bar{x}=0$), corresponding to an unbiased estimate.

Since all $a_1(l)$, $e^{\zeta_1(l)}$, $I(l)$ and $J(l)$ are independent, $E[I(l)]=E[J(l)]=0$, and $E[I(l_1)I(l_2)]=E[J(l_1)J(l_2)]=0$, for $l_1\neq l_2$, the cross-terms with odd-orders of $I(l)$ and $J(l)$ have zero expectation. Thus, from (7.37) and (7.39), we have

$$U=\frac{2P_R\xi_{11}}{L}\sum_{l=1}^{L}E\left[a_1^2(l)e^{2\zeta_1(l)-2\zeta_1(L)}\right]+\frac{1}{Lr_{11}^{-\gamma}}\sum_{l=1}^{L}E\left[e^{-2\zeta_1(L)}\left(I^2(l)+J^2(l)\right)\right]$$

$$=\frac{2P_R\xi_{11}}{L}\sum_{l=1}^{L}E\left[a_1^2(l)\right]E\left[e^{2\zeta_1(l)-2\zeta_1(L)}\right]+\frac{1}{Lr_{11}^{-\gamma}}\sum_{l=1}^{L}E\left[e^{-2\zeta_1(L)}\right]E\left[I^2(l)+J^2(l)\right]$$

$$=\frac{2P_R\xi_{11}}{L}\sum_{l=1}^{L}\exp\left(4\sigma_{\zeta_1}^2-4C_{\zeta_1}\left(\left(l-L\right)T\right)\right)+\frac{1}{r_{11}^{-\gamma}}e^{2\sigma_{\zeta_1}^2}\left(\sigma_I^2+\sigma_J^2\right)$$

$$= \frac{P_s \xi_{11}}{LR^{-\gamma}} \sum_{l=1}^{L} \exp\left(2\sigma_{\zeta_1}^2 - 4C_{\zeta_1}\left(\left(l-L\right)T\right)\right) + \frac{1}{r_{11}^{-\gamma}} e^{2\sigma_{\zeta_1}^2}\left(\sigma_I^2 + \sigma_J^2\right) \tag{7.43}$$

where $E\left[a_1^2(l)\right] = 1$, P_s is given by (7.25) and $C_{\zeta_1}(\tau)$ is the auto-covariance function of $\zeta(t)$, given by [31]

$$C_{\zeta_1}(\tau) = E\left[\zeta_1(t)\zeta_1(t+\tau)\right] - E^2\left[\zeta_1(t)\right]$$

$$= \sigma_{\zeta_1}^2 e^{-v|\tau|/D} \tag{7.44}$$

where v stands for the speed of a mobile, D represents shadowing correlation distance and has been measured as hundreds of meters for conventional terrestrial cells, and tens of meters for terrestrial microcells.

From (7.38) and (7.40), we have

$$V = \frac{4P_R^2 \xi_{11}^2}{L^2} \sum_{l_1=1}^{L} \sum_{l_2=1}^{L} E\left[a_1^2(l_1) a_1^2(l_2) e^{2\zeta_1(l_1) + 2\zeta_1(l_2) - 4\zeta_1(L)}\right]$$

$$+ \frac{8P_R \xi_{11}}{L^2 r_{11}^{-\gamma}} \sum_{l_1=1}^{L} \sum_{l_2=1}^{L} E\left\{a_1(l_1) a_1(l_2) e^{\zeta_1(l_1) + \zeta_1(l_2) - 4\zeta_1(L)} \left[I(l_1) I(l_2) + J(l_1) J(l_2)\right]\right\}$$

$$+ \frac{4}{L^2 r_{11}^{-2\gamma}} \sum_{l_1=1}^{L} \sum_{l_2=1}^{L} E\left[e^{-4\zeta_1(L)} I(l_1) I(l_2) J(l_1) J(l_2)\right]$$

$$+ \frac{1}{L^2 r_{11}^{-2\gamma}} \sum_{l_1=1}^{L} \sum_{l_2=1}^{L} E\left\{e^{-4\zeta_1(L)} \left[I^2(l_1) I^2(l_2) + 2I^2(l_1) J^2(l_2) + J^2(l_1) J^2(l_2)\right]\right\}$$

$$+ \frac{4P_R \xi_{11}}{L^2 r_{11}^{-\gamma}} \sum_{l_1=1}^{L} \sum_{l_2=1}^{L} E\left\{a_1^2(l_1) e^{2\zeta_1(l_1) - 4\zeta_1(L)} \left[I^2(l_2) + J^2(l_2)\right]\right\}$$

$$= \frac{4P_R^2 \xi_{11}^2}{L^2} \sum_{l_1=1}^{L} \sum_{l_2=1}^{L} E\left[a_1^2(l_1) a_1^2(l_2)\right] E\left[e^{2\zeta_1(l_1) + 2\zeta_1(l_2) - 4\zeta_1(L)}\right]$$

$$+\frac{8P_R\xi_{11}}{L^2 r_{11}^{-\gamma}}\sum_{l=1}^{L}E\left[e^{2\zeta_1(l)-4\zeta_1(L)}\right]E\left[I^2(l)+J^2(l)\right]$$

$$+\frac{4}{L^2 r_{11}^{-2\gamma}}\sum_{l=1}^{L}E\left[e^{-4\xi_1(L)}\right]E\left[I^2(l)\right]E\left[J^2(l)\right]$$

$$+\frac{1}{L^2 r_{11}^{-2\gamma}}\sum_{l_1=1}^{L}\sum_{l_2=1}^{L}E\left[e^{-4\zeta_1(L)}\right]E\left[I^2(l_1)I^2(l_2)+2I^2(l_1)J^2(l_2)+J^2(l_1)J^2(l_2)\right]$$

$$+\frac{4P_R\xi_{11}}{L r_{11}^{-\gamma}}\sum_{l=1}^{L}E\left[e^{2\zeta_1(l)-4\zeta_1(L)}\right]E\left[I^2(l)+J^2(l)\right] \tag{7.45}$$

where $\mathrm{E}\left[a_1^2(l_1)a_1^2(l_2)\right]$ is the autocorrelation function of $a_1^2(l)$, defined as

$$R_{a_1^2}(l_2-l_1)=\mathrm{E}\left[a_1^2(l_1)a_1^2(l_2)\right]$$

$$=1+\rho^2(l_2-l_1) \tag{7.46}$$

where $\rho(l_2-l_1)$ is the normalized autocovariance function of the Rayleigh process, given by [32]

$$\rho(l_2-l_1)=J_0\left(2\pi f_d\left|(l_2-l_1)T\right|\right) \tag{7.47}$$

where f_d is a maximum Doppler frequency or fading rate, given by $f_d=v/\lambda$, where λ is a carrier wavelength. $J_0(x)$ is a Bessel function of the first kind of zeroth order.

Since both $I(l)$ and $J(l)$ are Gaussian, we have

$$E\left[I^2(l_1)I^2(l_2)\right]=\begin{cases}3\sigma_I^4, & l_1=l_2\\ \sigma_I^4, & l_1\neq l_2\end{cases} \tag{7.48}$$

and

$$E\left[J_1^2(l_1)J_1^2(l_2)\right]=\begin{cases}3\sigma_J^4, & l_1=l_2\\ \sigma_J^4, & l_1\neq l_2\end{cases}. \tag{7.49}$$

Thus,

$$\sum_{l_1=1}^{L}\sum_{l_2=1}^{L}E\left[I^2(l_1)I^2(l_2)\right]=\sum_{l_1=1}^{L}\left\{E\left[I^4(l)\right]+\sum_{\substack{l_2=1\\ l_2\neq l_1}}^{L}E\left[I^2(l_1)I^2(l_2)\right]\right\}$$

$$=3L\sigma_I^4+\left(L^2-L\right)\sigma_I^4 \tag{7.50}$$

and

$$\sum_{l_1=1}^{L}\sum_{l_2=1}^{L}E\left[J_1^2(l_1)J_1^2(l_2)\right]=3L\sigma_J^4+\left(L^2-L\right)\sigma_J^4\,. \tag{7.51}$$

V in (7.45) can be simplified to

$$V=\frac{4P_R^2\xi_{11}^2}{L^2}\sum_{l_1=1}^{L}\sum_{l_2=1}^{L}\left[1+\rho^2(l_1-l_2)\right]$$

$$\cdot\exp\left\{12\sigma_{\zeta_1}^2+4C_{\zeta_1}\left[(l_1-l_2)T\right]-8C_{\zeta_1}\left[(l_1-L)T\right]-8C_{\zeta_1}\left[(l_2-L)T\right]\right\}$$

$$+\frac{8P_R\xi_{11}}{L^2r_{11}^{-\gamma}}\left(\sigma_I^2+\sigma_J^2\right)\sum_{l=1}^{L}\exp\left\{10\sigma_{\zeta_1}^2-8C_{\zeta_1}\left[(l-L)T\right]\right\}+\frac{4}{Lr_{11}^{-2\gamma}}\sigma_I^2\sigma_J^2e^{8\sigma_{\zeta_1}^2}$$

$$+\left(1+\frac{2}{L}\right)\frac{\left(\sigma_I^4+\sigma_J^4\right)}{r_{11}^{-2\gamma}}e^{8\sigma_{\zeta_1}^2}+\frac{2}{r_{11}^{-2\gamma}}\sigma_I^2\sigma_J^2e^{8\sigma_{\zeta_1}^2}$$

$$+\frac{4P_R\xi_{11}}{Lr_{11}^{-\gamma}}\left(\sigma_I^2+\sigma_J^2\right)\sum_{l=1}^{L}\exp\left\{10\sigma_{\zeta_1}^2-8C_{\zeta_1}\left[(l-L)T\right]\right\}$$

$$=\frac{P_s^2\xi_{11}^2}{L^2R^{-2\gamma}}\sum_{l_1=1}^{L}\sum_{l_2=1}^{L}\left[1+\rho^2(l_1-l_2)\right]$$

$$\cdot\exp\left\{8\sigma_{\zeta_1}^2+4C_{\zeta_1}\left[(l_1-l_2)T\right]-8C_{\zeta_1}\left[(l_1-L)T\right]-8C_{\zeta_1}\left[(l_2-L)T\right]\right\}$$

$$+\frac{2P_s\xi_{11}}{Lr_{11}^{-\gamma}R^{-\gamma}}\left(1+\frac{2}{L}\right)\left(\sigma_I^2+\sigma_J^2\right)\sum_{l=1}^{L}\exp\left\{8\sigma_{\zeta_1}^2-8C_{\zeta_1}\left[(l-L)T\right]\right\}$$

$$+\frac{2}{r_{11}^{-2\gamma}}\left(1+\frac{2}{L}\right)\sigma_I^2\sigma_J^2 e^{8\sigma_{\zeta 1}^2}+\frac{1}{r_{11}^{-2\gamma}}\left(1+\frac{2}{L}\right)\left(\sigma_I^4+\sigma_J^4\right)e^{8\sigma_{\zeta 1}^2}. \qquad (7.52)$$

7.4 NUMERICAL RESULTS

The numerical results of the standard deviation (7.42) of the power control error are shown for several cases. Unless noted otherwise, the following parameters are assumed: the path loss exponent $\gamma = 4$, the processing gain $N = 1000$, the number of active users per cell $K = 20$, the number of taps of the Wiener filter $M = 4$ on each side, the ratio of the interference bandwidth to the spread spectrum bandwidth $p = 0.2$, the ratio of the offset of the interference carrier frequency to half of the spread spectrum bandwidth $q = 0.2$, standard deviation of shadowing $\sigma_\zeta = 8$ dB, the information data rate $10Kb/s$, the speed of a mobile user $v = 100km/h$, the carrier frequency $f_0 = 2GHz$ (Dopplor frequency shift $f_d = v/\lambda = 185Hz$), the period for power estimation $L = 60$ (or 60 bits), and the shadowing correlation distance $D = 45m$.

Fig. 7.5 illustrates the standard deviation of power control error when optimum downlink power control is used as a function of r_{11}/R for various values of P_j/P_s. For comparison, the power control error without power control is also shown in the figure. Without downlink power control, assuming $\theta_{11} = \pi/6$, a base station transmits the same power to all the users wherever they are, so that a user close to the base station will receive more power than a user far from it. Therefore, when the distance ratio is small, i.e., $r_{11}/R < 0.3$, the received signal power of a mobile user is very large. However, narrowband interference and adjacent-cell interference are relatively stable. That is, signal to interference ratio (SIR) is very large. Thus, the power control error mainly results from the measurement error, which mainly depends on the measurement interval, mobile speed and shadowing correlation distance. However, when the distance ratio (r_{11}/R) increases,

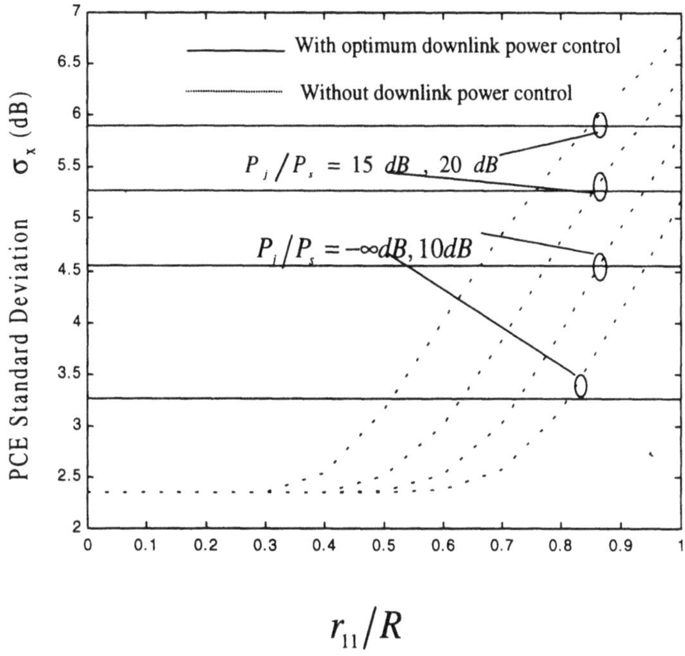

Fig. 7.5. Standard deviation of power control error versus relative distance

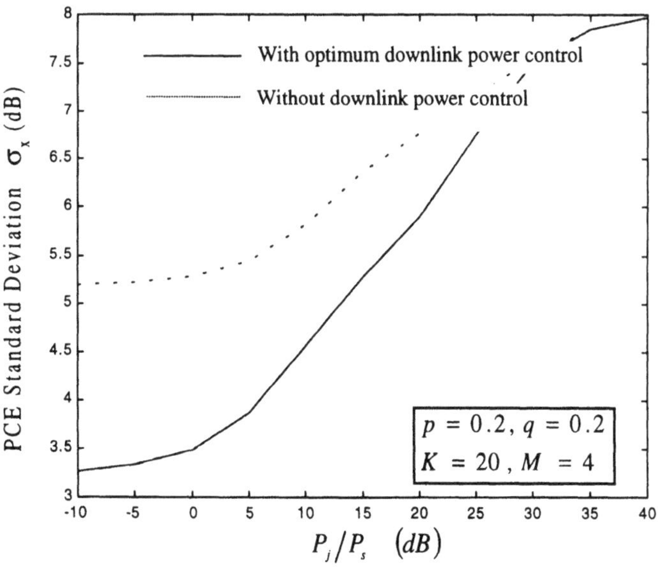

Fig. 7.6. Standard Deviation of power control error as a function of P_j/P_s

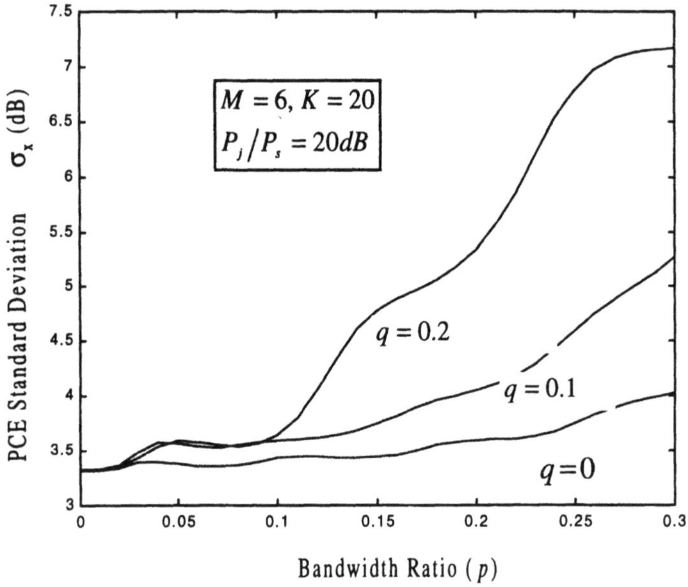

Fig. 7.7. Standard deviation of power control error as a function of the ratio (p) of the interference BW to the spread spectrum BW

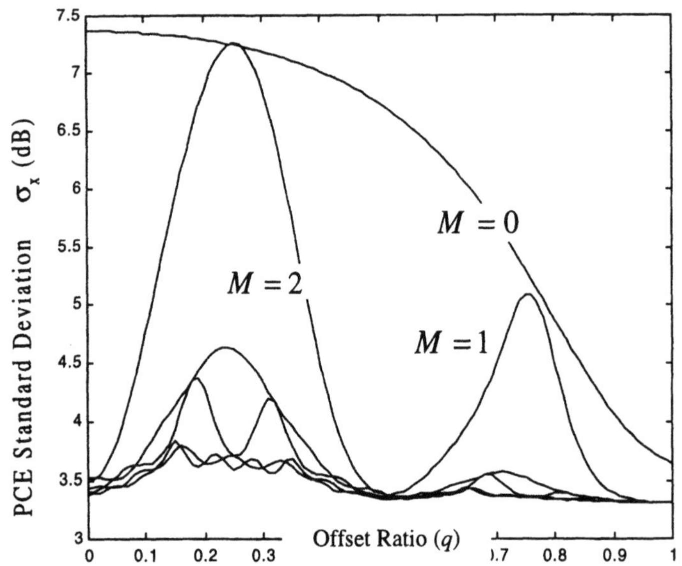

Fig. 7.8. Standard deviation of power control error as a function of the offset ratio (q) of the interference carrier frequency to the half spread spectrum

the received signal power of a mobile decreases. That is, the SIR decreases. In this case, the power control error is caused by both interference and measurement error. Therefore, the power control error standard deviation increases as the distance ratio increases. For a given value of P_j/P_s, when a mobile is located at one of vertices (i.e., $r_{11}/R = 1$), the power control error takes the maximum value, which is used for evaluation of system capacity. However, when power control is used, for a given value of P_j/P_s, the power control error keeps constant, irrespective of the relative distance, r_{11}/R. This is contrast to that without power control, where the power control error increases dramatically as r_{11}/R for large values of r_{11}/R. It can also be seen from Fig. 7.5 that in the presence of narrowband interference, the optimum downlink power control algorithm can reduce the power control error significantly. For $P_j/P_s = -\infty$ dB, 10dB, 15dB, 20dB, the power control error is 1.9dB, 1.3dB, 1.1dB, 0.9dB, respectively, less than that without power control.

Fig. 7.6 shows the power control error as a function of P_j/P_s both with and without power control. It can be seen that the power control algorithm performs very well for large range of P_j/P_s. However, when P_j/P_s is very large (i.e., $P_j/P_s > 30dB$), the power control error for both curves approaches 8dB (the maximum power control error). That is, the power control is ineffective when P_j/P_s is very large.

Fig. 7.7 is plotted for the power control error as a function of the bandwidth ratio (p) of interference bandwidth to the CDMA signal bandwidth for the offset ratio $q = 0, 0.1, 0.2$, respectively, when power control algorithm is used. It can be seen that the power control error does not change much for small bandwidth ratio p $(p \le 0.1)$, whereas it increases dramatically as q increases for large $p\,(p > 0.1)$.

Fig. 7.8 shows the power control error as a function of the offset ratio (q) for different values of the number (M) of each-side taps of the suppression filter. It can be seen that generally, power control error decreases as M increases for a given value of q and that power control error is more robust for large M.

Finally, Fig. 7.9 illustrates the power control error as a function of the number of active users (K) for various values of P_j/P_s. It can be seen that for a given value of power control error and given value of P_j/P_s, the

CDMA overlay system with optimum power control can support many more users than that without power control.

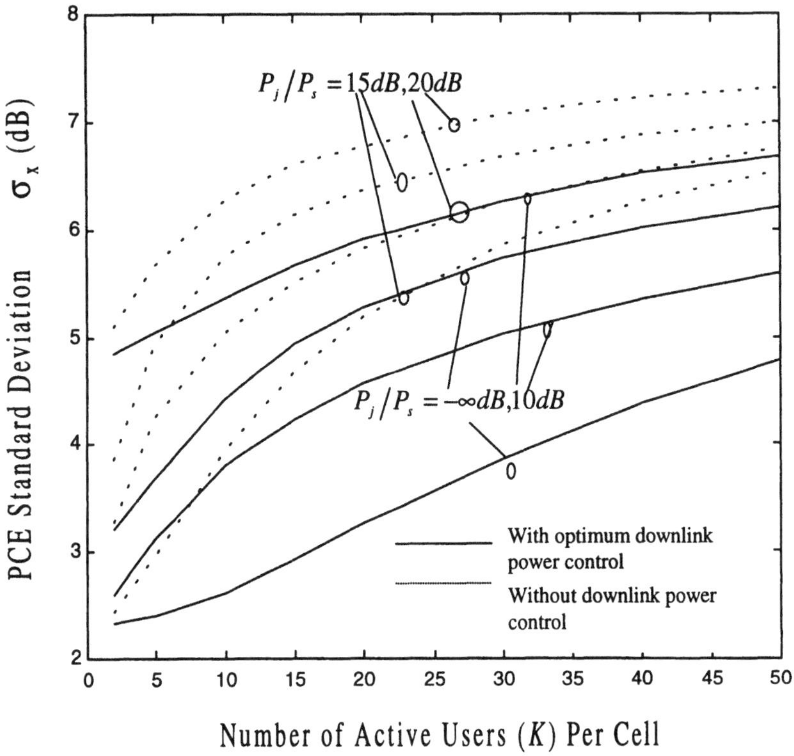

Fig. 7.9 Standard deviation of power control error as a function of active users K

7.5 CONCLUSIONS

The optimum open-loop power control error in cellular CDMA overlay systems has been studied. The following conclusions have been drawn:

- For wide range of the ratio of narrowband interference to signal power, the optimum power control algorithm performs well.
- In order for the power control error to be more robust against the offset ratio, more taps of the suppression filter should be used.
- The cellular CDMA overlay system with optimum open-loop power control can support many more users than that without power control.

REFERENCES

[1] R. L. Pickholtz, L. B. Milstein and D. L. Schilling, "Spread spectrum for mobile communications", *IEEE Trans. Veh. Technol.*, Vol.40, No.20, pp313-322, May 1991.

[2] R. Cameron and B. Woerner, "Performance analysis of CDMA with imperfect power control", *IEEE Trans. Commun.*, Vol.44, No.7, pp777-781, July 1996.

[3] W. C. Y. Lee, *Mobile Communications Design Fundamentals*. New York: Wiley, 1993.

[4] Qiang Wu, Wei-Ling Wu and Jiong-Pan Zhou, "Centralised power control in CDMA cellular mobile systems", *Electron. Lett.*, Vol.33, No.2, 16th, Jan. 1997.

[5] S. A. Grandhi, R. Vijayan, D. J. Goodman and J. Zander, "Centralized power control in cellular radio system ", *IEEE Trans. Veh. Technol.*, Vol.42, No.4, pp466-468, Nov. 1993.

[6] J. Zander, "Distributed cochannel interference control in cellular radio systems", *IEEE Trans. Veh. Technol.*, Vol.41, No.3, pp305-311, Aug. 1992.

[7] S. Ariyavisitakul, L. F. Chang, "Signal and interference statistics of a CDMA system with feedback power control", *IEEE Trans. On Commun.* vol. 41, pp.1626-1634, Nov. 1993.

[8] S. Ariyavisitakul, "SIR based power control in CDMA system", in *Proc. IEEE GLOBECOM*, Orlando FL, pp. 868-873, Dec. 1992.

[9] S. Ariyavisitakul, "Signal and interference statistics of a CDMA system with feedback power control – Part II", *IEEE Trans. On Commun.* vol. 42, pp.597-605, Feb/Mar/April, 1994.

[10] S. D. Fina, "Error probability analysis for CDMA systems with closed loop power contro," To appear in IEEE Transactions on Communications.

[11] S. A. Grandhi, R. Vijayan, D. J. Goodman, "Distributed cochannel interference control in cellular radio system ", *IEEE Trans. On Commun.* vol. 42, pp. 226-228, Feb, 1994.

[12] A. M. Monk and L. B. Milstein, "Open-loop power control error in a land mobile satellite system", *IEEE J. Select. Areas Commun.*, Vol.13, No.2, pp205-212, Feb. 1995.

[13] A. Chockalingam, P. Dietrich, L. B. Milstein and R. R. Rao, "Performance of closed-loop power control in DS-CDMA cellular systems", *IEEE Trans. On Veh. Technol.*, vol. 47, pp. 774-789, Aug. 1998.

[14] C. J. Chang, J. H. Lee and F. C. Ren, "Design of power control mechanism with PCM realization for the uplink of a DS-CDMA

cellular mobile radio system", *IEEE Trans. On Veh. Technol.*, vol. 45, pp. 522-530, Aug. 1996.

[15] M. L. Sim, E. Gunawan and B. H. Soong, and C. B. Soh, "Performance study of closed-loop power control algorithm for a cellular CDMA system", *IEEE Trans. On Veh. Technol*, vol. 48, pp. 911-921, May 1999.

[16] Y. J. Yang and J. F. Chang, "A strength-and-SIR combined adaptive power control for CDMA mobile radio channels", *IEEE Trans. On Veh. Technol.*, vol.48, No.6, pp1996-2004, Nov. 1999.

[17] D. L. Schilling, "Wireless communications going into the 21st century", *IEEE Trans. On Veh. Technol.*, vol. 43, pp. 645-651, Aug. 1994.

[18] C. C. Lee and R. Steele, "Closed-loop power control in CDMA systems", *IEE Proc. Commun.*, vol.143, No.4, pp231-239, Aug.1996.

[19] P. R. Chang and B. C. Wang, "Adaptive fuzzy power control for CDMA mobile radio system", *IEEE Trans. Veh. Technol.*, Vol. 45, pp225-236, Feb. 1996.

[20] P. R. Chang and B. C. Wang, "Adaptive fuzzy proportional integral power control for a cellular CDMA with time delay", *IEEE J. Select. Areas Commun.*, Vol14, pp1818-1829, Sept. 1996.

[21] A. M. Viterbi and A. J. Viterbi, "Erlang capacity of a power controlled CDMA system", *IEEE J. Select. Areas Commun.*, Vol.11, No.6, Aug. 1993.

[22] T. K. Woo, "Trellis-based power control for direct sequence CDMA in a Rayleigh fading environment", *IEE Proc. Commun.* Vl146, No.4, pp258-264, Aug. 1999

[23] R. R. Gejji, "Forward-link power control in CDMA cellular systems", *IEEE Trans. Veh. Technol.*, Vol.41, No.4, pp532-536, Nov. 1992.

[24] D. Kim, "Downlink power allocation and adjustment for CDMA cellular systems", *IEEE Commun. Letter*, vol.1, No.4, July 1997

[25] A. J. Goldsmith, L. J. Greenstein and G. J. Foschini, "Error statics of real-time power measurements in cellular channels with multipath and shadowing", *IEEE Trans. Veh. Technol.*, Vol.43, No.3, pp439-446, Aug. 1994.

[26] A. Yu, "A study of power control in CDMA overlay," Ph.D. Thesis, University of Hong Kong, 2000.

[27] R. Padovani, "Reverse link performance of IS-95 based cellular system", *IEEE Pers. Commun.*, vol.1, No.3, third quarter, pp28-34, 1994.

[28] J. M. Holtzman, "CDMA power control for wireless Networks", in Third Genearation Wireless Information Networks, S. Nanda and D. J. Goodman, Eds. Boston, MA: Kluwer Academic, pp299-311, 1992.

[29] J. Wang, L. B. Milstein, "CDMA overlay situations for microcellular mobile communications," *IEEE Trans. Commun.*, vol.43, NO.2/3/4, Feb/ Mar/April, 1995.

[30] M. B. Pursley, "Performance evaluation for phase-ceded spread spectrum multiple access communications, part I: system analysis," IEEE *Trans. Commun.*, vol.25, pp295-299, Aug., 1977.

[31] M. Gudmundson, "Correlation model for shadow fading in mobile radio systems," *Electr. Letts.*, vol.27, pp. 2145-2146, Nov. 1991.

[32] R. H. Clarke, "A statistical theory of mobile radio reception," *Bell Syst. Tech. J.*, vol. 47, pp. 957-1000, July/Aug. 1968.

[33] K. S. Gilhousen, I. M. Jacobs, R. Padovani, A. J. Viterbi, L. A. Weaver, Jr., and C. E. Wheatley III, "On the capacity of a cellular CDMA system," *IEEE Trans. On Veh. Technol.*, vol. 40, pp. 303-312, May 1991.

[34] J. Wang and A. Yu, "Open loop power control error in cellular CDMA overlay systems," IEEE Journal on Selected Areas in Communications, July 2001.

[35] A. Yu, "A study of power control in CDMA overlay," Ph.D. Thesis, University of Hong Kong, April 2000.

Chapter 8

Closed-loop Power Control

Abstract In this chapter, we investigate the effect of power control error of a closed-loop power controlled CDMA system with imperfect power estimation. By use of a simplified log-linear power control model, the power control error of a strength-based power control algorithm is studied in the presence of narrowband interference. Analytical and simulation results show that for normal load (K) of CDMA systems when the ratio of the narrowband interference power to signal power ratio is less than 10dB, the power control error caused by narrowband interference is very small, whereas when the ratio is large (larger than 10dB), the power control error increases dramatically.

Keywords Closed-Loop Power Control, Narrowband Interference

8.1 INTRODUCTION

Power control is one of the most crucial issues in the design of CDMA systems, due to the well known sensitivity of the spread spectrum technology to "near-far" effects [1]. Terrestrial mobile access systems require very accurate power control schemes, especially in the uplink. Power control aims at balancing the multiple access interference (MAI) for the users within the coverage area of a base station. This is made by monitoring the signal energy of users and avoiding that the users with large received power cause an excessive increase of the overall MAI. Terrestrial mobile radios may experience round-trip delays, between the mobile and the base stations, which turn out to be smaller than the channel time coherence. Therefore, the feedback from the base station can be used to implement a closed-loop power control (CLPC) scheme that compensates for the channel multipath fading. It is known that CLPC is effective against multipath fading if the power command update rate is significantly higher than 10 times the maximum fading rate [2].

Closed-loop power control has mainly two types of mechanisms: strength-based and SIR (signal to interference ratio)-based power control mechanisms. The strength-based mechanism [2-4] is based on the original idea of power control — to keep the powers of all signals received at the base station as nearly the same as possible so as to mitigate the near-far effect. The SIR-based mechanism [5-6] is based on the objective of directly controlling the communication quality or the system performance according to the SIR received at the base station. In this chapter, a strength-based power control mechanism is adopted.

Closed loop power control has been widely studied, and the capacity of a CDMA system is found to greatly depend on the quality of power control [1]. In practice applications, there always exists power control error, which usually results from the factors such as the accuracy of power measurement at a base station, the dynamic ranges of the transmit power of mobiles and the loop delay. In the strength-based power control algorithm, the base station makes an estimate of the received power by averaging the signal power at the outputs of the correlator during bit sampling instants. However, in much published work on power control loop, it is assumed that base stations generate the power control commands based on perfect estimates of the received signal power. In fact, the square law estimation of the received signal power at the output of the correlator generates power estimation error because of the existence of noise components such as MAI (multi-access interference) and the narrowband interference of our study. This chapter provides analytical expressions of the power control error and error probability of CDMA system when narrowband interference is present.

8.2 SYSTEM MODEL

The channel is modeled as frequency nonselective with Rayleigh statistics, and log-normal distributed shadowing. However, results of this chapter can be extended to frequency selective channels. The receiver is a conventional coherent receiver with a suppression filter for narrowband interference rejection [7]. A closed-loop power control algorithm is adopted to compensate for the channel fading and shadowing. Compared to the Rayleigh fading process, the power loss due to propagation distance and shadowing is relatively slow and can be assumed to be "tracked out" perfectly by the closed-loop power control algorithm. Neglecting channel noise, the received signal at the base station can be written as

$$r(t)=\sum_{k=1}^{K}\sqrt{2P_k(t)}b_k(t-\tau_k)c_k(t-\tau_k)\cos(2\pi f_0 t+\phi_k)+j(t) \quad (8.1)$$

where K is the number of active users, $P_k(t)$ represents the received signal power of the kth user. In the case of perfect power control, $P_k(t)$ is a constant. However, in a practical power control system, power control errors occur. $P_k(t)$ is a random process due to the adaptation of power control. From [8], the *pdf* of $P_k(t)$ can be assumed to be log-normally distributed, i.e.

$$P_k(t)=10^{\zeta(t)/10} \quad (8.2)$$

where $\zeta(t)$ is normally distributed with mean $m_{P_{k,dB}}$ and variance $\sigma^2_{P_{k,dB}}$. In (8.1), $b_k(t)$ and $c_k(t)$ represent the kth-user binary data sequence with duration T and the spreading sequence with chip duration T_c, respectively. τ_k and ϕ_k are the corresponding time delay and phase, respectively. $j(t)$ represents the narrowband interference, which is modelled as a zero-mean Gaussian random process with a double-sided power spectral density of $\eta_j/2$ over the bandwidth of B_j, and is given by (7.5). The power of $j(t)$ is given by $P_j=\eta_j B_j$. The auto-correlation function of $j(t)$ is given by (7.7). $p=B_jT_c/2$ is defined as the ratio of the interference bandwidth to the spread spectrum bandwidth, $q=\Delta T_c$ is defined as the ratio of the offset (Δ) of the interference carrier frequency to the half $(1/T_c)$ of the spread spectrum bandwidth.

The interference suppression filter is assumed to be a double-sided Wiener filter with M taps on each side. Its impulse response is $\sum_{m=-M}^{M} \alpha_m \delta(t - mT_c)$, where $\alpha_0 = 1$ and $\alpha_m = \alpha_{-m}$. Therefore, the signal at the output of the suppression filter is given by

$$r_f(t) = \sum_{m=-M}^{M} \alpha_m r(t - mT_c). \tag{8.3}$$

Assuming the user 1 is the reference user and $\tau_1 = \phi_1 = 0$, the output of despreader/demodulator at time *iT*, is given by

$$Z(iT) = \frac{1}{T} \int_{(i-1)T}^{iT} r_f(t) \cdot c_1(t) \sqrt{2} \cos(2\pi f_0 t) dt \tag{8.4}$$

where $\sqrt{2}\cos(2\pi f_0 t)$ and $c_1(t)$ are the recovered carrier and the spreading sequence of the reference user. Since high frequency terms are removed by the lowpass filter following the mixer of the demodulator, the above expression reduces to

$$\begin{aligned} Z(iT) &= \sqrt{P_1(iT)} b_1(iT) + I(iT) + J(iT) \\ &= \sqrt{P_1(i)} b_1(i) + I(i) + J(i). \end{aligned} \tag{8.5}$$

Note that, in the following, "iT" is replaced by "i" for simple notation.

The first term on the right hand side of (8.5) is the desired component. The second term is the multi-access interference from the zero-th tape of the suppression filter. The multi-access interference from other taps is relatively small and is neglected here like [7]. Thus, $I(i)$ is given by

$$I(i) = \sum_{k=2}^{K} \sqrt{P_k(i)} I_k(i) \cos(\phi_k) \tag{8.6}$$

where

$$I_k(i) = \frac{1}{T} \int_{(i-1)T}^{iT} b_k(t - \tau_k) c_k(t - \tau_k) c_1(t) dt. \tag{8.7}$$

When τ_k is uniformly distributed in [0, T], the variance of $I_k(i)$ is given by [9], i.e. $E[I_k^2(i)] = 2/(3N)$. According to the central-limit theorem, when K

is large, $I(i)$ can be approximated as a zero-mean Gaussian random variable with the variance given by

$$\sigma_I^2 = \frac{1}{2}\sum_{k=2}^{K} E[P_k(i)]E[I_k^2(i)]. \tag{8.8}$$

Assuming that all the users adopt the same power control algorithm, the distribution of $P_k(i)$ can be assumed to be the same as $P_1(i)$. Thus,

$$\sigma_I^2 = \frac{K-1}{3N} E[P_1(i)]. \tag{8.9}$$

The third term on the right hand side of (8.5) is the narrowband interference component and is given by

$$J(i) = \frac{1}{T}\sum_{m=-M}^{M} \alpha_m \int_{(i-1)T}^{iT} j(t - mT_c)c_1(t)\sqrt{2}\cos(2\pi f_0 t)dt. \tag{8.10}$$

The variance of the narrowband interference is given by (7.17)

$$\begin{aligned}\sigma_J^2 &= \frac{P_j}{\sqrt{2}N}\sum_{m_1=-M}^{M}\alpha_{m_1}\sum_{m_2=-M}^{M}\alpha_{m_2}Q(m_1, m_2)\\ &= P_j\hat{\sigma}_J^2\end{aligned} \tag{8.11}$$

where

$$\hat{\sigma}_J^2 = \frac{1}{\sqrt{2}N}\sum_{m_1=-M}^{M}\alpha_{m_1}\sum_{m_2=-M}^{M}\alpha_{m_2}Q(m_1, m_2), \tag{8.12}$$

and

$$\begin{aligned}&Q(m_1, m_2)\\ &= \int_{-1}^{1}(1-|\tau|)\text{sinc}\{2\pi[\tau-(m_1-m_2)]p\}\cos\{2\pi[\tau-(m_1-m_2)]q\}d\tau.\end{aligned} \tag{8.13}$$

8.3 POWER CONTROL MODEL

A simplified log-linear closed-loop power control model is adopted as shown in Figure 8.1. All values of inputs and outputs of the figure are marked in decibels. The blocks of the figure represent linear filters, with either a fixed delay or filter impulse response. The model only shows the instantaneous powers at various points in the system model and not the actual signals. In the figure, $P^{t}_{1,dB}(i)$ stands for the transmit power of the reference user. $a^{2}_{1,dB}(i)$ represents propagation loss introduced by the channel and changes very slowly compared to the bit rate and, hence, is considered constant during bit duration. From [3], the correlation function of $a^{2}_{1,dB}(i)$ is given by

$$R_{a_1^2}(i-j)=E\left[a^{2}_{1,dB}(i)a^{2}_{1,dB}(j)\right]=\left(\frac{10}{\ln 10}\right)^2\left[\rho^{2}_{i,j}\sum_{n=1}^{\infty}\frac{\rho_{ij}^{2(n-1)}}{n^2}+C^2\right] \tag{8.14}$$

where C is the Euler Gamma constant and is approximately equal to 0.577216. $\rho_{i,j}$ is the normalised autocovariance function of the Rayleigh process $a_1(i)$, given by

$$\rho_{i,j}=J_0\left[2\pi f_d(i-j)T\right] \tag{8.15}$$

where f_d is a maximum Doppler frequency or fading rate, given by $f_d=v/\lambda$, and λ is a carrier wavelength. $J_0(x)$ is a Bessel function of the first kind of the zero-th order. The standard deviation of $a^{2}_{1,dB}(i)$ is then given by

$$\begin{aligned}\sigma_{a_1^2}&=\sqrt{R_{a_1^2}(0)-m^2_{a_1^2}}\\&=\sqrt{100\pi^2/\left(6\ln^2 10\right)}\\&\approx 5.6dB\end{aligned} \tag{8.16}$$

where $m_{a_1^2}=-C$.

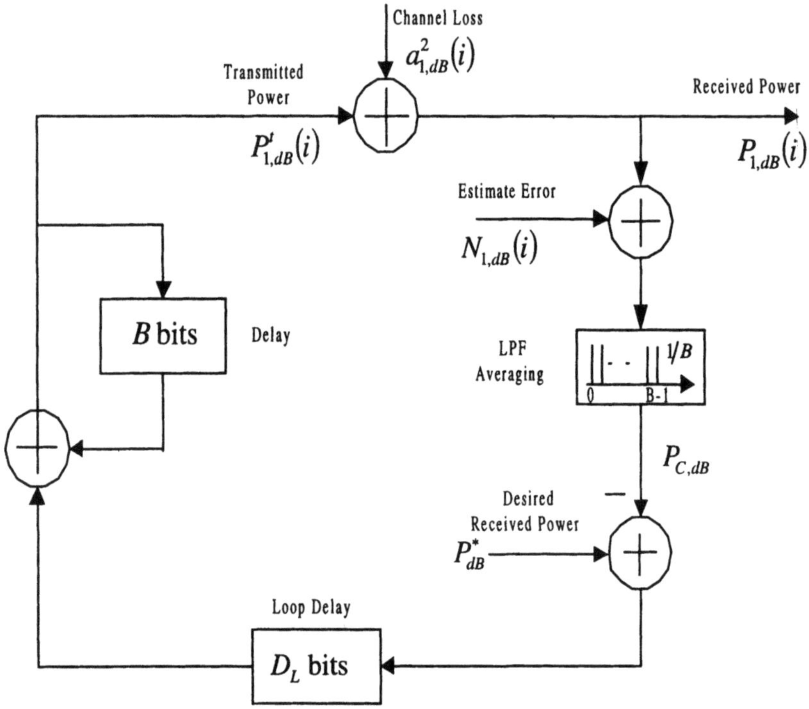

Fig. 8.1. Simplified log-linear power control model [12].

$P_{1,dB}(i)$ represents the signal power of the *i*th bit of the reference user at the base station. $P_{1,dB}(i)$ has to be measured accurately at the base station, in order to achieve good closed-loop power control. However, because the multiple access interference and narrowband interference are also present in the system, the estimate of the received power contains an error, which is defined by the random process $N_{1,dB}(i)$.

From (8.5), the signal power of the demodulator output at *i*th bit is given by

$$S(i) = Z^2(i)$$

$$= \left(\sqrt{P_1(i)}b_1(i) + I(i) + J(i)\right)^2$$

$$= P_1(i) + X(i) \tag{8.17}$$

where

$$X(i)=2\sqrt{P_1(i)}b_1(i)[I(i)+J(i)]+[I(i)+J(i)]^2$$

$$=2\sqrt{P_1(i)}b_1(i)Y(i)+Y^2(i) \tag{8.18}$$

where $Y(i)=I(i)+J(i)$ is Gaussian distributed with variance $\sigma_Y^2=\sigma_I^2+\sigma_J^2$, where σ_I^2 and σ_J^2 are given by (8.9) and (8.11), respectively.

The series expansion of log function of (8.17) is given by [10]

$$10\log_{10} S(i)=10\log_{10}(P_1(i)+X(i))$$

$$=10\log_{10} P_1(i)+\frac{20}{\ln 10}\left[\frac{X(i)}{2P_1(i)+X(i)}+\frac{1}{3}\left(\frac{X(i)}{2P_1(i)+X(i)}\right)^3\right.$$

$$\left.+\cdots+\frac{1}{2n+1}\left(\frac{X(i)}{2P_1(i)+X(i)}\right)^{2n+1}+\cdots\right]$$

$$=10\log_{10} P_1(i)+\frac{20}{\ln 10}\sum_{n=0}^{\infty}\frac{1}{2n+1}\left[\frac{X(i)}{2P_1(i)+X(i)}\right]^{2n+1}$$

$$=P_{1,dB}(i)+N_{1,dB}(i) \tag{8.19}$$

where $N_{1,dB}(i)$ is the estimation error and is given by

$$N_{1,dB}(i)=\frac{20}{\ln 10}\sum_{n=0}^{\infty}\frac{1}{2n+1}\left[\frac{X(i)}{2P_1(i)+X(i)}\right]^{2n+1}$$

$$=\frac{20}{\ln 10}\sum_{n=0}^{\infty}\frac{1}{2n+1}\left[\frac{2\sqrt{P_1(i)}b_1(i)Y(i)+Y^2(i)}{2P_1(i)+2\sqrt{P_1(i)}b_1(i)Y(i)+Y^2(i)}\right]^{2n+1}$$

$$= \frac{20}{\ln 10} \sum_{n=0}^{\infty} \frac{1}{2n+1} \left[\frac{\left(\sqrt{P_1(i)} b_1(i) + Y(i)\right)^2 - P_1(i)}{\left(\sqrt{P_1(i)} b_1(i) + Y(i)\right)^2 + P_1(i)} \right]^{2n+1} \tag{8.20}$$

Since $Y(i)$ is Gaussian with variance of σ_Y^2, the mean of $N_{1,dB}(i)$, conditioned on $P_1(i)$ and $b_1(i)$, is given by

$$m_{N_{1,dB}(i)|P_1(i),b_1(i)} = E\left[N_{1,dB}(i)|P_1(i), b_1(i)\right]$$

$$= \frac{20}{\sqrt{2\pi}\sigma_Y \ln 10} \int_{-\infty}^{\infty} \sum_{n=0}^{\infty} \frac{1}{2n+1} \left[\frac{\left(\sqrt{P_1(i)} + b_1(i) y\right)^2 - P_1(i)}{\left(\sqrt{P_1(i)} + b_1(i) y\right)^2 + P_1(i)} \right]^{2n+1} e^{-y^2/2\sigma_Y^2} dy \tag{8.21}$$

Assuming that $b_1(i)$ takes values from ± 1 with equal probability and averaging over $b_1(i)$, (8.21) reduces to

$$m_{N_{1,dB}(i)|P_1(i)} = E\left[N_{1,dB}(i)|P_1(i)\right]$$

$$= \frac{10}{\sqrt{2\pi}\sigma_Y \ln 10} \int_{-\infty}^{\infty} \sum_{n=0}^{\infty} \frac{1}{2n+1} \left[\frac{\left(\sqrt{P_1(i)} + y\right)^2 - P_1(i)}{\left(\sqrt{P_1(i)} + y\right)^2 + P_1(i)} \right]^{2n+1} e^{-y^2/2\sigma_Y^2} dy$$

$$+ \frac{10}{\sqrt{2\pi}\sigma_Y \ln 10} \int_{-\infty}^{\infty} \sum_{n=0}^{\infty} \frac{1}{2n+1} \left[\frac{\left(\sqrt{P_1(i)} - y\right)^2 - P_1(i)}{\left(\sqrt{P_1(i)} - y\right)^2 + P_1(i)} \right]^{2n+1} e^{-y^2/2\sigma_Y^2} dy$$

$$= \frac{20}{\sqrt{2\pi}\sigma_Y \ln 10} \int_{-\infty}^{\infty} \sum_{n=0}^{\infty} \frac{1}{2n+1} \left[\frac{\left(\sqrt{P_1(i)} + y\right)^2 - P_1(i)}{\left(\sqrt{P_1(i)} + y\right)^2 + P_1(i)} \right]^{2n+1} e^{-y^2/2\sigma_Y^2} dy$$

$$= \frac{20}{\sqrt{2\pi}\sigma_Y \ln 10} \int_{-\infty}^{\infty} \sum_{n=0}^{\infty} \frac{1}{2n+1} \left[\frac{x^2 - P_1(i)}{x^2 + P_1(i)} \right]^{2n+1} e^{-\left(x - \sqrt{P_1(i)}\right)^2/2\sigma_Y^2} dx$$

$$= \frac{40}{\sqrt{2\pi}\sigma_Y \ln 10} \sum_{n=0}^{\infty} \frac{1}{2n+1} \int_0^{\infty} \left[\frac{x^2 - P_1(i)}{x^2 + P_1(i)} \right]^{2n+1} e^{-\left(x^2 + P_1(i)\right)/2\sigma_Y^2} ch\left(\frac{x\sqrt{P_1(i)}}{\sigma_Y^2} \right) dx \tag{8.22}$$

where $ch(x) = \left(e^{x} + e^{-x}\right)/2$.

Similarly, the variance of $N_{1,dB}(i)$, conditioned on $P_1(i)$ is given by

$$\sigma^2_{N_{1,dB}(i)|P_1(i)} = E\left[N^2_{1,dB}(i)\big|_{P_1(i)}\right] - m^2_{N_{1,dB}(i)|P_1(i)}$$

$$= \frac{400}{\sqrt{2\pi}\sigma_Y (\ln 10)^2} \int_{-\infty}^{\infty} \sum_{n_1=0}^{\infty}\sum_{n_2=0}^{\infty} \frac{1}{(2n_1+1)(2n_2+1)}$$

$$\times \left[\frac{\left(\sqrt{P_1(i)}b_1(i) + y\right)^2 - P_1(i)}{\left(\sqrt{P_1(i)}b_1(i) + y\right)^2 + P_1(i)}\right]^{2n_1+2n_2+2} e^{-y^2/2\sigma_Y^2} dy - \frac{800}{\pi\sigma_Y^2 (\ln 10)^2}$$

$$\cdot \left[\sum_{n=0}^{\infty} \frac{1}{2n+1} \int_0^{\infty} \left[\frac{x^2 - P_1(i)}{x^2 + P_1(i)}\right]^{2n+1} e^{-\left(x^2 + P_1(i)\right)/2\sigma_Y^2} ch\left(\frac{x\sqrt{P_1(i)}}{\sigma_Y^2}\right) dx\right]^2$$

$$= \frac{400}{\sqrt{2\pi}\sigma_Y (\ln 10)^2} \sum_{n_1=0}^{\infty}\sum_{n_2=0}^{\infty} \frac{1}{(2n_1+1)(2n_2+1)}$$

$$\times \int_{-\infty}^{\infty} \left[\frac{x^2 - P_1(i)}{x^2 + P_1(i)}\right]^{2n_1+2n_2+2} e^{-\left(x^2 + P_1(i)\right)/2\sigma_Y^2} ch\left(\frac{x\sqrt{P_1(i)}}{\sigma_Y^2}\right) dx$$

$$- \frac{800}{\pi\sigma_Y^2 (\ln 10)^2} \left[\sum_{n=0}^{\infty} \frac{1}{2n+1} \int_0^{\infty} \left[\frac{x^2 - P_1(i)}{x^2 + P_1(i)}\right]^{2n+1} e^{-\left(x^2 + P_1(i)\right)/2\sigma_Y^2} ch\left(\frac{x\sqrt{P_1(i)}}{\sigma_Y^2}\right) dx\right]^2$$

(8.23)

Since $P_1(i)$ is log-normally distributed, i.e., the *pdf* of $P_1(i)$ is given by [11]

$$f_{P_1(i)}(p) = \frac{10}{\sqrt{2\pi}\sigma_{P_{1,dB}}\, p \ln 10} \exp\left(-\frac{\left(10\log p - m_{P_{1,dB}}\right)^2}{2\sigma^2_{P_{1,dB}}}\right), \quad p > 0$$

(8.24)

the mean and variance of $N_{1,dB}(i)$ are given, respectively, by

$$m_{N_{1,dB}(i)} = \frac{10}{\sqrt{2\pi}\sigma_{P_{1,dB}} \ln 10} \int_0^\infty m_{N_{1,dB}(i)|P_1(i)=p} \frac{1}{p} e^{-\left(10\log(p)-m_{P_{1,dB}}\right)^2/2\sigma_{P_{1,dB}}^2} dp \tag{8.25}$$

and

$$\sigma^2_{N_{1,dB}(i)} = \frac{10}{\sqrt{2\pi}\sigma_{P_{1,dB}} \ln 10} \int_0^\infty \sigma^2_{N_{1,dB}(i)|P_1(i)=p} \frac{1}{p} e^{-\left(10\log(p)-m_{P_{1,dB}}\right)^2/2\sigma_{P_{1,dB}}^2} dp \tag{8.26}$$

where $m_{P_{1,dB}}$ and $\sigma^2_{P_{1,dB}}$ will be given by (8.33) and (8.34) later.

The low pass averaging filter averages the input samples $P_{1,dB}(i)+N_{1,dB}(i)$ over B bits, as shown in Fig.8.1 and its output, $P_{c,dB}$, is compared to a desired power threshold P^*_{dB}. The value of $P_{c,dB} - P^*_{dB}$ is sent back to the reference user for its transmit power adjustment. When the value is positive (negative), the reference user needs to decrease (increase) its transmit power. Due to round-trip propagation and processing, the closed-loop power control process involves D_L bits delay. Finally, the mobile reference adjusts its transmit power every B bits.

In order to derive the mean and variance of $P_{1,dB}(i)$, the following transfer function [3] are useful. The transfer function of the averaging filter is

$$H_f(\omega) = \frac{1}{B}\frac{\sin(\omega B/2)}{\sin(\omega/2)} e^{-j\omega[(B-1)/2]}. \tag{8.27}$$

The transfer function between $P_{1,dB}(i)$ and $a^2_{1,dB}(i)$ is

$$H_{P_1,a_1^2}(\omega) = \frac{1-e^{-j\omega B}}{1-e^{-j\omega B} + H_f(\omega)e^{-j\omega D_L}}. \tag{8.28}$$

The transfer function between $P_{1,dB}(i)$ and P^*_{dB} is given by

$$H_{P_1,P^*}(\omega) = \frac{1}{1-e^{-j\omega B} + H_f(\omega)e^{-j\omega D_L}}, \tag{8.29}$$

and the transfer function between $P_{1,dB}(i)$ and $N_{1,dB}(i)$ is given by

$$H_{P_1,N_1}(\omega)=\frac{-H_f(\omega)e^{-j\omega D_L}}{1-e^{-j\omega B}+H_f(\omega)e^{-j\omega D_L}}. \tag{8.30}$$

Since the system is linear, $P_{1,dB}(i)$ is the sum of three components $P_{1,dB}^{a_1^2}(i)$, $P_{1,dB}^{P^*}(i)$ and $P_{1,dB}^{N_1}(i)$, due to the three inputs $a_{1,dB}^2(i)$, P_{dB}^*, and $N_{1,dB}(i)$, respectively. In other words, $P_{1,dB}(i)=P_{1,dB}^{a_1^2}(i)+P_{1,dB}^{P^*}(i)+P_{1,dB}^{N_1}(i)$. Therefore, the mean of $P_{1,dB}(i)$ is given by

$$m_{P_{1,dB}(i)}=E\left[P_{1,dB}^{a_1^2}(i)\right]+E\left[P_{1,dB}^{N_1}(i)\right]+E\left[P_{1,dB}^{P^*}(i)\right]. \tag{8.31}$$

Defining $h_{P_1,a_1^2}(j)$, $h_{P_1,N_1}(j)$, $h_{P_1,P^*}(j)$ as the discrete inverse Fourier transforms of $H_{P_1,a_1^2}(\omega)$, $H_{P_1,N_1}(\omega)$ and $H_{P_1,P^*}(\omega)$, respectively, one obtains

$$m_{P_{1,dB}(i)}=$$

$$\sum_{j=-\infty}^{\infty}E\left[a_{1,dB}^2(i-j)\right]h_{P_1,a_1^2}(j)+\sum_{j=-\infty}^{\infty}E\left[N_{1,dB}(i-j)\right]h_{P_1,N_1}(j)+P_{dB}^*\sum_{j=-\infty}^{\infty}h_{P_1,P^*}(j) \tag{8.32}$$

Assuming that in the main window of $h_{P_1,a_1^2}(j)$, $E\left[a_{1,dB}^2(i-j)\right]$ is constant, on obtains

$$m_{P_{1,dB}(i)}=E\left[a_{1,dB}^2(i)\right]\sum_{j=-\infty}^{\infty}h_{P_1,a_1^2}(j)+E\left[N_{1,dB}(i)\right]\sum_{j=-\infty}^{\infty}h_{P_1,N_1}(j)+P_{dB}^*\sum_{j=-\infty}^{\infty}h_{P_1,P^*}(j)$$

$$=E\left[a_{1,dB}^2(i)\right]H_{P_1,a^2}(0)+E\left[N_{1,dB}(i)\right]H_{P_1,N_1}(0)+P_{dB}^*H_{P_1,P^*}(0)$$

$$=-E\left[N_{1,dB}(i)\right]$$

$$=m_{N_{1,dB}(i)} \tag{8.33}$$

where $H_{P_1,a_1^2}(0)=H_{P_1,P^*}(0)=0$ and $H_{P_1,N_1}(0)=-1$.

Statistically, due to power control, $a_{1,dB}^2(i)$, and $N_{1,dB}(i)$ can be considered independent. It is assumed that $P_{dB}^*(i)=0$. Defining $S_{a_1^2}(\omega)$ and

$S_{N_1}(\omega)$ are the power spectral densities of $a_{1,dB}^2(i)$ and $N_{1,dB}(i)$, respectively, the variance of $P_{1,dB}(i)$ is given by

$$\sigma^2_{P_{1,dB}(i)} = E\left[P^2_{1,dB}(i)\right] - m^2_{P_{1,dB}(i)}$$

$$= E\left[\left(P^{a_1^2}_{1,dB}(i)\right)^2\right] + E\left[\left(P^{N_1}_{1,dB}(i)\right)^2\right] + E\left[\left(P^{P^*}_{1,dB}(i)\right)^2\right] - m^2_{N_{1,dB}(i)}$$

$$= \frac{1}{2\pi}\int_{-\infty}^{\infty}\left[S_{a_1^2}(\omega)\left|H_{P_1,a_1^2}(\omega)\right|^2 + S_{N_1}(\omega)\left|H_{P_1,N_1}(\omega)\right|^2\right]d\omega - m^2_{N_{1,dB}(i)}$$

$$= \frac{1}{2\pi}\int_{-\infty}^{\infty}\sum_{n=-\infty}^{\infty} R_{a_1^2}(n)e^{-j\omega n}\left|H_{P_1,a_1^2}(\omega)\right|^2 d\omega$$

$$+ \frac{1}{2\pi}\sum_{n=-\infty}^{\infty} R_{N_1}(n)e^{-j\omega n}\int_{-\infty}^{\infty}\left|H_{P_1,N_1}(\omega)\right|^2 d\omega - m^2_{N_{1,dB}(i)}$$

$$= \frac{1}{2\pi}\sum_{n=-\infty}^{\infty} R_{a_1^2}(n)\int_{-\infty}^{\infty} e^{-j\omega n}\left|H_{P_1,a_1^2}(\omega)\right|^2 d\omega + \frac{1}{2\pi}R_{N_1}(0)\int_{-\infty}^{\infty}\left|H_{P_1,N_1}(\omega)\right|^2 d\omega - m^2_{N_{1,dB}(i)}$$

$$= \frac{1}{2\pi}\sum_{n=-\infty}^{\infty} R_{a_1^2}(n)\int_{-\infty}^{\infty} e^{-j\omega n}\left|H_{P_1,a_1^2}(\omega)\right|^2 d\omega + \frac{1}{2\pi}\left[\mathrm{var}\left[N_{1,dB}(i)\right] + \left[E\left(N_{1,dB}(i)\right)\right]^2\right]$$

$$\times\int_{-\infty}^{\infty}\left|H_{P_1,N_1}(\omega)\right|^2 d\omega - m^2_{N_{1,dB}(i)}$$

$$= \frac{1}{2\pi}\sum_{n=-\infty}^{\infty} R_{a_1^2}(n)\int_{-\infty}^{\infty} e^{-j\omega n}\left|H_{P_1,a_1^2}(\omega)\right|^2 d\omega + \frac{1}{2\pi}\left[\sigma^2_{N_{1,dB}(i)} + m^2_{N_{1,dB}(i)}\right]$$

$$\times\int_{-\infty}^{\infty}\left|H_{P_1,N_1}(\omega)\right|^2 d\omega - m^2_{N_{1,dB}(i)} \qquad (8.34)$$

where $R_{a_1^2}(n)$ is given by (8.14), $R_{N_1}(n)$ stands for the autocorrelation function of $N_1(i)$ and $R_{N_1}(n)=0$ for $n \neq 0$. Note that substituting (8.33) and (8.34) into (8.25) and (8.26), we can obtain $m_{N_{1,dB}(i)}$ and $\sigma^2_{N_{1,dB}(i)}$ by computer search.

8.4 SYSTEM PERFORMANCE

From (8.5), since the decision variable is Gaussian, the conditional error probability, conditioned on $P_1(i)$, is given by

$$P\left(e\middle|_{P_1(i)}\right)=Q\left(\sqrt{\frac{P_1(i)}{\frac{K-1}{3N}E[P_1(i)]+\sigma_J^2}}\right) \tag{8.35}$$

where

$$Q(x)=\int_x^{\infty}\frac{1}{\sqrt{2\pi}}e^{-x^2/2}dx\,.$$

Assuming that $\xi(i)=\zeta(i)\ln 10/10$, one obtains

$$\begin{aligned}E[P_1(i)]&=E\left[10^{\zeta(i)/10}\right]\\&=E\left[e^{\xi(i)}\right]\\&=\exp\left(\bar{\xi}(i)+\sigma_{\xi(i)}^2/2\right)\\&=\exp\left[m_{P_{1,dB}}\ln 10/10+\left(\sigma_{P_{1,dB}}\ln 10/10\right)^2/2\right].\end{aligned} \tag{8.36}$$

Therefore,

$$P\left(e\middle|_{P_1(i)}\right)=Q\left(\sqrt{\frac{P_1(i)}{\frac{K-1}{3N}e^{m_{P_{1,dB}}\ln 10/10+\left(\sigma_{P_{1,dB}}\ln 10/10\right)^2/2}+\sigma_J^2}}\right). \tag{8.37}$$

The average error probability is given by

$$P(e)=\int_0^{\infty}Q\left(\sqrt{\frac{x}{\frac{K-1}{3N}e^{m_{P_{1,dB}}\ln 10/10+\left(\sigma_{P_{1,dB}}\ln 10/10\right)^2/2}+\sigma_J^2}}\right)f_{P_1(i)}(x)dx$$

$$= \int_0^\infty Q\left(\sqrt{\frac{x}{\frac{K-1}{3N}e^{m_{P_1,dB}\ln 10/10+\left(\sigma_{P_1,dB}\ln 10/10\right)^2/2}+P_j\hat{\sigma}_J^2}}\right)$$

$$\times\frac{10}{\sqrt{2\pi}\sigma_{P_1,dB}\,x\ln 10}\exp\left(-\frac{\left(10\log x-m_{P_1,dB}\right)^2}{2\sigma_{P_1,dB}^2}\right)dx \qquad (8.38)$$

where $\hat{\sigma}_J^2$, $m_{P_1,dB}$ and $\sigma_{P_1,dB}$ are given by (8.12), (8.33) and (8.34), respectively.

3.5 NUMERICAL RESULTS

Equations (8.33) and (8.34) give rise to the power control error expressions of the simplified log-linear model. It is shown that the power control error is a function of various parameters: fading rate, LPF averaging interval, loop delay, number of users and the variance of narrowband interference. Since the effect of fading rate and loop delay has been widely studied by others, our investigations mainly focus on the power control errors caused by the MAI and narrowband interference. Unless noted otherwise, fading rate, LPF averaging interval and loop delay are fixed, i.e. the Doppler frequency shift is $f_d = 27\,Hz$ (corresponding to $15km/h$ mobile speed at 2GHz carrier frequency), the averaging interval is $B = 10$ bits, the loop delay is $D_L = 1$ bit. The other parameters are assumed as: the threshold is $P_{dB}^* = 0dB$, the processing gain is $N = 250$, the number of taps of Wiener filter is $M = 4$ on each side, the ratio of the interference bandwidth to the spread spectrum bandwidth is $p = 0.2$, the ratio of the offset of the interference carrier frequency to half of the spread spectrum bandwidth is $q = 0.2$, the information data rate is $r_b = 8Kb/s$, the number of active users is $K = 20$.

Fig. 8.2 shows the standard deviation of received signal power versus update rate $(1/BT)$ when there is no power estimation error. In the case of perfect power estimation, i.e., $N_{1,dB}(i) = 0$, the mean of the received signal power (see (8.33)) keeps constant (0dB), and the standard deviation of the received signal power only changes with the variation of update rate (e.g.

update rate 800Hz corresponding to $B = 10$, as $r_b = 8Kb/s$) and loop delay D_L. It is seen from the figure that the standard deviation of the power control error increases when update rate decrease for a fixed D_L. In the case where the update rate is 0Hz (no power control is employed), the standard deviation of the power control error is given by $\sigma_{P_{1,dB}} = 5.6$ dB (see (8.16)).

Fig. 8.3 and Fig. 8.4 illustrate the mean and variance of received signal power as a function of the ratio of interference power to desired signal power P_j/P^*. It can be seen that when $P_j/P^* = -\infty$, the power control error (mean and variance) results from the MAI. When P_j/P^* increases from $-\infty$ to around 10dB, the power control error only increases slightly. This is because narrowband interference is small with respect to the MAI for normal load ($K = 10$ to 40). However, when P_j/P^* increases from around 10dB to a large value (30dB), the power control error increases dramatically. In this case, the narrowband interference dominates the MAI. It can also be seen that the difference between analytic and simulation results is very small when P_j/P^* is small. However, the difference in power control error increases when P_j/P^* increases. Therefore, the simplified log-linear power

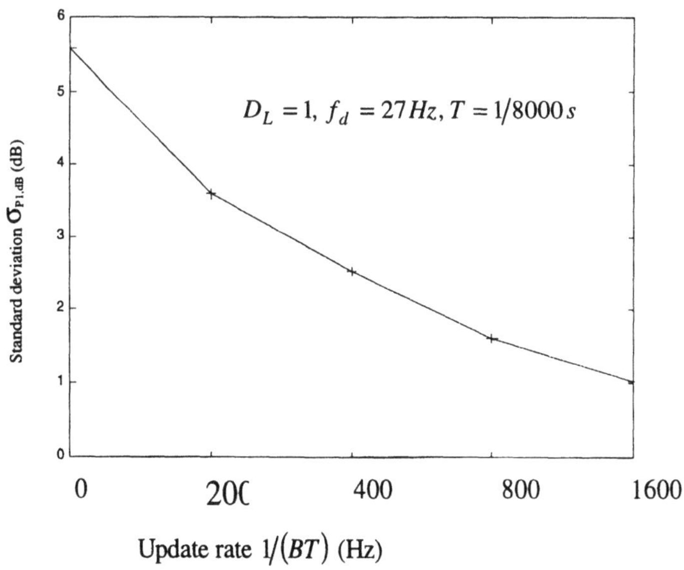

Fig. 8.2. The standard deviation of received signal power versus power control update rate when there is no power estimation error

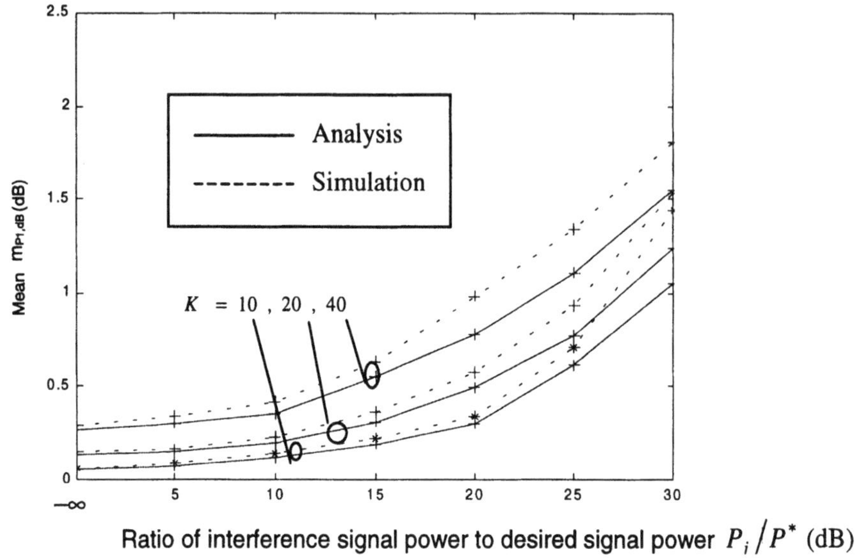

Fig. 8.3. The mean of received signal power as a function of the ratio of interference power to desired signal power P_j/P^*

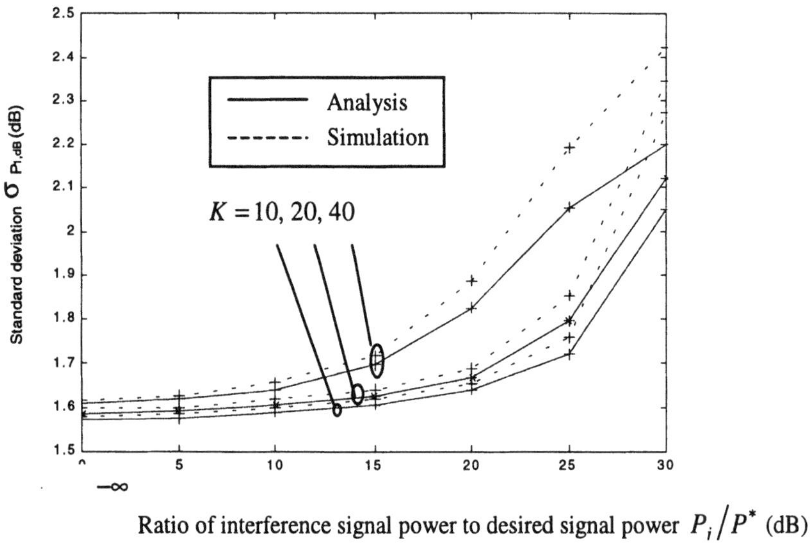

Fig. 8.4. The standard deviation of received signal power as a function of the ratio of interference power to desired signal power P_j/P^*

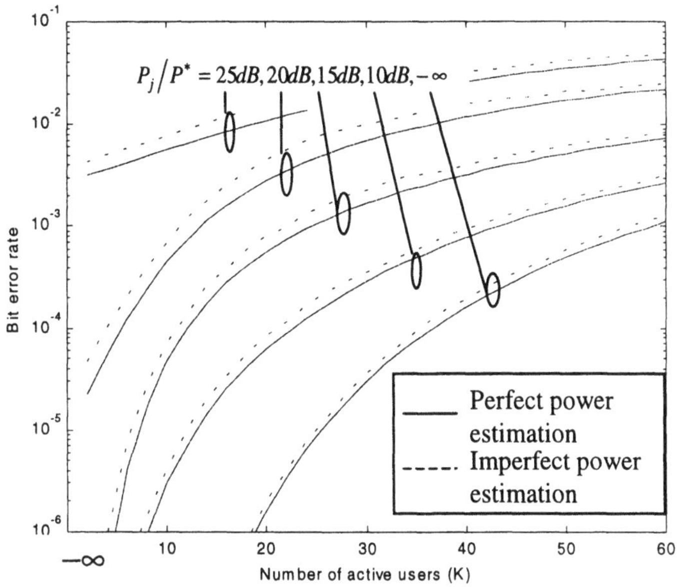

Fig. 8.5. BER comparison with both perfect and imperfect power estimation

control model is effective for not-too-large narrowband interference.

Fig. 8.5 shows the BER comparison of perfect power estimation with that of imperfect power estimation for different values of P_j/P^*. It is shown that for the fixed BER of 10^{-3}, there is capacity loss of 2~4 users caused by the imperfect power estimation.

8.6 CONCLUSIONS

The power control error of CDMA system in the presence of narrowband interference has been studied. The following conclusions have been drawn:

- For normal load (K) of CDMA systems when the ratio of the narrowband interference power to signal power ratio is less than 10dB, the power control error caused by narrowband interference is very small, whereas when the ratio is large (larger than 10dB), the power control error increases dramatically.
- The simplified log-linear power control model is effective for not-too-large narrowband interference.
- For the fixed BER of 10^{-3}, there is capacity loss of 2~4 users caused by the imperfect power estimation.

REFERENCES

[1] K. S. Gilhousen, I. M. Jacobs, R. Padovani, A. J. Viterbi, L. A. Weaver, Jr., and C. E. Wheatley III, "On the capacity of a cellular CDMA system," *IEEE Trans. On Veh. Technol.*, vol. 40, pp. 303-312, May 1991.

[2] S. Ariyavisitakul, L. F. Chang, "Signal and interference statistics of a CDMA system with feedback power control", *IEEE Trans. On Commun.* vol. 41, pp.1626-1634, Nov. 1993.

[3] A. Chockalingam, P. Dietrich, L. B. Milstein and R. R. Rao, "Performance of closed-loop power control in DS-CDMA cellular systems", *IEEE Trans. On Veh. Technol.*, vol. 47, pp. 774-789, Aug. 1998.

[4] M. L. Sim, E. Gunawan and B. H. Soong and C. B. Soh, "Performance study of closed-loop power control algorithm for a cellular mobile radio system", *IEEE Trans. Veh. Technol.*, Vol.48, pp911-921, May 1999.

[5] C. J. Chang, J. H. Lee and F. C. Ren, "Design of power control mechanism with PCM realization for the uplink of a DS-CDMA cellular mobile radio system", *IEEE Trans. On Veh. Technol.*, vol. 45, pp. 522-530, Aug. 1996.

[6] S. Ariyavisitakul, "Signal and interference statistics of a CDMA system with feedback power control – Part II", *IEEE Trans. On Commun.* vol. 42, pp.597-605, Feb/Mar/April, 1994.

[7] J. Wang, L. B. Milstein, "CDMA overlay situations for microcellular mobile communications," *IEEE Trans. Commun.*, vol.43, NO.2/3/4, Feb/Mar/April, 1995.

[8] A. J. Viterbi, A. M. Viterbi and E. Zehavi, "Performance of power-controlled wideband terrestrial digital communication," *IEEE Trans. Commun.*, vol.41, NO. 4, Apr. 1995.

[9] M. B. Pursley, "Performance evaluation for phase-ceded spread spectrum multiple access communications, part I: system analysis," IEEE *Trans. Commun.*, vol.25, pp295-299, Aug., 1977.

[10] J. W. Harris, and H. Stocker, *Handbook of mathematics and computational science*. Springer, NY, 1998.

[11] W. C. Y. Lee, *Mobile Communications Design Fundamentals*. New York: Wiley, 1993.

[12] A. Yu and J. Wang, "Closed loop power control in CDMA overlay systems," Submitted to IEEE Transactions on Communications.

Chapter 9

UPLINK MULTICARRIER DS/SFH-CDMA

Abstract In this chapter, uplink multicarrier direct-sequence/slow-frequency-hopping (MC DS/SFH) CDMA systems are proposed, in which multiple carriers are modulated by the same DS waveform and hopped in frequency according to a random hopping pattern. The receiver dehops the received signal with the same pattern, provides RAKE receivers for each carrier and combines the outputs with a maximal ratio combiner (MRC). The performance of the proposed system is investigated over a frequency selective Rayleigh fading channel and compared to that of the MC DS-CDMA systems. It is shown that for the same diversity order, the MC DS/SFH-CDMA systems are superior in reducing multiple access interference (MAI) while preserving the good capability of narrowband interference suppression, when the system parameters are selected properly.

Keywords CDMA, Multicarrier Systems, Narrowband Interference suppression.

9.1 LITERATURE REVIEW

The most commonly used forms of spread spectrum fall into three categories, Direct-sequence (DS), in which a high-rate pseudonoise (PN) code (termed as signature sequence) is used to spread the data spectrum before carrier modulation; Frequency-hopped (FH), in which a pseudorandom code (termed as frequency-hopping pattern) is used to control the frequency of the carrier which the data sequence is to modulate; Multicarrier (MC), in which multiple orthogonal carriers are transmitted simultaneously.

Much literature has addressed on the possibility and performance of CDMA overlay. DS systems have attracted most of the attentions (see Chapters 5-6). While DS has inherent NBI suppression capability, researches have shown that extra interference rejection techniques are indispensable to allow satisfactory performance when the interferer is so powerful that the processing gain alone is insufficient. This is especially true for commercial applications that presuppose a spread spectrum system constrained in power so as not to disrupt existing narrowband systems. The methods in both time and transform domains are described extensively in Chapters 5-6.

Multicarrier CDMA has drawn a lot of concern in recent years. Various schemes of MC CDMA systems have been proposed, which can be categorized into two groups. One is a combination of orthogonal frequency division multiplexing (OFDM) and CDMA, in which the spreading sequence is serial-to-parallel converted and each chip is transmitted by a different carrier [1-2]. The other is a parallel transmission of narrowband DS waveforms in frequency domain, in which the total system bandwidth is divided into multiple subchannels, the number of which is typically much less than the processing gain [3-5]. While initially proposed as a new way in realizing wideband CDMA which requires low chip rate and is robust to multipath fading, it exhibits desirable capability in narrowband suppression [5]. Since the NBI only overlays parts of spectrum, the decision can be made based on the remaining 'clean' subchannels. Thus, there is no need for either a suppression filter at the receiver or a notch filter at the transmitter.

In the MC DS-CDMA system proposed in [5], the division of the channel is determined solely by its own characteristics, that is, the subchannel width is chosen to be equal to or less than the coherence bandwidth of the channel, ensuring independent and flat fading on all of the carriers. As a result, the system performance may vary considerably when the interference location and bandwidth changes. For example, as shown in Fig.8 of [5], the BER when two subchannels are affected is about six times higher than that when only one subchannel is affected, although the bandwidth and power of the interference remain the same.

SFH techniques can be introduced to relieve such dependence. One scheme is to transmit the signal only in a subset of the subchannels in one hopping interval, and hop to a different subset in the next interval. An obvious advantage of this scheme is the reduction of co-channel users, since different users have different hopping patterns. However, the diversity order is also reduced. This scheme is useful for uplink due to multiple antennas are necessary for diversity. Another scheme for downlink is to divide each subchannel into multiple frequency slots, and transmit the signal within only one slot in each subchannel, and hop to a different slot in the next interval. This approach reduces the number of co-channel users while maintaining the same diversity order as the MC DS system, but the spreading gain is sacrificed. This scheme is very useful for downlink since no antenna diversity is needed.

In this and next chapter, we provide analytical results for the two schemes of MC DS/SFH systems over a frequency-selective Rayleigh fading channel, and compare their performance to that of the MC DS system with the same bandwidth and data rate, with and without narrowband interference. According to our best knowledge, this is also the first attempt to provide closed form solutions to the bit error rate (BER) of maxim ratio combining (MRC) in Rayleigh fading channels, when the average SNR of different diversity branches are not all identical or all distinct. The results are also validated through Monte Carlo simulations.

As discussed in, MC DS-CDMA system has shown ingenious capability in suppressing narrowband interference. Adaptive algorithm has also been proposed to exploit the correlation between different carriers to cancel MAI and make the MC DS system robust to the near-far problem [6]. However, the algorithm needs a few symbol intervals to converge, and is not applicable when the channel fading is fast and signals on different carriers lost correlation.

An alternative resort is the hybrid DS/SFH technique [7], in which the signal occupies only one subchannel at any instant and is kept hopping in frequency. Since different users employ independent hopping patterns, the average number of co-channel users is reduced by a factor equal to the number of total subchannels, and the near-far problem is alleviated. Combining the good features of both MC and SFH techniques, MC DS/SFH-CDMA systems are expected to show good capability in combating both narrowband and multi-access interference. In this chapter, the first (unlink) scheme of MC DS/SFH-CDMA systems is investigated, which use one subset of subchannels to transmit the signal during each hopping interval, and change to a different subset in the next interval. While reducing the average co-channel user number, it also avoids the NBI during some hopping intervals, and then improves the average performance.

9.2 CHANNEL AND INTERFERENCE MODELS

The mobile radio channel between the subscribers and the base station is assumed to be slowly Rayleigh fading and frequency-selective, which is characterized by two parameters [9]. One is the coherence time (Δt), which is usually greater than the bit duration for slowly fading, hence we can assume a near constant fading characteristic during each bit transmission. The other is the coherence bandwidth $(\Delta f)_c$, which is measured by the reciprocal of the multipath spread T_m, i.e., $(\Delta f)_c \approx 1/T_m$. If the signal bandwidth exceeds the coherence bandwidth, the channel is said to be frequency-selective.

For a single carrier system with total bandwidth B_T, like a pure DS system with chip duration T_s ($B_T = 2/T_s$), the channel for the kth user can be approximated by a Tapped-Delay-Line model [9] with lowpass equivalent impulse response given by

$$h_k(t) = \sum_{l=0}^{L_T-1} \beta_{kl} e^{j\mu_{kl}} \delta(t - lT_s - \tau_k), \tag{9.1}$$

where L_T is the number of resolvable paths and given by

$$L_T = \lfloor T_m/T_s \rfloor + 1 = \lfloor T_m B_T/2 \rfloor + 1 \tag{9.2}$$

The amplitude gains $\{\beta_{kl}\}$ are Rayleigh distributed and the phases $\{\mu_{kl}\}$ are uniformly distributed in $[-\pi, \pi)$. τ_k is the relative delay of the first path signal of the kth user to that of the reference user.

In the proposed MC DS/SFH system, B_T is divided into M equi-width subchannels with bandwidth $B_M = B_T/M$ and central frequency

$$f_m = f_c + \frac{B_T}{M}\left[m - (M+1)/2\right], \quad m = 1, \ldots, M, \tag{9.3}$$

where f_c is the central frequency of the total system bandwidth. We adopt the same assumption as that in [5] to choose M, that is, the following two conditions are to be satisfied:

1. Each carrier of the system has only one resolvable path, i.e., $T_m/(MT_s) \le 1$;
2. All carriers are subject to independent fading, i.e., $B_M \ge (\Delta f)_c$.

To ensure the inequalities above, we choose $M = L_T$, which also maintain the same diversity order as the pure DS system and make a fair comparison. Then the lowpass equivalent impulse response of the *m*th subchannel for the *k*th user is then given by

$$h_{k,m}(t) = \beta_{k,m} e^{j\mu_{k,m}} \delta(t-\tau_k), \tag{9.4}$$

where $\{\beta_{k,m}\}$ are i.i.d. Rayleigh distributed, with $E\left[\beta_{k,m}^2\right] = 2\rho$ for different k and m, and the phases $\{\mu_{k,m}\}$ are uniformly distributed in $[-\pi, \pi)$. τ_k is randomly distributed in $[0, T)$, where T is the bit interval.

The narrowband interference is characterized as band-limited white Gaussian noise with power spectrum density (psd) function given by

$$S_J(f) = \begin{cases} \dfrac{\eta_J}{2}, & f_J - \dfrac{B_J}{2} \le |f| \le f_J + \dfrac{B_J}{2} \\ 0, & else \end{cases} \tag{9.5}$$

To simplify the expressions, the following parameters are defined:

— The ratio of the interference bandwidth to the system bandwidth

$$p = \frac{B_J}{B_T} \tag{9.6}$$

— The ratio of the offset of the interference central frequency to half of the system bandwidth

$$\delta = \frac{f_J - f_c}{B_T/2} \tag{9.7}$$

— The ratio of the interference power to the signal power

$$J/S = \frac{P_J}{E_b/T} \tag{9.8}$$

where $P_J = \eta_J B_J$ is the interference power, E_b is the total received signal energy per bit.

In time domain, the interference can then be written as

$$J(t) = \text{Re}\{[J_c(t) + jJ_s(t)]\exp(j2\pi f_J t)\}, \tag{9.9}$$

where both $J_c(t)$ and $J_s(t)$ are independent lowpass processes with $E[J_c^2(t)] = E[J_s^2(t)] = P_J$. The autocorrelation function of $J(t)$ is then given by

$$R_J(\tau) = P_J \text{sinc}(\pi B_J \tau)\cos(2\pi f_J \tau) \ , \tag{9.10}$$

where $\text{sinc}(x) = \sin(x)/x$.

9.3 SYSTEM MODEL

The transmitter for the *k*th user is shown in Fig. 9.1, where $b_k(t)$ is the data sequence with bit rate $R_b = 1/T$, $a_k(t)$ is the random signature sequence with chip duration $T_c = T/N$, where N is the processing gain of one carrier. The bit and chip waveforms are all rectangular, and the spread signal is bandlimited by its null-to-null bandwidth $2/T_c = B_M$. The total processing gain is $N_T = B_T/(2/T) = MN$.

The hopping pattern $f_k(t)$ takes values randomly from the central frequencies of the first Q subchannels, $\{f_1, f_2, \cdots, f_Q\}$ at any hop interval. The narrowband DS waveform modulates L carriers ($f_k(t) + lB_H$, $l = 0,\ldots,L-1$) simultaneously, while the relationship $M = LQ$ is satisfied. The frequency separation between the nearest two carriers, B_H, is equal to QB_M, so the signal spectrums of two users do not overlap if they take different hopping frequencies.

The transmitted signal of the *k*th user can be written as

$$S_k(t) = \text{Re}\left[\sqrt{2P}\sum_{l=0}^{L-1} C_{kl}(t)\right], \tag{9.11a}$$

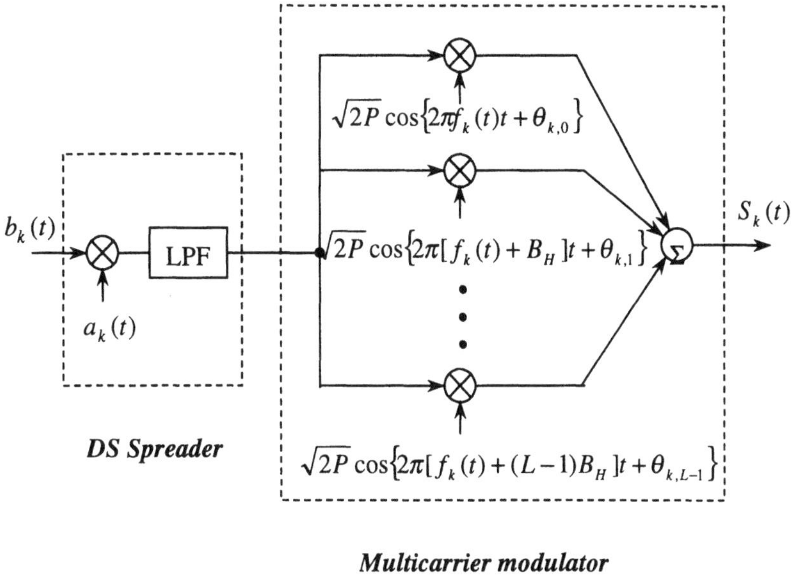

Figure 9.1 The uplink transmitter structure of the MC DS/SFH system.

where P is the transmitted power of each carrier and

$$C_{kl}(t) = b_k(t)a_k(t)\exp\{j2\pi[f_k(t)+lB_H]t + j\theta_{k,l}\}, \tag{9.11b}$$

where $\theta_{k,l}$ is the random phase introduced.

Note that when $Q = 1$, the system is identical to the MC DS system. The power spectrum of the transmitted signals of the MC DS and the MC DS/SFH systems are shown in Fig. 9.2, where the number of total subchannels $S = 4$.

Since MC DS/SFH system employs fewer carriers than the MC DS system, Q antennas are employed to maintain the same diversity order. The received signal by antenna v can be written as

$$r^{(v)}(t) = \mathrm{Re}\left[\sqrt{2P}\sum_{k=1}^{K}\sum_{l=0}^{L-1} C_{kl}(t) * h^{(v)}_{k,k_q+lQ}(t)\right] + n^{(v)}(t) + J^{(v)}(t)$$

$$= \mathrm{Re}\left\{\sqrt{2P}\sum_{k=1}^{K}\sum_{l=0}^{L-1}\beta^{(v)}_{k,k_q+lQ}\exp\left[j\mu^{(v)}_{k,k_q+lQ}\right]C_{km}(t-\tau_k)\right\} + n^{(v)}(t) + J^{(v)}(t)$$

(9.12)

where * denotes convolution and the superscript $^{(v)}$ denotes 'for the vth antenna'. K is the number of active users, while k_q is determined by the instant value of the hopping pattern ($f_k(t) = f_{k_q}$). $n(t)$ is the additive white Gaussian noise with one-sided spectral power density of N_0.

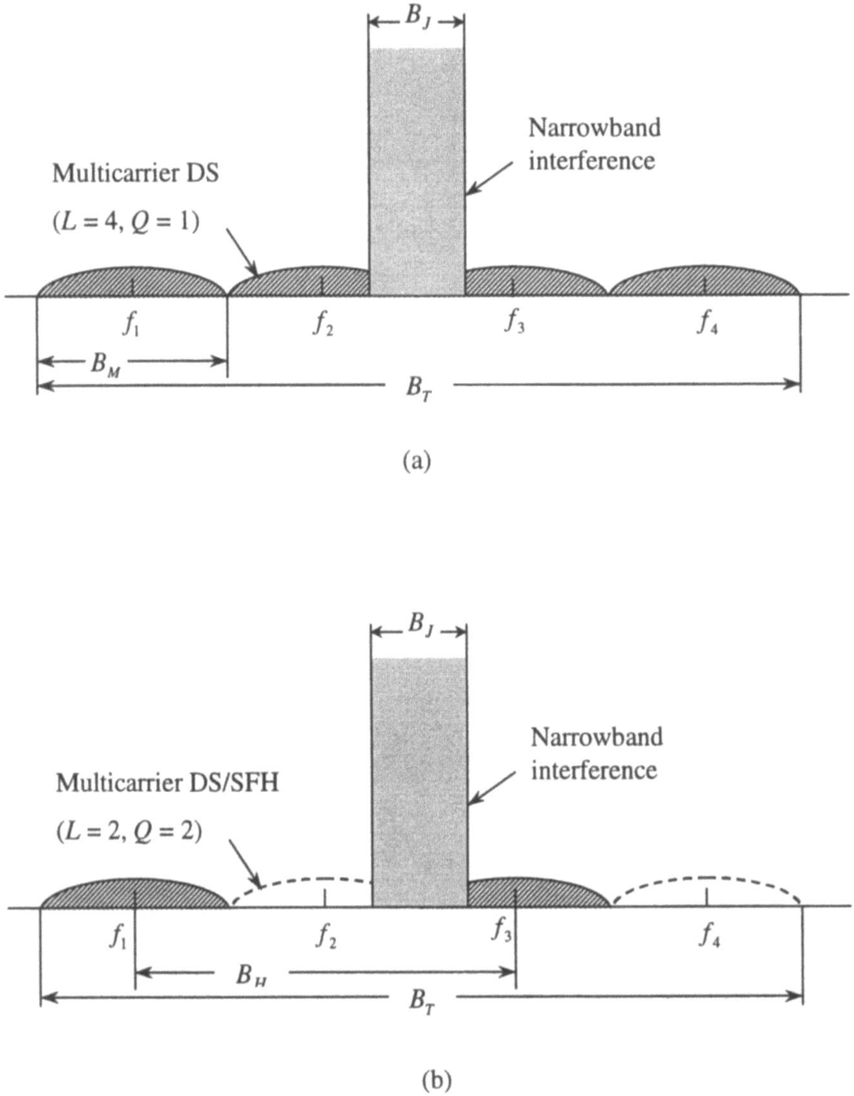

Figure 9.2 Power spectrum of transmitted signals of (a) MC DS (b) MC DS/SFH systems and the narrowband interference [12].

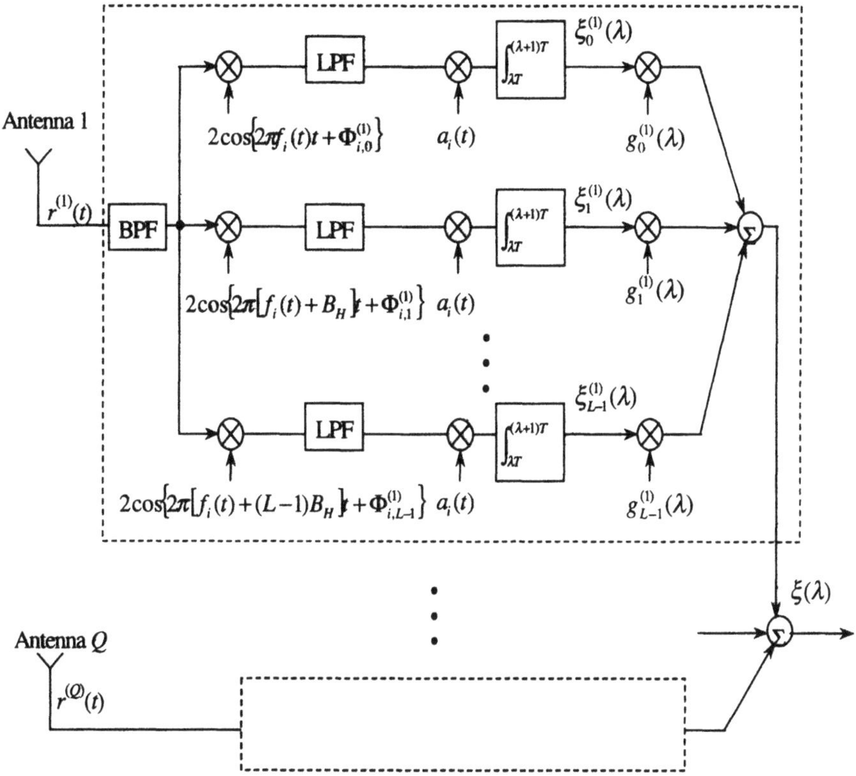

Figure 9.3 The block diagram of the receiver for user *i*.

The receiver for the reference user (the *i*th user) is shown in Fig.9.3. The received signal of each antenna is first mixed with L local carriers controlled by the hopping pattern of the reference user. An ideal lowpass filter (LPF) with frequency response nonzero only in $[-B_M/2, B_M/2]$ is employed for each carrier to eliminate the NBI components out of the subchannel. The outputs are correlated with $a_i(t)$ in one bit interval and combined with a maximal ratio combiner (MRC) with diversity order $LQ = M$, to yield the final decision variable.

9.4 PERFORMANCE ANALYSIS

Perfect carrier and bit synchronization are assumed, as well as perfect power control and channel estimation. Therefore, the BER performance presented below should be considered as the lower bounds of the practical systems.

In the presence of narrowband interference, the performance of a SFH system depends on the instantaneous value of its hopping frequency. Conditioned on $f_i(\lambda T) = f_q$, the output of the lth correlator for the vth antenna in the λth symbol interval is given by

$$\xi_l^{(v)}(\lambda) = \int_{\lambda T}^{(\lambda+1)T} \left\{2r^{(v)}(t)\cos\left[2\pi f_{q+lQ}t + \Phi_{i,l}^{(v)}\right] * h(t)\right\} \cdot a_i(t)dt\,, \tag{9.13}$$

where $\Phi_{i,l}^{(v)} = \theta_{i,l} + \mu_{i,q+lQ}^{(v)}$ and $h(t)$ denotes the impulse response of the ideal LPF.

Neglecting the high frequency terms, (9.13) can be written as

$$\xi_l^{(v)}(\lambda) = D_l^{(v)} + N_l^{(v)} + J_l^{(v)} + I_l^{(v)} \quad , \tag{9.14}$$

where

$D_l^{(v)} = \sqrt{2PT}\beta_{i,q+lQ}^{(v)} b_i^{(\lambda)}$ is the desired signal component, where $b_i^{(\lambda)}$ stands for the λth bit of the reference user with $b_i^{(\lambda)}$ = 1 or -1.

$N_l^{(v)}$ is the noise term and given by

$$N_l^{(v)} = \int_{\lambda T}^{(\lambda+1)T} \left\{2n^{(v)}(t)\cos\left[2\pi f_{q+lQ}t + \Phi_{i,l}^{(v)}\right] * h(t)\right\} \cdot a_i(t)dt\,, \tag{9.15}$$

It is easy to show that $N_l^{(v)}$ is a Gaussian random variable with zero mean and variance $\sigma_N^2 = N_0 T$.

$J_l^{(v)}$ is due to the narrowband interference can be written as

$$J_l^{(v)} = \int_{\lambda T}^{(\lambda+1)T} \hat{J}_l(t) \cdot a_i(t)dt, \tag{9.16a}$$

where

$$\hat{J}_l(t) = 2J^{(v)}(t)\cos\left[2\pi f_{q+lQ}t + \Phi_{i,l}^{(v)}\right] * h(t), \tag{9.16b}$$

The power density spectrum of $\hat{J}_l(t)$ is given by

$$\begin{aligned} S_{J_l}(f) &= \left[S_J(f + f_{q+lQ}) + S_J(f - f_{q+lQ})\right]H^2(f) \\ &= S_l(f) + S_l(-f) \end{aligned}, \tag{9.17a}$$

where $H(f) = 1$ when $|f| \le B_M/2$ and $H(f) = 0$ otherwise, and

$$S_l(f) = S_J(f + f_{q+lQ})H^2(f) = \begin{cases} \eta_J/2, & f_l \le f \le f_h \\ 0, & else \end{cases}, \tag{9.17b}$$

where

$$\begin{cases} f_h = \min\left[f_J + B_J/2 - f_{q+lQ}, \quad B_M/2\right] \\ f_l = \max\left[f_J - B_J/2 - f_{q+lQ}, \quad -B_M/2\right] \end{cases} \tag{9.17c}$$

The autocorrelation function of $\hat{J}_l(t)$ is then given by the inverse Fourier transform of $S_{J_l}(f)$,

$$\hat{R}_{J_l}(\tau) = E\left[\hat{J}_l(t)\hat{J}_l(t+\tau)\right] = \eta_J \frac{\sin(\pi B_{J_l}\tau)}{\pi\tau}\cos(2\pi\Delta_{J_l}\tau), \tag{9.18a}$$

where

$$\begin{cases} B_{J_l} = \max(f_h - f_l, \quad 0) \\ \Delta_{J_l} = (f_h + f_l)/2 \end{cases} \tag{9.18b}$$

Since $J^{(v)}(t)$ is a Gaussian process, $J_l^{(v)}$ is also a zero mean Gaussian random variable. Its variance conditioned on q, $\sigma_l^2(q)$, is given by

$$\sigma_l^2(q) = \int_{\lambda T}^{(\lambda+1)T}\int_{\lambda T}^{(\lambda+1)T} E\left[\hat{J}_l(t_1)\hat{J}_l(t_2)\right]\cdot E\left[a_i(t_1)a_i(t_2)\right]dt_1 dt_2$$

$$= \int_{\lambda T}^{(\lambda+1)T}\int_{\lambda T}^{(\lambda+1)T} \hat{R}_{J_l}(t_1 - t_2)R_i(t_1 - t_2)dt_1 dt_2$$

$$= \int_{-T}^{T} \hat{R}_{J_l}(\tau)R_i(\tau)(T - |\tau|)d\tau \quad , \tag{9.19a}$$

where $R_i(\tau)$ is the autocorrelation function of $a_i(t)$ and triangular in ($-T_c$, T_c).

Substituting $\hat{R}_{J_l}(\tau)$ by (9.18a), one obtains

$$\sigma_l^2(q) = 2\int_0^{T_c} \eta_J \frac{\sin(\pi B_{J_l}\tau)}{\pi\tau}\cos(2\pi\Delta_{J_l}\tau)(1 - \tau/T_c)(T - \tau)d\tau$$

$$= 2\eta_J T\int_0^1 \frac{\sin(\pi B_{J_l}T_c\tau)}{\pi\tau}\cos(2\pi\Delta_{J_l}T_c\tau)(1-\tau)(1-\tau/N)d\tau \quad . \tag{9.19b}$$

Replacing T_c with $2M/B_T$, and B_{J_l}, Δ_{J_l} with (9.18b), (9.19b) can be rewritten as

$$\sigma_l^2(q) = P_J T^2 \mathrm{E}(l,q) \ , \tag{9.20a}$$

where

$$\mathrm{E}(l,q) = \frac{1}{pN_T} U\begin{pmatrix} \min\left[M(\delta + p) - 2(q + lQ) + M + 1, \quad 1\right] \ , \\ \max\left[M(\delta - p) - 2(q + lQ) + M + 1, \quad -1\right] \end{pmatrix} \tag{9.20b}$$

where

$$U(u,v)=\begin{cases}\int_0^1 \frac{\sin(2\pi u\tau)-\sin(2\pi v\tau)}{2\pi\tau}(1-\tau)(1-\tau/N)d\tau & u>v\\ 0 & u\le v\end{cases}$$

(9.20c)

$I_l^{(v)}=\sum_{k=1}^{K_h} I_{k,l}^{(v)}$ is the total MAI term, where K_h is the number of non-reference users that collide with the reference user in frequency ($f_k(\lambda T)=f_q$). Assuming the hopping patterns of different users are mutually independent, K_h has a binomial distribution with probability density function (psd) given by [7-8] for slow frequency hopping

$$P_H(K_h)=\binom{K-1}{K_h}\cdot\left(\frac{1}{Q}\right)^{K_h}\left(1-\frac{1}{Q}\right)^{K-K_h-1}, \tag{9.21}$$

Note that Q is equal to 1 for the MC DS system, therefore, its $P_H(K_h)$ is nonzero only for $K_h = K$-1.

$I_{k,l}^{(v)}$ is due to the signal of the kth user ($k\neq i$). Since the transmitted signal is already limited in bandwidth, it passes through the LPF without any distortion or attenuation. Therefore, $I_{k,l}^{(v)}$ can be written as

$$\begin{aligned}I_{k,l}^{(v)}&=\sqrt{2P}\beta_{k,q+lQ}^{(v)}\cos\left(\Phi_{k,l}^{(v)}-\Phi_{k,l}^{(v)}\right)\int_{\lambda T}^{(\lambda+1)T} b_k(t-\tau_k)a_k(t-\tau_k)a_i(t)dt\\ &=\sqrt{2P}\beta_{k,q+lQ}^{(v)}\cos\left(\Phi_{k,l}^{(v)}-\Phi_{k,l}^{(v)}\right)\cdot\left[b_k^{(\lambda-1)}R_{ki}(\tau_k)+b_k^{(\lambda)}\hat{R}_{ki}(\tau_k)\right]\end{aligned},$$

(9.22)

where $\Phi_{k,l}^{(v)}=\theta_{k,l}+\mu_{k,q+lQ}^{(v)}-2\pi f_k(\lambda T)\tau_k$, $b_k^{(\lambda-1)}$ and $b_k^{(\lambda)}$ stand for the (λ-1)th and the λth bit of the kth user, respectively, which are assumed to be uncorrelated and take values randomly from $\{+1,-1\}$. The partial crosscorrelation functions $R_{ki}(\tau)$ and $\hat{R}_{ki}(\tau)$ are defined as

$$\begin{cases} R_{ki}(\tau) = \int_0^{\tau} a_k(t-\tau)a_i(t)dt \\ \hat{R}_{ki}(\tau) = \int_{\tau}^{T} a_k(t-\tau)a_i(t)dt \end{cases} \tag{9.23}$$

It is easy to prove that $\{ I_{k,l}^{(v)} \}$ are identically distributed but uncorrelated for different k. When K_h is large, one can use the Gaussian approximation [10], and treat $I_l^{(v)}$ as a random Gaussian variable with variance, conditioned on K_h given by

$$\sigma_I^2(K_h) = \sum_{k=1}^{K_h} P \cdot E\left[\left(\beta_{k,q+lQ}^{(v)}\right)^2\right] \cdot E\left[R_{ki}^2(\tau_k) + \hat{R}_{ki}^2(\tau_k)\right]$$

$$= 4K_h P\rho T^2/(3N) \quad , \tag{9.24}$$

where $E\left[R_{ki}^2(\tau_k)\right] = E\left[\hat{R}_{ki}^2(\tau_k)\right] = T^2/(3N)$ and $E\left[\left(\beta_{k,q+lQ}^{(v)}\right)^2\right] = 2\rho$.

The SNR of $\xi_l^{(v)}(\lambda)$, conditioned on $\beta_{i,q+lQ}^{(v)}$, K_h and q is then defined as

$$\gamma_l^{(v)}\Big|_{\beta_{i,q+lQ}^{(v)}, K_h, q} = \frac{E^2\left[\xi_l^{(v)}(\lambda) \mid \beta_{i,q+lQ}^{(v)}\right]}{2\,\mathrm{var}[\xi_l^{(v)}(\lambda) \mid K_h, q]}$$

$$= \frac{2PT^2\left(\beta_{i,q+lQ}^{(v)}\right)^2}{2[\sigma_I^2(K_h) + \sigma_l^2(q) + \sigma_N^2]}$$

$$= \frac{\left(\beta_{i,q+lQ}^{(v)}\right)^2}{2\rho}\left[\frac{2K_h}{3N} + \frac{P_J T}{E_c}\mathrm{E}(l,q) + \left(\frac{E_c}{N_0}\right)^{-1}\right]^{-1} , \tag{9.25}$$

where $E_c = 2\rho PT$ is the average received signal energy per carrier.

By definition, $\{ N_l^{(v)} \}$ in (9.15) are i.i.d. It is also easy to verify that $\{ I_l^{(v)} \}$ and $\{ J_l^{(v)} \}$ are mutually uncorrelated. Therefore, $\{ \xi_l^{(v)}(\lambda) \}$ are uncorrelated for different l and v. To maximize the SNR of the final decision variable, $\xi_l^{(v)}(\lambda)$ is weighted by a gain [5]

$$g_l^{(v)}(\lambda) = \frac{E\left[b_i^{(\lambda)}\xi_l^{(v)}(\lambda) \mid \beta_{i,q+lQ}^{(v)}\right]}{\mathrm{var}[\xi_l^{(v)}(\lambda) \mid K_h, q]}, \tag{9.26}$$

where we assume that the fading is slow enough that the conditional mean and variance can be estimated accurately. The final decision variable is given by

$$\xi(\lambda) = \sum_{v=1}^{Q}\sum_{l=0}^{L-1} g_l^{(v)}(\lambda)\xi_l^{(v)}(\lambda). \tag{9.27}$$

The instantaneous SNR of $\xi(\lambda)$, conditioned on $\{\beta_{i,q+lQ}^{(v)}\}$, K_h and q is given by

$$\gamma_b\Big|_{\{\beta_{i,q+lQ}^{(v)}\}K_h,q} = \frac{E^2\left[\xi(\lambda) \mid \{\beta_{i,q+lQ}^{(v)}\}\right]}{2\,\mathrm{var}[\xi(\lambda) \mid K_h, q]} = \sum_{v=1}^{Q}\sum_{l=0}^{L-1}\gamma_l^{(v)}\Big|_{\beta_{i,q+lQ}^{(v)},K_h,q}$$

$$= \sum_{v=1}^{Q}\sum_{l=0}^{L-1}\frac{\left(\beta_{i,q+lQ}^{(v)}\right)^2}{2\rho}\left[\frac{2K_h}{3N} + M\cdot J/S\cdot \mathrm{E}(l,q) + M\left(\frac{E_b}{N_0}\right)^{-1}\right]^{-1}, \tag{9.28}$$

where $E_b = ME_c$ is the total received signal energy per bit.

The conditional BER as a function of γ_b takes the simple form

$$P_e(\gamma_b) = \frac{1}{\sqrt{2\pi}}\int_{\sqrt{2\gamma_b}}^{\infty} e^{-x^2/2}dx, \tag{9.29}$$

To get the final average BER P_e, we first find $P_e\big|_{K_h,q}$, the BER conditioned on K_h and q, which can be obtained by averaging $P_e(\gamma_b)$ over the statistics of $\{\beta_{i,q+lQ}^{(v)}\}$. The detailed derivation of $P_e\big|_{K_h,q}$ is shown in Appendix A. Assuming all hopping frequencies appear with equal probability, the final average BER is given by

$$P_e = \frac{1}{Q}\sum_{q=1}^{Q}\left[\sum_{K_h=0}^{K-1} P_e\big|_{K_h,q} \cdot P_H(K_h)\right], \tag{9.30}$$

where $P_H(K_h)$ is the pdf of K_h and given by (9.21), and $P_e\big|_{K_h,q}$ is given by (9A.6) of Appendix 9A when all diversity branches have the same SNR or (9A.9) when the SNRs of different branches are not identical.

In case K_h is small, the BER obtained by Gaussian approximation may be not accurate enough. However, small values of K_h appear with small probability and small conditional BER. Therefore, their contributions to the final average BER are not significant. In the numerical section, the results of Gaussian approximation are verified by Monte Carlo simulation, whose model is described in Appendix 9B.

9.5 NUMERICAL RESULTS

In this section, we compare two systems with the same bandwidth and data rate: the MC DS system (N = 127, L = 4, Q =1) and the MC DS/SFH system (N = 127, L = 2, Q = 2). To make a fair comparison, both systems have the same order of diversity (M = 4), the same processing gain (N_T = MN) and the same total average received energy (E_b = ME_c). Pseudonoise (PN) sequences applied in Monte Carlo simulations are Gold sequences of period 127 generated by the polynomial 41567 (octal representation). The initial loadings are given in [11] to generate a class of codes known as Auto-Optimal with Least Sidelobe Energy (AO/LSE).

In the absence of narrowband interference, the average SNRs of all carriers are identical, and $P_e\big|_{K_h,q}$ for both systems can be given by

$$P_e\big|_{K_h,q} = \mu^4 \sum_{i=0}^{3}\binom{3+i}{i}(1-\mu)^i, \tag{9.31}$$

where $\mu = \left[1-\sqrt{r_0/(1+r_0)}\right]/2$, while r_0 is the common SNR given by

$$r_0 = \left[2K_h/(3N) + 4(E_b/N_0)^{-1}\right]^{-1}, \tag{9.32}$$

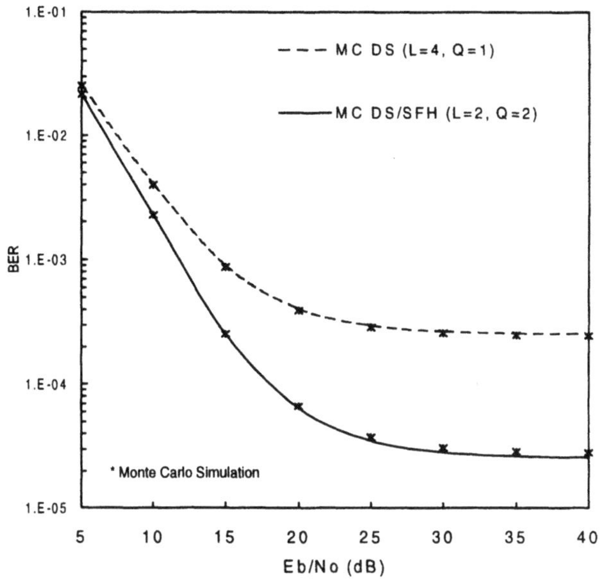

Figure 9.4 BER performance with respect to E_b/N_0 in the absence of narrowband interference (K = 50, J/S = - ∞ dB).

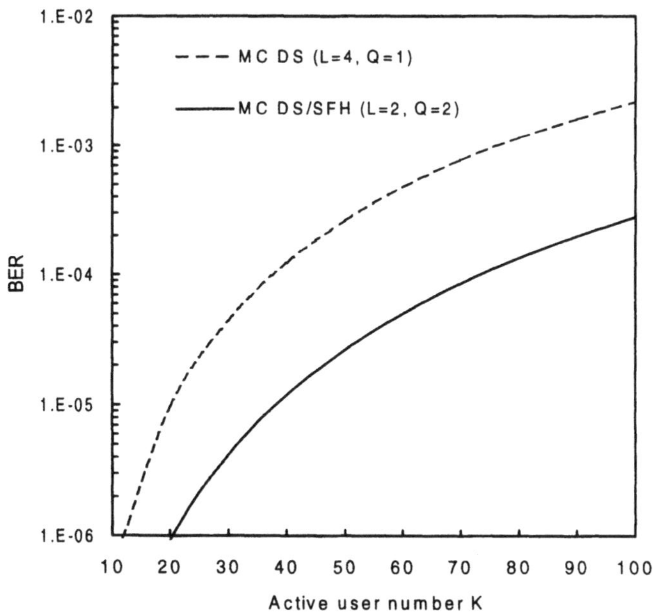

Figure 9.5 BER performance versus active user number K (J/S = - ∞ dB, $E_b/N_0 \rightarrow \infty$).

The BER of all the systems with respect to E_b/N_0 is shown in Fig. 9.4, obtained by both Gaussian approximation and Monte Carlo simulation. It is shown that the MC DS/SFH system outperforms the MC DS system noticeably, which is not surprising because the average hits number, $E[K_h]$ in the MC DS/SFH system is only $1/Q$ of that in the MC DS system, and average SNR per carrier is increased accordingly. However, the improvement in performance is at the expense of increased complexity, since extra antennas are required to maintain the same diversity order. The Gaussian approximation results match the simulation results quite well, although being a little optimistic with large E_b/N_0. It is also shown that the Gaussian approximation is more accurate for the MC DS systems, which have a larger value of average hits number.

In Fig. 9.5 The asymptotic BERs ($E_b/N_0 \to \infty$) of the MC DS/SFH and the MC DS systems are plotted for different number of active users. It is seen that the advantage of the MC DS/SFH system becomes more obvious when the system load is large.

The performance of the two systems in the presence of NBI is of more concern. We consider the case that the interference bandwidth B_J is equal to one subchannel ($p = 0.25$) and the interference is located at the center of the system bandwidth ($\delta = 0$). The average SNRs are no longer identical for all carriers. In the MC DS system, the two middle carriers are overlaid by the NBI, with average SNR given by

$$r_1 = \left\{ 2K_h/(3N) + 4\left[\frac{J/S \cdot U(1,0)}{0.25N_T} + \left(E_b/N_0\right)^{-1} \right] \right\}^{-1}, \tag{9.33}$$

while the SNRs of the other two carriers free of interference are equal to r_0. For the MC DS/SFH system, one carrier is affected and the other one is clean at any instant, so there are also two affected diversity branches and two clean ones, by means of two antennas. Therefore, $P_e|_{K_h,q}$ takes the same formula in the two systems and is given by (9A.11)

$$P_e|_{K_h,q} = A_{10}^2\mu_0^2(3-2\mu_0) + 2A_{10}^2A_{01}\mu_0 + 2A_{01}^2A_{10}\mu_1 + A_{01}^2\mu_1^2(3-2\mu_1), \tag{9.34}$$

where $A_{ij} = r_j/(r_j - r_i)$, $\mu_i = \left(1-\sqrt{r_i/(1+r_i)}\right)/2$, r_0 and r_1 are given by (9.32) and (9.33), respectively.

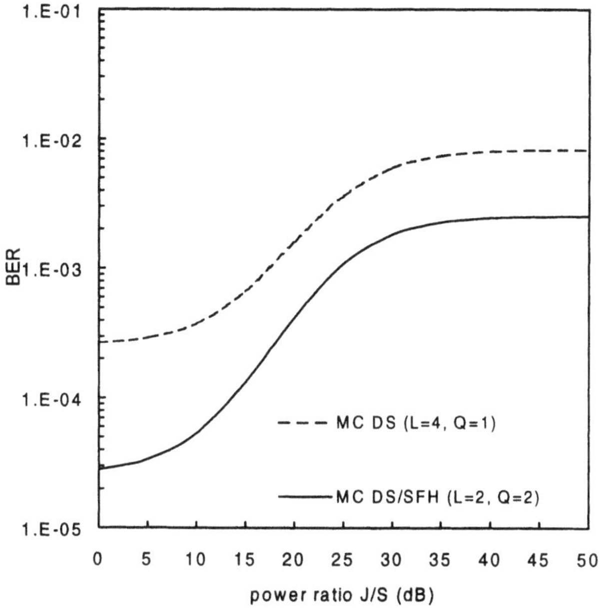

Figure 9.6 BER performance versus the ratio of interference power to signal power J/S (K = 50, p = 0.25, $\delta = 0$, $E_b/N_0 \rightarrow \infty$)

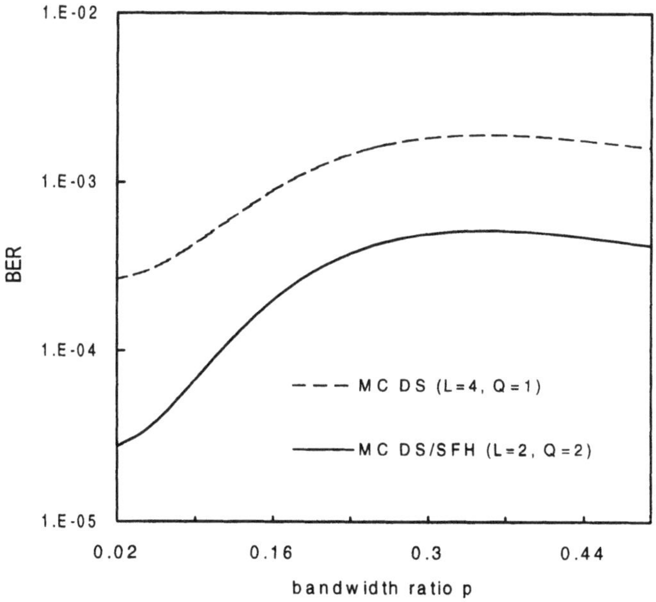

Figure 9.7 BER performance versus the ratio of interference bandwidth to system bandwidth p (K = 50, J/S = 20dB, $\delta = 0$, $E_b/N_0 \rightarrow \infty$)

In Fig. 9.6 the BER performance is plotted versus *J/S*, the ratio of interference power to signal power. It is shown that the MC DS/SFH system outperforms the MC DS system at any interference power level. From (9.34) one can see that given K_h, the two systems have identical BER, that is, they perform almost the same in rejecting NBI. However, the MC DS/SFH system possesses an advantage because it undergoes less MAI.

Fig. 9.7 illustrates the effects of the interference bandwidth on their performance (K = 50, *J/S* = 20 dB, $E_b/N_0 \to \infty$). It is seen that the performance of both systems degrades as the interference bandwidth increases, but the MC DS/SFH system keeps its advantage.

Finally, the performance with respect to the offset ratio of the interference central frequency is shown in Fig. 9.8 ($B_J = B_M$, *J/S* = 50dB). It is seen that the performance of the MC DS system varies significantly for different interference center frequencies, while that of the MC DS/SFH system is much more stable. Both systems perform best when the interference is contained in one subchannel, where the MC DS system has three unaffected diversity branches and the MC DS/SFH system has only two in its worst case. Therefore, the former performs better when active user number is small (e.g., K = 10). However, when user number is large and MAI is comparable to NBI, the MC DS/SFH system still performs better (e.g., K = 50). In designing communication systems, the exact locations of the narrowband interference are usually unknown so the worst-case performance is of more interest. In this sense, the MC DS/SFH system gives a better performance than the MC DS system.

9.6 CONCLUSIONS

The uplink scheme of MC DS/SFH-CDMA system is investigated in a multipath Rayleigh fading channel. Its performance is compared with that of the MC DS-CDMA system, with or without the presence of NBI. The results show that the MC DS/SFH system can suppress the NBI as effectively as the MC DS system while undergoing less MAI, and as a result, achieve better performance in most cases. However, its implementation complexity is higher than the MC DS system because it needs extra antennas to maintain the same diversity order.

It is also found that Gaussian approximation is practically accurate enough to evaluate the BER performance in a Rayleigh fading channel, though it is a little optimistic for high SNR situations.

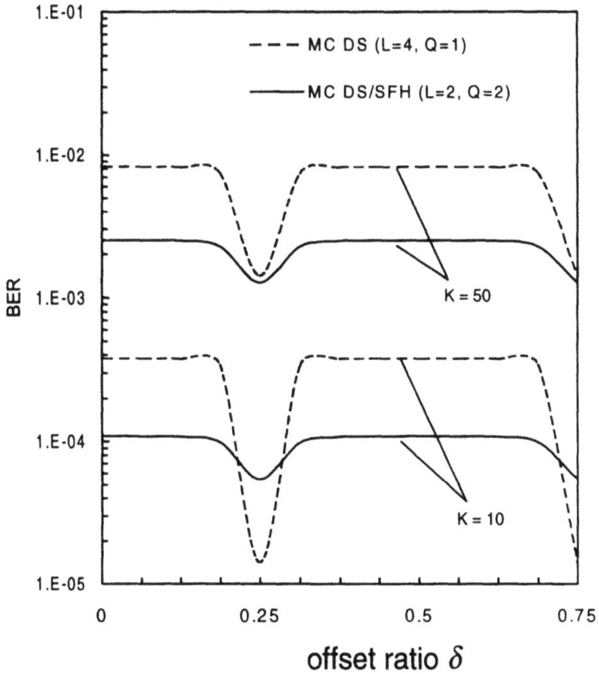

Figure 9.8 BER with respect to the ratio of the offset of the interference central frequency to half of the system bandwidth δ (J/S = 50dB, $B_J = B_M$, $E_b/N_0 \to \infty$)

Appendix 9A

The derivation of BER conditioned on K_h and q

In this appendix, we derive the closed form solutions to the BER of maxim ratio combining in Rayleigh fading channels, when the average SNR of different diversity branches are not all identical or all distinct.

The conditional SNR of the dth diversity branch can be generally written as

$$\gamma_d = \frac{\beta_d^2}{2\rho}\overline{\gamma}_d \tag{9A.1}$$

where $E\left[\beta_d^2\right] = 2\rho$ and $\overline{\gamma}_d$ is the average SNR conditioned on K_h and q. Since β_d is Rayleigh distributed, γ_d has a chi-square distribution with characteristic function given by

$$\psi_d(ju) = \frac{1}{1 - ju\bar{\gamma}_d}\ , \tag{9A.2}$$

For MRC of *M*th order, the SNR of the final decision variable, γ_b, is given by

$$\gamma_b = \sum_{d=1}^{M} \gamma_d \tag{9A.3}$$

Because $\{\gamma_d\}$ are mutually independent, the characteristic function of γ_b is simply the product of all $\psi_d(ju)$, i.e.,

$$\psi_b(ju) = \prod_{d=1}^{M} \frac{1}{1 - ju\bar{\gamma}_d} \tag{9A.4}$$

The conditional BER, conditioned on K_h and q is then given by

$$P_e\big|_{K_h,q} = \int_0^\infty P_e(\gamma_b) f_\gamma(\gamma_b | K_h, q) d\gamma_b\ , \tag{9A.5}$$

where $P_e(\gamma_b)$ is defined by (9.29) and $f_\gamma(\gamma_b|K_h,q)$ is the inverse Fourier transform of $\psi_b(ju)$.

When all diversity branches have identical average SNR equal to *r*, i.e., $\psi_b(ju) = \frac{1}{(1 - jur)^M}$, (9A.5) has a closed form solution [9]

$$P(M,r) = \mu^M \sum_{i=0}^{M-1} \binom{M-1+i}{i} (1-\mu)^i\ . \tag{9A.6}$$

where $\mu = \left[1 - \sqrt{r/(1+r)}\right]/2$.

When the SNRs of different branches are not identical (in the presence of narrowband interference), $\psi_b(ju)$ can be expressed by a partial fraction expansion, i.e.,

$$\psi_b(ju) = \frac{1}{\prod_i (1 - jur_i)^{D_i}} = \sum_i \sum_{j=1}^{D_i} \frac{B_{i,j}}{(1 - jur_i)^j}\ , \tag{9A.7}$$

where $\sum_i D_i = M$ and

$$B_{i,j} = \frac{(-r_i)^{(j-D_i)}}{(D_i - j)!} \left\{ \frac{d^{(D_i-j)}}{dx^{(D_i-j)}} \left[(1 - r_i x)^{D_i} \psi_b(x) \right] \right\} \Bigg|_{x=1/r_i} \tag{9A.8}$$

where $\frac{d^k}{dx^k}[f(x)]$ represents the kth derivative of $f(x)$. The corresponding probability of bit error can be written as

$$P_e\big|_{K_h,q} = \sum_i \sum_{d=1}^{D_i} B_{i,d} P(d, r_i) \tag{9A.9}$$

For example, when $M = 4$, two branches have average SNR r_0 and the other two have average SNR r_1, i.e.,

$$\psi_b(ju) = \frac{1}{(1 - jur_0)^2 (1 - jur_1)^2}, \tag{9A.10}$$

one has

$$P_e\big|_{K_h,q} = A_{10}{}^2 P(2, r_0) + 2A_{10}{}^2 A_{01} P(1, r_0) + 2A_{01}{}^2 A_{10} P(1, r_1) + A_{01}{}^2 P(2, r_1)$$

$$= A_{10}^2 \mu_0^2 (3 - 2\mu_0) + 2A_{10}^2 A_{01} \mu_0 + 2A_{01}^2 A_{10} \mu_1 + A_{01}^2 \mu_1^2 (3 - 2\mu_1)$$

(9A.11)

where $A_{ij} = r_j/(r_j - r_i)$ $(i \neq j, r_i \neq r_j)$, $\mu_i = \left[1 - \sqrt{r_i/(1 + r_i)}\right]/2$.

In the case that three branches have identical average SNR r_0 and one has average SNR r_1, i.e.,

$$\psi_b(ju) = \frac{1}{(1 - jur_0)^3 (1 - jur_1)} \tag{9A.12}$$

one has

$$P_e\big|_{K_h,q} = A_{10} \sum_{d=1}^{3} A_{01}{}^{d-1} P(d, r_0) + A_{01}^3 P(1, r_1)$$

$$= A_{10}\mu_0 + 3A_{10}A_{01}\mu_0^2 + (6A_{10}A_{01}^2 - 2A_{10}A_{01})\mu_0^3$$

$$-8A_{10}A_{01}^2\mu_0^4 + 3A_{10}A_{01}^2\mu_0^5 + A_{01}^3\mu_1 \tag{9A.13}$$

where $A_{ij} = r_j/(r_j - r_i)$ $(i \neq j, r_i \neq r_j)$, $\mu_i = \left[1-\sqrt{r_i/(1+r_i)}\right]/2$.

Appendix 9B

The Monte Carlo simulation model

In this appendix we provide a simulation model to evaluate the BER of the MC DS/SFH system in the presence of multi-access interference, via Monte Carlo methods.

The MAI term from the *k*th non-reference user in the output of the *d*th diversity branch can be written as

$$I_{k,d} = \sqrt{2P}\beta_{k,d}\cos\left(\Phi_{k,d} - \Phi_{i,d}\right)\cdot\left[b_k^{(\lambda-1)}R_{ki}(\tau_k) + b_k^{(\lambda)}\hat{R}_{ki}(\tau_k)\right], \quad (9B.1)$$

where P is the transmitted signal power, $\beta_{k,d}$ is Rayleigh distributed with $E\left[\beta_{k,d}^2\right] = 2\rho$, $\Phi_{k,d}$ is a random phase, $b_k^{(\lambda-1)}$ and $b_k^{(\lambda)}$ are assumed to be uncorrelated and take values randomly from $\{+1,-1\}$. τ_k is assumed to be uniformly distributed in $[0,T)$.

For rectangular chips and $0 \le nT_c \le \tau \le (n+1)T_c \le T$, as shown in [10], the partial crosscorrelation functions $R_{ki}(\tau)$ and $\hat{R}_{ki}(\tau)$ are given by

$$R_{ki}(\tau) = C_{ki}(n-N)T_c + \left[C_{ki}(n+1-N) - C_{ki}(n-N)\right](\tau - nT_c)$$

$$\hat{R}_{ki}(\tau) = C_{ki}(n)T_c + \left[C_{ki}(n+1) - C_{ki}(n)\right](\tau - nT_c) \quad (9B.2)$$

where the discrete aperiodic cross-correlation term $C_{ki}(.)$ is defined as

$$C_{ki}(n) = \begin{cases} \sum_{j=0}^{N-1-n} a_k^{(j)} a_i^{(j+n)} & 0 \le n \le N-1 \\ \sum_{j=0}^{N-1+n} a_k^{(j-n)} a_i^{(j)} & -(N-1) \le n \le 0 \\ 0 & else \end{cases} \quad (9B.3)$$

Once the code sequences are obtained, $R_{ki}(\tau)$ and $\hat{R}_{ki}(\tau)$ are determined by τ. Note that $I_{k,d}$ is a zero-mean Gaussian random variable if τ_k is given, which follows from the fact that $\beta_{k,d}\cos\left[\Phi_{k,d}-\Phi_{i,d}\right]$ is a zero-mean Gaussian variable with variance equal to ρ, therefore, $b_k^{(\lambda-1)}\beta_{k,d}\cos\left[\Phi_{k,d}-\Phi_{i,d}\right]$ and $b_k^{(\lambda)}\beta_{k,d}\cos\left[\Phi_{k,d}-\Phi_{i,d}\right]$ are also zero-mean Gaussian variables with the same variance. The conditional variance of $I_{k,d}$ is given by

$$\sigma_k^2 = 2\rho P\left[R_{ki}^2(\tau_k)+\hat{R}_{ki}^2(\tau_k)\right] \tag{9B.4}$$

It is also obvious that $\{I_{k,d}\}$ are all uncorrelated since $E\left\{\cos\left[\Phi_{k_1,d}-\Phi_{k_2,d}\right]\right\}=0$ for $k_1 \neq k_2$. Therefore, the output of the dth diversity branch, $\xi_d(\lambda)$, is a Gaussian random variable conditioned on $\beta_{i,d}$, $\{\tau_k\}$ and the number of hits, K_h, which has a binomial distribution given by (9.21). The conditional SNR of the output of the dth diversity branch is then given by

$$\gamma_d\Big|_{\beta_{i,d},\{\tau_k\},K_h} = \frac{E^2\left[\xi_m(\lambda)\,|\,\beta_{i,d}\right]}{2\,\mathrm{var}\left[\xi_m(\lambda)\,|\,\{\tau_k\},K_h\right]}$$

$$= PT^2\left(\beta_{i,d}\right)^2\left[\sum_{k=1}^{K_h}\sigma_k^2+\sigma_J^2(d)\right]^{-1}$$

$$= \alpha_{i,d}\left\{T^{-2}\sum_{k=1}^{K_h}\left[R_{ki}^2(\tau_k)+\hat{R}_{ki}^2(\tau_k)\right]+D\frac{\sigma_J^2(d)}{E_bT}\right\}^{-1} \tag{9B.5}$$

where $\alpha_{i,d}=\beta_{i,d}^2/2\rho$ is a normalized chi-square distributed random variable, $\sigma_J^2(d)$ denotes the variance of the NBI and noise, D is the order of diversity.

The instantaneous SNR of the final decision variable, conditioned on $\{\alpha_{i,d}\}$, $\{\tau_k\}$ and K_h, is given by

$$\gamma_b\Big|_{\{\beta_{i,d}\},\{\tau_k\},K_h} = \sum_{d=1}^{D} \gamma_d\Big|_{\beta_{i,d},\{\tau_k\},K_h}$$

$$= \sum_{d=1}^{D} \alpha_{i,d} \left\{ T^{-2} \sum_{k=1}^{K_h} \left[R_{ki}^2(\tau_k) + \hat{R}_{ki}^2(\tau_k) \right] + D \frac{\sigma_J^2(d)}{E_b T} \right\}^{-1} \tag{9B.6}$$

The average BER P_e is obtained by averaging the conditional BER over the statistics of γ_b, that is,

$$P_e = \int_0^{\infty} P_e(\gamma_b) f_\gamma(\gamma_b) d\gamma_b \ , \tag{9B.7}$$

where $f_\gamma(\gamma_b)$ represents the probability density function of γ_b. It is difficult to get the analytical results, therefore Monte Carlo integration over $\{\alpha_{i,d}\}$, $\{\tau_k\}$ and K_h is performed to get P_e.

REFERENCES

[1] E. A. Sourour and M. Nakagawa, "Performance of Orthogonal Multicarrier CDMA in a Multipath Fading Channel," *IEEE Transactions on Communications*, vol. 44, pp. 356-367, March 1996.

[2] X. Gui and T. S. Ng, "Performance of Asynchronous Orthogonal Multicarrier CDMA System in Frequency Selective Fading Channel," *IEEE Transactions on Communications*, vol. 47, pp. 1084-1091, July 1999.

[3] D. Lee and L. B. Milstein, "Comparison of Multicarrier DS CDMA Broadcast Systems in a Multipath Fading Channel," *IEEE Transactions on Communications*, vol. 47, pp. 1897 -1904, December 1999.

[4] D. N. Rowitch and L. B. Milstein, "Convolutionally coded multicarrier DS-CDMA systems in a multipath fading channel. II. Narrow-band interference suppression," *IEEE Transactions on Communications*, vol. 47, pp. 1729 -1736, November 1999.

[5] S. Kondo and L. B. Milstein, "Performance of Multicarrier DS CDMA Systems," *IEEE Transactions on Communications*, vol. COM-44, pp. 238-246, February 1996.

[6] T. M. Lok, T. F. Wong and J. S. Lehnert, "Blind Adaptive Signal Reception for MC-CDMA Systems in Rayleigh Fading Channels," *IEEE Transactions on Communications*, vol. COM-47, pp. 464-465, March 1999.

[7] J. Wang and M. Moeneclaey "Hybrid DS/SFH-SSMA with predetection diversity and coding over indoor radio multipath Rician fading channels," IEEE Transactions on Communications, vol. COM-40, pp. 1654-1662, October 1992.

[8] J. Wang and M. Moeneclaey "Hybrid DS/SFH spread spectrum multiple access with predetection diversity and coding for indoor radio," IEEE Journal on Selected Areas in Communications, vol. JSAC-10, pp. 705-713, May 1992.

[9] J. G. Proakis, *Digital Communication*, 3rd edition, McGraw-Hill, 1995.

[10] M. B. Pursley, "Performance Evaluation for Phase-Coded Spread Spectrum Multiple Access Communication – Part I: System Analysis," *IEEE Transactions on Communications*, vol. 25, pp. 795-799, August 1977.

[11] H. F. A. Roefs and M. B. Pursley, "Correlation parameters of Random Sequences and Maximal length Sequences for Spread-Spectrum Multiple-Access Communications," *IEEE Transactions on Communications*, vol. 27, pp. 1797-1604, October 1979.

[12] J. Wang and H. Huang, "Multicarrier DS/SFH-CDMA systems", Submitted to IEEE Transactions on Vehicular Technology.

[13] J. Wang and H. Huang, "MC DS/SFH-CDMA systems for overlay situations", Submitted to IEEE Journal on Selected Areas in Communications.

[14] H. Huang, "Multicarrier DS/SFH-CDMA systems", MPhil Thesis, University of Hong Kong, April 2000.

Chapter 10

DOWNLINK MULTICARRIER DS/SFH-CDMA

Abstract In this chapter, we introduce the slow frequency hopping (SFH) technique to the multicarrier (MC) CDMA systems for downlink overlay situations. Using lower chip rate, which results in a narrower spectrum for each carrier and hopping the signal in frequency, the MC DS/SFH system achieves better performance than the MC DS CDMA system in most cases, especially when the bandwidth of the narrowband interference is narrower than one subchannel. It also exhibits a stable performance against the variations of the narrowband interference location and bandwidth. When there is no narrowband interference, the two systems perform approximately the same. The evaluation is performed over a frequency selective Rayleigh fading channel, with both Gaussian approximation and Monte Carlo simulation.

Keywords CDMA, Multicarrier Systems, Frequency Hopping, Narrowband Interference Suppression.

10.1 INTRODUCTION

The performance of the uplink scheme of MC DS/SFH system has been investigated in Chapter 9. The results show that for the same diversity order, it is effective in reducing MAI while preserving the good capability of NBI suppression. However, antenna diversity, which increases the implementation complexity, is indispensable to combat fading and ensure satisfactory performance.

In this chapter we consider downlink scheme of MC DS/SFH system, which avoids this disadvantage and then is suitable for downlink transmission. The same number of carriers as the MC DS system is employed, but each carrier is modulated by a narrower DS waveform so that only a slot of each subchannel is occupied at any instant. The whole system can be view as a juxtaposition of M narrowband hybrid DS/SFH systems, which employ a uniform hopping pattern.

10.2 SYSTEM MODEL

The division of the channel is consistent to that in uplink scheme, and the channel model is given by (9.4). However, in this scheme, each subchannel is further divided into Q frequency slots with bandwidth B_s = B_M/Q, where Q is the number of available hopping frequencies. The M carriers of the kth user are uniformly hopped by the hopping pattern, $f_k(t)$, which takes values randomly from the frequency set

$$\left\{ f^{(q)} = \frac{B_M}{Q}\left[q-(Q+1)/2\right], \quad q=1,\ldots,Q \right\} \tag{10.1}$$

in each hopping interval and is independent to the hopping pattern of other users.

The transmitter for the kth user is shown in Fig. 10.1, where $b_k(t)$ is the data sequence with bit duration T and bit rate $R_b = 1/T$, $a_k(t)$ is the random signature sequence with chip duration $T_c = T/N$, where N is the spreading gain of one carrier. The bit and chip waveforms are all rectangular, and the spread signal is bandlimited by its null-to-null bandwidth $2/T_c$ = B_s, ensuring that there is no overlap between transmitted signals of two users if they take different hopping frequencies. The total processing gain is

$N_T = B_T/(2/T) = MQN$. The transmitted signal of the kth user can be written as

$$S_k(t) = \mathrm{Re}\left[\sqrt{2P}\sum_{m=1}^{M} C_{km}(t)\right], \tag{10.2a}$$

where P is the transmitted power of each carrier and

$$C_{km}(t) = b_k(t)a_k(t)\exp\{j2\pi[f_m + f_k(t)]t + j\theta_{k,m}\}, \tag{10.2b}$$

f_m is the center frequency of the mth subchannel and $\theta_{k,m}$ is the random phase introduced. Note that when $Q = 1$, the system is identical to the MC DS-CDMA system. The power spectrum of the transmitted signal is shown in Fig. 10.2, where the number of subchannels $M = 4$.

The received signal can be written as:

$$r(t) = \mathrm{Re}\left[\sqrt{2P}\sum_{k=1}^{K}\sum_{m=1}^{M} C_{km}(t) * h_{k,m}(t)\right] + J(t) + n(t)$$

$$= \mathrm{Re}\left[\sqrt{2P}\sum_{k=1}^{K}\sum_{m=1}^{M} \beta_{k,m} e^{j\mu_{k,m}} C_{km}(t-\tau_k)\right] + J(t) + n(t), \tag{10.3}$$

where $J(t)$ is the narrowband interference defined by (9.5-9.8) and $n(t)$ is the additive white Gaussian noise with one-sided spectral power density of N_0.

The receiver for the ith user (the reference user) is shown in Fig. 10.3, which consists of M correlators and a maximal ratio combiner. It differs to the receiver employed in uplink scheme of the MC DS/SFH system in that the frequency response of the LPF is nonzero only in [$-B_s/2$, $B_s/2$] to eliminate the NBI components out of the current slot.

10.3 PERFORMANCE EVALUATION

In the presence of NBI, the performance of a SFH system depends on the instantaneous value of its hopping frequency. Conditioned on

$f_i(\lambda T) = f^{(q)}$, the output of the correlator for the mth carrier in the λth symbol interval is given by

$$\xi_m(\lambda) = \int_{\lambda T}^{(\lambda+1)T} \left\{ 2r(t)\cos\left[2\pi\left(f_m + f^{(q)}\right)t + \Phi_{i,m}\right] * h(t) \right\} a_i(t)dt\,, \tag{10.4}$$

where $\Phi_{i,m} = \theta_{i,m} + \mu_{i,m}$ and $h(t)$ is the impulse response of the ideal LPF.

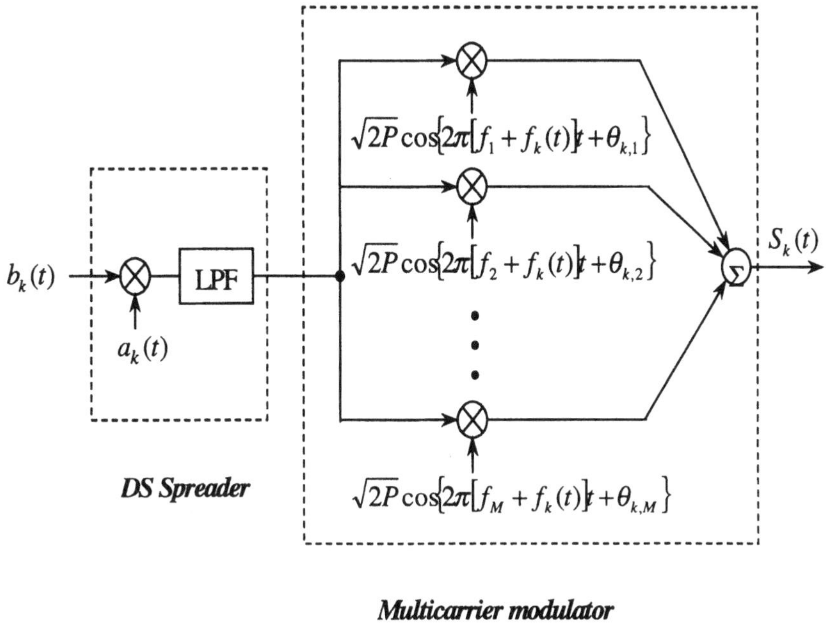

Figure 10.1 The downlink transmitter structure of the MC DS/SFH system.

Neglecting the high frequency terms, (10.4) reduces to

$$\xi_m(\lambda) = D_m + N_m + J_m + I_m\ , \tag{10.5}$$

where

$D_m = \sqrt{2PT}\beta_{i,m}b_i^{(\lambda)}$ is the desired signal component, where $b_i^{(\lambda)}$ stands for the λth bit of the reference user with $b_i^{(\lambda)} = 1$ or -1.

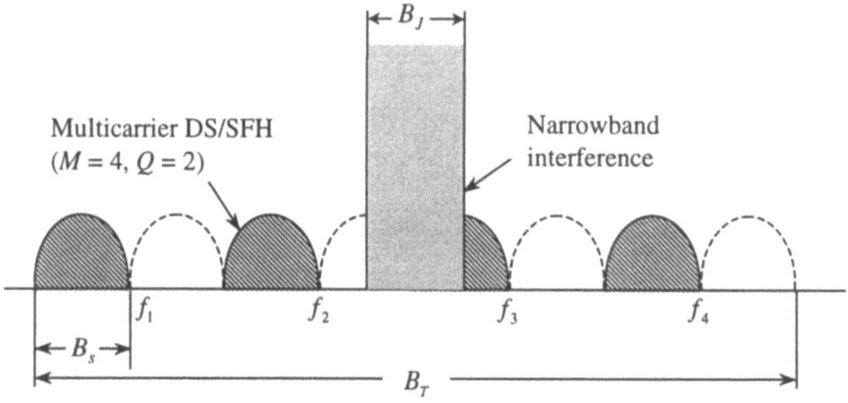

Figure 10.2 Power spectrum of transmitted signal of the MC DS/SFH system and the narrowband interference ([13] in Chapter 9).

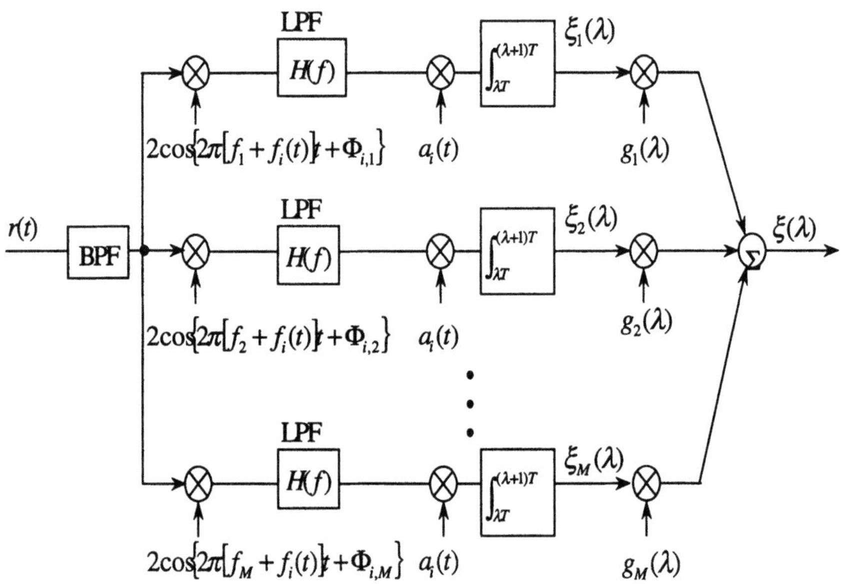

Figure 10.3 The block diagram of the receiver for user *i*.

N_m is the noise term and given by

$$N_m = \int_{\lambda T}^{(\lambda+1)T} \left\{2n(t)\cos\left[2\pi\left(f_m + f^{(q)}\right)t + \Phi_{i,m}\right] * h(t)\right\} a_i(t)dt \tag{10.6}$$

It is easy to shown that N_m is a Gaussian random variable with zero mean and variance $\sigma_N^2 = N_0 T$.

J_m is due to the narrowband interference and can be written as

$$J_m = \int_{\lambda T}^{(\lambda+1)T} \hat{J}_m(t) a_i(t)dt \ , \tag{10.7a}$$

where

$$\hat{J}_m(t) = 2J(t)\cos\left[2\pi(f_m + f^{(q)})t + \Phi_{i,m}\right] * h(t) \, , \tag{10.7b}$$

Derived in a similar way as (9.17-9.18), the autocorrelation function of $\hat{J}_m(t)$ is then given by

$$\hat{R}_{J_m}(\tau) = E\left[\hat{J}_m(t)\hat{J}_m(t+\tau)\right]$$

$$= \eta_J \frac{\sin\left[\pi\tau \cdot \max\left(f_h - f_l, \ 0\right)\right]}{\pi\tau} \cos\left[\pi\tau\left(f_h + f_l\right)\right] \ , \tag{10.8}$$

where

$$\begin{cases} f_h = \min\left[f_J + B_J/2 - f_m - f^{(q)}, \quad B_s/2\right] \\ f_l = \max\left[f_J - B_J/2 - f_m - f^{(q)}, \quad -B_s/2\right] \end{cases} \tag{10.9}$$

Since $J(t)$ is a Gaussian process, J_m is also a zero mean Gaussian random variable. Its variance conditioned on q, $\sigma_m^2(q)$, is given by

$$\sigma_m^2(q) = \int_{-T}^{T} \hat{R}_{J_m}(\tau) R_i(\tau)(T - |\tau|)d\tau \, , \tag{10.10}$$

where $R_i(\tau)$ is the autocorrelation function of $a_i(t)$ and triangular in $(-T_c, T_c)$. Substituting $\hat{R}_{J_m}(\tau)$ with (10.8), B_{J_m} and Δ_{J_m} with (10.9), and T_c with $2/B_s$, (10.10) can be rewritten as

$$\sigma_m^2(q) = P_J T^2 \mathrm{X}(m,q) \ , \tag{10.11a}$$

where

$$\mathrm{X}(m,q) = \frac{1}{pN_T} U\begin{pmatrix} \min[MQ(\delta+p) - Q(2m-M-2) - 2q+1,\ 1]\ , \\ \max[MQ(\delta-p) - Q(2m-M-2) - 2q+1,\ -1] \end{pmatrix} \tag{10.11b}$$

where $U(u,v)$ is defined by (9.20c).

$I_m = \sum_{k=1}^{K_h} I_{k,m}$ is the total MAI term, where K_h is the number of non-reference users that collide with the reference user in frequency ($f_k(\lambda T) = f^{(q)}$) and has a binomial distribution with psd given by (9.21). $I_{k,m}$ is due to the signal of the kth user ($k \neq i$) and given by

$$I_{k,m} = \sqrt{2P}\beta_{k,m} \cos(\Phi_{k,m} - \Phi_{i,m}) \cdot \left[b_k^{(\lambda-1)} R_{ki}(\tau_k) + b_k^{(\lambda)} \hat{R}_{ki}(\tau_k)\right] \ , \tag{10.12}$$

where $\Phi_{k,m} = \theta_{k,m} + \mu_{k,m} - 2\pi f_k(\lambda T)\tau_k$, $b_k^{(\lambda-1)}$ and $b_k^{(\lambda)}$ stand for the (λ-1)th and the λth bit of the kth user, respectively. The partial crosscorrelation functions $R_{ki}(\tau)$ and $\hat{R}_{ki}(\tau)$ are defined by (9.23).

It is also easy to prove that $\{I_{k,m}\}$ are identically distributed but uncorrelated for different k. When K_h is large, one can use the Gaussian approximation and treat I_m as a random Gaussian variable. Its variance, conditioned on K_h, is given by

$$\sigma_I^2(K_h) = \sum_{k=1}^{K_h} P \cdot E[\beta_{k,m}^2] \cdot E[R_{ki}^2(\tau_k) + \hat{R}_{ki}^2(\tau_k)] = 4K_h P\rho T^2/(3N) \tag{10.13}$$

where $E[R_{ki}^2(\tau_k)] = E[\hat{R}_{ki}^2(\tau_k)] = T^2/(3N)$ and $E[\beta_{k,m}^2] = 2\rho$.

The SNR of $\xi_m(\lambda)$, conditioned on $\beta_{i,m}$, K_h and q, is defined as

$$\gamma_m\Big|_{\beta_{i,m},K_h,q} = \frac{E^2[\xi_m(\lambda)\mid\beta_{i,m}]}{2\operatorname{var}[\xi_m(\lambda)\mid K_h,q]} = \frac{2PT^2\beta_{i,m}^2}{2[\sigma_I^2(K_h)+\sigma_m^2(q)+\sigma_N^2]}$$

$$= \frac{\beta_{i,m}^2}{2\rho}\left[2K_h/(3N) + \frac{P_J T^2 \mathrm{X}(m,q)+N_0 T}{E_b T/M}\right]^{-1}$$

$$= \frac{\beta_{i,m}^2}{2\rho}\left[2K_h/(3N) + M\cdot J/S\cdot \mathrm{X}(m,q) + M\left(E_b/N_0\right)^{-1}\right]^{-1} \qquad ,(10.14)$$

where $E_b = 2M\rho PT$ is the total received signal energy per bit.

By definition, $\{N_m\}$ in (10.6) are i.i.d. It is also easy to verify that $\{I_m\}$ and $\{J_m\}$ are uncorrelated for different m. Therefore, $\{\xi_m(\lambda)\}$ are mutually uncorrelated. To maximize the SNR of the final decision variable, $\xi_m(\lambda)$ is weighted by

$$g_m(\lambda) = \frac{E[b_i^{(\lambda)}\xi_m(\lambda)\mid\beta_{i,m}]}{\operatorname{var}[\xi_m(\lambda)\mid K_h,q]}, \qquad (10.15)$$

where we assume that the fading is slow enough that the conditional mean and variance of $\xi_m(\lambda)$ can be accurately estimated. The final decision variable is then given by

$$\xi(\lambda) = \sum_{m=1}^{M} g_m(\lambda)\xi_m(\lambda). \qquad (10.16)$$

and its instantaneous SNR, conditioned on $\{\beta_{i,m}\}$, K_h and q, is given by

$$\gamma_b\Big|_{\{\beta_{i,m}\},K_h,q} = \frac{E^2[\xi(\lambda)\mid\{\beta_{i,m}\}]}{2\operatorname{var}[\xi(\lambda)\mid K_h,q]} = \sum_{m=1}^{M}\gamma_m\Big|_{\beta_{i,m},K_h,q}$$

$$= \sum_{m=1}^{M}\frac{\beta_{i,m}^2}{2\rho}\left[2K_h/(3N) + M\cdot J/S\cdot\mathrm{X}(m,q) + M\left(E_b/N_0\right)^{-1}\right]^{-1},$$

(10.7)

To get the final average BER P_e, we follow the same approach as in chapter 9, that is, first find the BER $P_e\big|_{K_h,q}$, conditioned on K_h and q by averaging $P_e(\gamma_b)$ over the statistics of $\{\beta_{i,m}\}$, then average $P_e\big|_{K_h,q}$ over the statistics of K_h and q.

10.4. NUMERICAL RESULTS

In this section, we consider the MC DS/SFH system ($M = 4$, $N = 63$, $Q = 2$), and compare its performance to that of the MC DS system ($M = 4$, $N = 127$, $Q = 1$). $K = 50$ is assumed if not specified otherwise. A set of sequences of period 63 is generated by the polynomial 13055 to perform the Monte Carlo simulation.

In the absence of NBI, the average SNRs of all carriers are identical, and $P_e\big|_{K_h,q}$ for both systems can be given by

$$P_e\big|_{K_h,q} = \mu^4 \sum_{i=0}^{3} \binom{3+i}{i} (1-\mu)^i , \tag{10.18}$$

where $\mu = \left[1 - \sqrt{r_0/(1+r_0)}\right]/2$, while r_0 is the common SNR given by

$$r_0 = \left[2K_h/(3N) + 4(E_b/N_0)^{-1}\right]^{-1}, \tag{10.19}$$

The BERs against E_b/N_0 of the two systems are given in Fig. 10.4, obtained by both Gaussian approximation and Monte Carlo simulation. It is shown that the two systems have approximately the same performance, although the BER of the MC DS/SFH system is slightly higher when E_b/N_0 is large. This is predictable because r_0 is a concave function of K_h, therefore, the averaged SNR per carrier

$$E[r_0] \le \left[\frac{2E[K_h]}{3N} + 4\left(\frac{E_b}{N_0}\right)^{-1}\right]^{-1} = \left[\frac{2(K-1)}{3N_T/4} + 4\left(\frac{E_b}{N_0}\right)^{-1}\right]^{-1}, \tag{10.20}$$

while the right hand of the inequality is the average SNR per carrier for the MC DS system.

In Fig. 10.5 the asymptotic BERs ($E_b/N_0 \to \infty$) of the MC DS/SFH and the MC DS systems are plotted for different system loads. It is shown that the gap between the two systems diminishes as the number of active user increases.

Next, we compare their performance in the presence of NBI with bandwidth B_J equal to one subchannel ($p = 0.25$) and central frequency equal to f_c ($\delta = 0$). The BER of the MC DS system has already given by (9.32)-(9.34). For the MC DS/SFH system, only one of its carrier is overlaid by the NBI, with average SNR given by

$$r_1 = \left\{ 2K_h/(3N) + 4\left[\frac{J/S \cdot U(1,-1)}{0.25N_T} + \left(E_b/N_0\right)^{-1} \right] \right\}^{-1} \qquad (10.21)$$

and the SNRs of the other three carriers free of interference are equal to r_0. Therefore, its $P_e\big|_{K_h,q}$ takes the formula given by (9A.13)

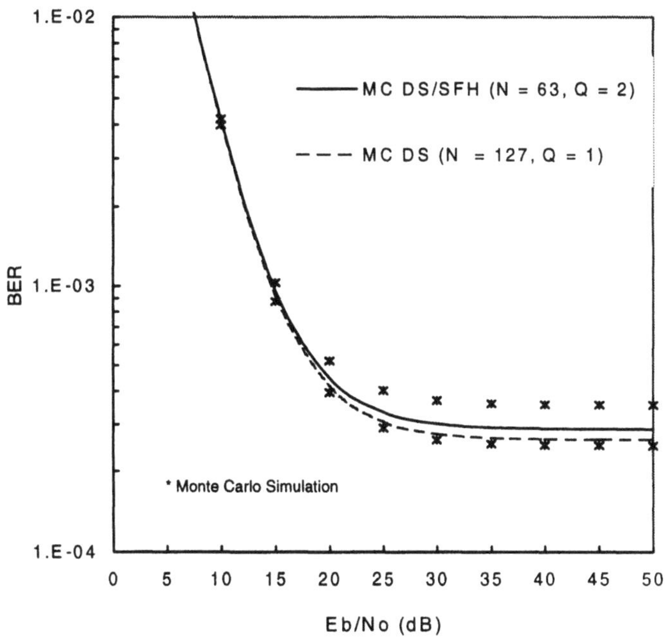

Figure 10.4 BER performance with respect to E_b/N_0 in the absence of narrowband interference ($K = 50$, $J/S = -\infty$ dB).

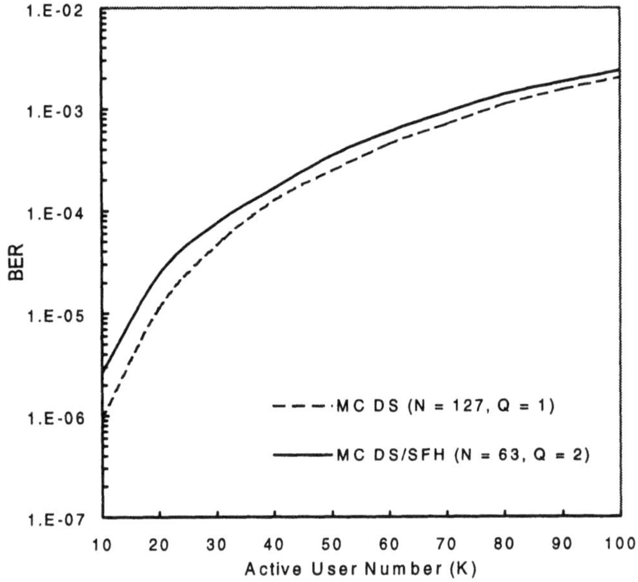

Figure 10.5 Comparison of the MC DS/SFH system and the MC DS system for different system loads (*J/S* = - ∞ dB, $E_b/N_0 \rightarrow \infty$).

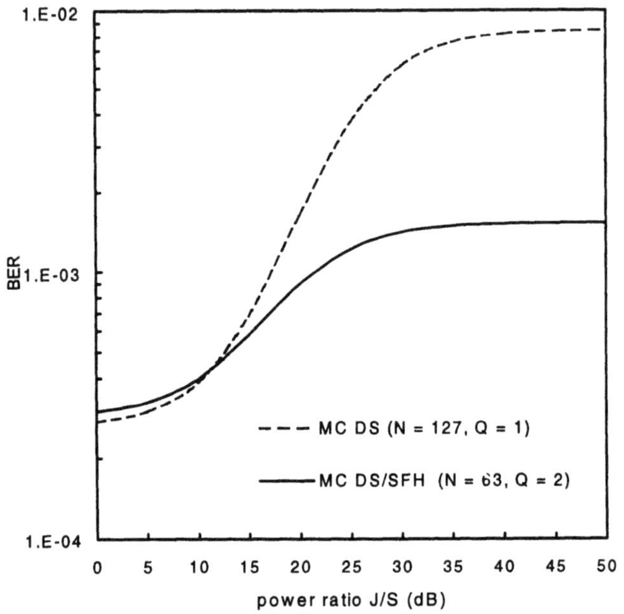

Figure 10.6 BER performance versus interference power to signal power ratio *J/S* (K = 50, p = 0.25, $\delta = 0$, $E_b/N_0 \rightarrow \infty$)

$$P_e\Big|_{K_h,q} = A_{10}\mu_0 + 3A_{10}A_{01}\mu_0^2 + (6A_{10}A_{01}^2 - 2A_{10}A_{01})\mu_0^3$$

$$- 8A_{10}A_{01}^2\mu_0^4 + 3A_{10}A_{01}^2\mu_0^5 + A_{01}^3\mu_1 \quad , \tag{10.22}$$

where $A_{ij} = r_j/(r_j - r_i)$, $\mu_i = \left(1 - \sqrt{r_i/(1+r_i)}\right)/2$, r_0 and r_1 are given by (10.19) and (10.21), respectively.

In Fig. 10.6 the BERs of both systems are plotted against *J/S*. It is seen that when *J/S* is low, the performance of the two systems is approximately the same, however, the MC DS/SFH system outperforms the MC DS system significantly when *J/S* is high. The reason is obvious that the MC DS/SFH system benefits from a higher 'effective' diversity order since the affected carriers contribute little to the final decision.

Fig. 10.7 plots the performance of the two systems for different NBI bandwidth (K = 50, *J/S* = 30 dB, $\delta = 0$). It is shown that the MC DS/SFH system keeps its advantage as long as the NBI bandwidth is less than on subchannel ($p \leq 0.25$) so that the interference does not overlay two carriers simultaneously. However, as B_J exceeds B_M, its advantage diminishes rapidly because it no longer has more clean carriers than the MC DS system. Once the NBI fully covers the two middle subchannels ($p = 0.5$), the MC DS/SFH system lost its advantages completely.

Fig. 10.8 illustrates the performance of the two systems with respect to the offset ratio of the interference central frequency (K = 50, *J/S* = 30 dB, p = 0.25). It is shown that the performance of both systems varies periodically with the NBI central frequency. The MC DS/SFH system is shown to have approximately the same performance as the MC DS system when the NBI is largely contained in one subchannel, because both systems have the same number of clean carriers. In other cases, the performance of the MC DS/SFH system is better and does not fluctuate as violently as the MC DS system for different interference locations.

When the interference bandwidth is narrower than one subchannel, this scheme of MC DS/SFH system exhibits greater advantages. In Fig. 10.9 the BERs of the two systems are plotted as a function of δ when the interference is only half as wide as one subchannel (*J/S* = 30 dB, p = 0.125). It is shown that the MC DS/SFH system outperforms the MC DS system for almost all occasions and exhibits an impressive robustness against the variation of the interference location.

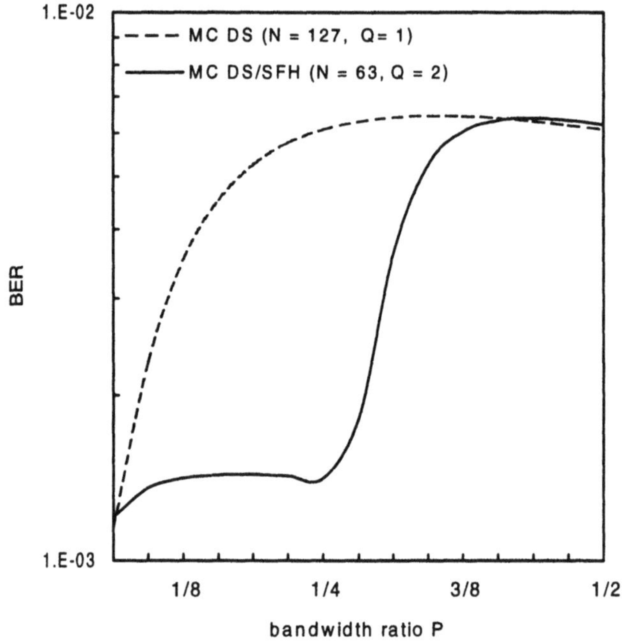

Figure 10.7 BER performance versus interference bandwidth to system bandwidth ratio (K = 50, J/S = 30dB, $E_b/N_0 \rightarrow \infty$)

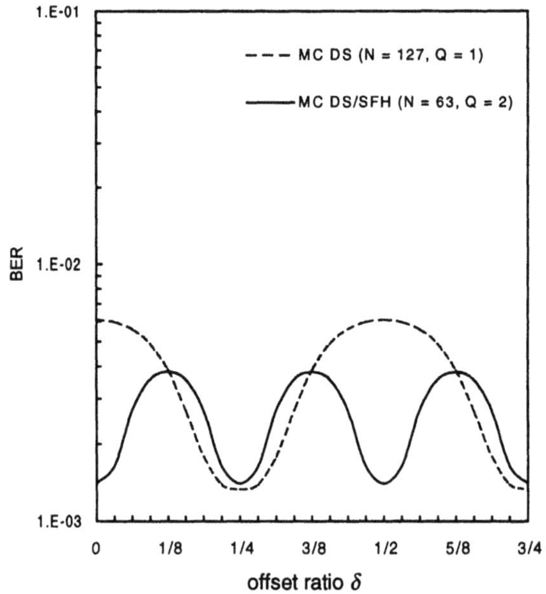

Figure 10.8 Asymptotic BER with respect to central frequency offset ratio (K = 50, J/S = 30dB, $B_J = B_M$, $E_b/N_0 \rightarrow \infty$)

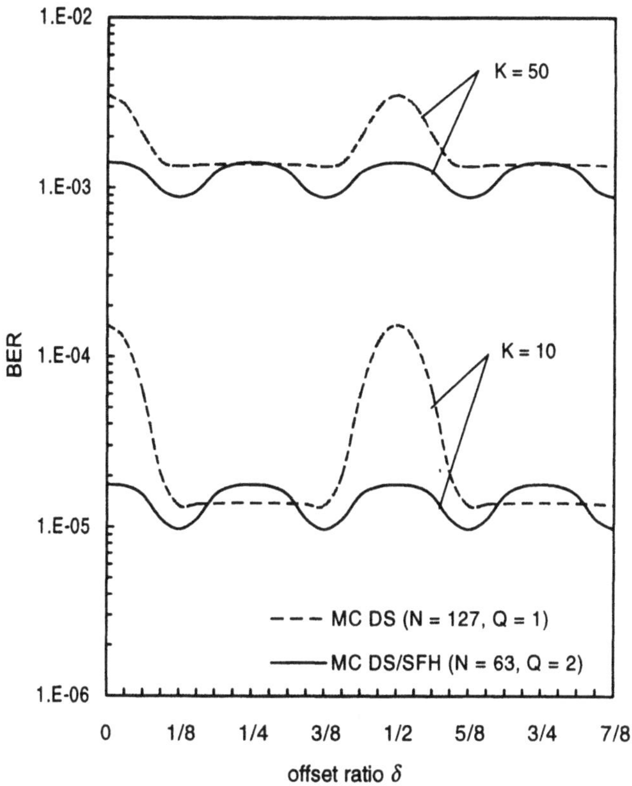

Figure 10.9 Asymptotic BER with respect to central frequency offset ratio (J/S = 30dB, $B_J = B_M/2$, $E_b/N_0 \to \infty$)

10.5 CONCLUSION

The downlink scheme of MC DS/SFH CDMA system is investigated in a multipath Rayleigh fading AWGN channel. Its performance is compared with that of the MC DS system, with and without the presence of NBI. The result shows that this scheme of MC DS/SFH system, with a lower chip rate, can achieve approximately the same capability as the MC DS system in MAI suppression. However, it offers better and more robust performance than the MC DS system in the presence of NBI, especially when the interference bandwidth is narrower than one subchannel, which suggests that SFH technique can make a more efficient use of the 'clean' bandwidth.

Chapter 11

WIRELESS FREQUENCY HOPPING

Abstract Frequency hopping is an alternative technique for wireless mobile communications. This chapter is concerned with the bit error rate (BER) performance of a cellular one-hop/symbol frequency hopping multiple access (FHMA) system operating through a multipath Rayleigh fading channel. *M*-ary FSK (MFSK) modulation with non-coherent square-law envelope demodulation and Reed-Solomon coding is considered. The multiple access and adjacent cell interference of the cellular FHMA system has been studied. In order to illustrate how sensitive the both systems are to the near/far problem, performance of the FHMA system is compared with that of a DS-CDMA system for an equal system bandwidth. Also, this chapter investigates the effect of the values of frequency reuse factor (*N*) on the system capacity of the cellular frequency hopping system.

Keywords Frequency hopping, CDMA, multipath fading channels

11.1 OVERVIEW

In the past twenty years, stimulated by their considerable commercial applications, spread-spectrum techniques have captured substantial attention. The market potential of spread-spectrum systems arises from their many significant advantages over the conventional techniques, which include the huge capacity to accommodate multiple users and the capability to combat variety of intended interference. A spread-spectrum system can be implemented by using two techniques: direct sequence and frequency hopping. In the context of mobile cellular radio, discussions are traditionally focused on DS system, leaving little room for frequency hopping. This situation is attributed to the fact that DS is much easier to implement than frequency hopping and especially, the fast frequency hopping. However, as VLSI technologies evolve, fast hopping frequency synthesizers will be feasible at relatively lower cost. In fact, slow frequency hopping has been installed in the second generation of cellular mobile systems such as GSM. The frequency hopping possesses some advantages not shared by direct sequence; for example, less severe near-far problem, no need for phase synchronization and a larger available frequency bandwidth, which implies a larger spread-spectrum processing gain..

Typically, a frequency hopping transmitter consists of a data modulator, a PN code generator and a frequency synthesizer capable of responding to the coded output from the PN code generator (Fig.11.1). A common data modulation for FH systems is M-ary frequency shift keyed, where one M-ary symbol is represented by one of $M = 2^{\lambda}$ tones, where M is order of MFSK modulation and λ represents the number of bits per symbol transmitted. Finally, the output of modulation is mixed with frequency-hopping tone of frequency f_h.

In the receivers of FH systems (Fig.11.2), the frequency hopping is removed by mixing with a local oscillator signal, which is hopping synchronously with the receiver signal. Because of the difficulty of building truly coherent frequency synthesizers, common frequency-hop spread spectrum systems use noncoherent data modulation scheme. The data demodulation of MFSK systems consists of a bank of M matched filters and quadratic detectors. The output of each filter/detector is sampled every T_S second for slow FH systems. In a fast FH system, the L hopping chips are sampled in the duration of a symbol and need be combined to decide which one of the M channel outputs is largest. The combination of hopping samples can operate in several different modes. One mode is simply to calculate the sum of all L hopping samples. The other modes would be to calculate the likelihood of each data modulation output symbol as nonlinear functions of the total hopping samples and to choose the largest. Some combining nonlinear calculations even utilize the side-information of signal and channel noise. Each of these possible-operating modes performs differently and has different complexity. The spread spectrum system designer must choose the appropriate modes of operation that best solve the particular problem.

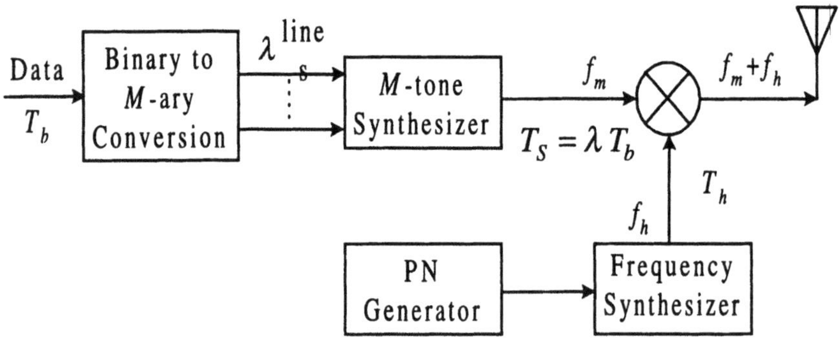

Fig. 11.1 Block diagram for a frequency-hopping MFSK transmitter

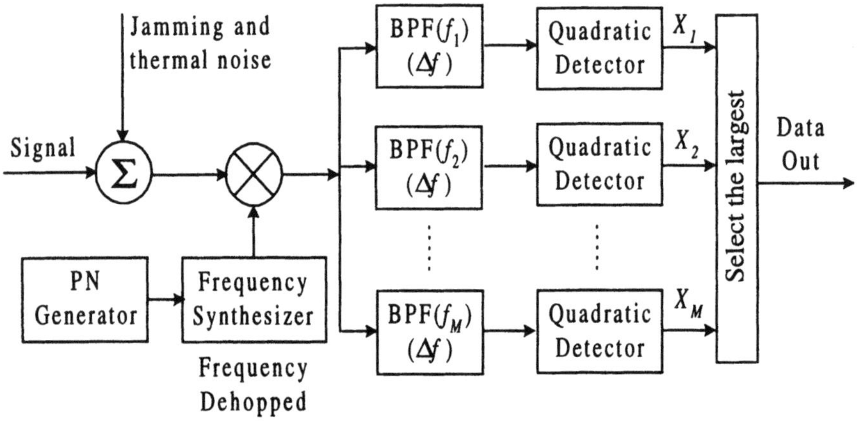

Fig. 11.2 Block diagram for a noncoherent frequency-hopping MFSK receiver [6].

The performance of FHMA over Rayleigh fading channels with single cell was extensively analyzed in the past years. [1] proposed a multi-hop per symbol FHMA system for indoor wireless communications. Continuing the previous research on performance analysis of single-cell FHMA systems, we consider multi-cell FH systems. It has been shown in [1] that in a multi-hop per symbol FHMA system, performance improves when the number of hops per symbol increases due to multi-hop diversity. However, the multi-hop system is very complicated in implementation due to the use of high-speed frequency synthesizers. On the other hand, in a slow frequency hopping system, a very large buffer is required to realize interhop interleaving techniques (whereby each bit of a codeword is transmitted during a hop to reduce the effect of hits form non-reference users), which is used in conjunction with error correcting coding. Therefore, one hop per symbol is proposed in our study. The bit error rate (BER) performance of one hop per symbol FHMA systems employing non-coherent square-law envelope detection and Reed-Solomon coding is studied for a cellular mobile channel. The hopping instants of the various users are not aligned in time. For FH systems it is natural to consider MFSK (M = 2, 4, 8, ...) as a candidate modulation scheme. Reed-Solomon codes are preferable in non-binary systems because of their good burst error correcting capability [2].

11.2 SYSTEM MODELS

The transmitter model of a frequency hopping system consists of a serial-to-parallel converter, a Reed-Solomon encoder, an interleaver, an MFSK modulator and a frequency hopper. The transmitted signal of the kth user in the FHMA system takes the following form

$$S_k(t) = \mathrm{Re}\{\sqrt{2\widehat{P}_k}\exp\{j2\pi[f_0 + f_{c_k} + f_k(t) + b_k(t)\Delta]t + j\alpha_k(t)\}\}, \quad (11.1)$$

where Re{•} stands for real part, $\widehat{P}_k$ is the transmitted power of the kth user. f_0 denotes the carrier frequency, which is common to all users. f_{c_k} is the frequency of the c_kth cell, which belongs to the set $\{F_1, ..., F_N\}$, whose frequency reuse factor is N. $f_k(t)$ is the hopping pattern of the kth user and takes on a constant value $f_k^{(\lambda)}$ during the λth hop, which belongs to the set $\{v_1, ..., v_q\}$ of q not necessarily equally spaced frequencies with minimum spacing W. It is assumed that $\{f_k^{(\lambda)}\}$ is a first-order Markov sequence, so that two consecutive hopping frequencies are always different. It is assumed that the duration of a single hopping interval (dwell time) equals one symbol duration (i.e. one hop per symbol). $\alpha_k(t)$ is the phase introduced by the MFSK modulator and takes on a constant value $\alpha_k^{(\lambda)}$ during λth coded symbol. $b_k(t)$ is the interleaved sequence of Reed-Solomon coded symbols; the λth symbol of $b_k(t)$ has amplitude $b_k^{(\lambda)}$, taking values from {±1, ±3, ...,

$\pm(M-1)\}$, where M is restricted to be a power of 2 (one symbol corresponds to $\log_2 M$ bits). The coded symbol rate and coded symbol duration are given by $R_s = R_b / \log_2 M$ and $T_s = 1/R_s = T_b \log_2 M$, respectively where R_b and T_b denote the rate and duration of the coded bits ($T_b = 1/R_b$). The information bit rate (i.e. the rate of source bits) is given by $R_i = R_b m/n$, where m/n is the rate of the RS code. During a given hop interval, one of M possible signals is transmitted. The M signals are sinusoidal tones of duration T_s, with frequency spacing of 2Δ. In order that these tones be orthogonal when aligned in time, $2\Delta T_s$ must be integer: $2\Delta T_s = n_0$. However, multi-access interference signals are not time aligned with the useful signal, partially due to multipath propagation. Assuming a misalignment, which is uniformly distributed over a symbol interval, the RMS correlation between adjacent tones equals $(\sqrt{2}\pi n_0)^{-1}$. Subsequently, it will be assumed that n_o is sufficiently large to ignore the correlation between tones at different frequencies. The resulting cellular system bandwidth is given by

$$B = 2\Delta MqN = MqN(n_0 / T_s) = MqNn_0 / (T_b \log_2 M) = MqNn_0 (n/m) / (T_i \log_2 M), \tag{11.2}$$

where T_i denotes the duration of the source (uncoded) bits.

It is assumed that the cellular channel between the kth user and the corresponding receiver at the base station of the cell of interest is a multipath Rayleigh fading channel [3]. The multipath Rayleigh fading channel between the kth user and receiver of interest (namely the receiver in the base station of what we refer to as the first cell) is modeled by the complex lowpass equivalent impulse response

$$h_k(t) = \frac{1}{(d_{1,k})^{\gamma/2}} \sum_{l=1}^{L} \beta_{kl} \exp(j\mu_{kl}) \delta(t - \tau_{kl}), \tag{11.3}$$

where $d_{1,k} (d_{1,k} \neq 0)$ is the distance between the kth user and the base station of the first cell, γ is the propagation path loss exponent. The random gain β_{kl} and random phase μ_{kl} of the fading component of the lth path of the kth user have a Rayleigh distribution with $E[\beta_{kl}^2] = 2\rho_{kl} = 2\rho$ for all k and l, and a uniform distribution in $[0, 2\pi]$, respectively. The path delay, τ_{kl}, is uniformly distributed in $[0, T_s]$. We assume that there are L paths associated with each user. The gains, delays, and phases of different paths and/or of different users are all statistically independent. Also, the channel introduces additive white Gaussian noise $n(t)$, with two-sided power spectral density $N_o/2$. Hence, the received signal can be represented as

$$r(t) = \mathrm{Re}\{\sum_{k=1}^{CK}\sqrt{2P_k\xi(\gamma,c_k,k)}\sum_{l=1}^{L}\beta_{kl}\exp[j\psi_{kl}(t)]$$
$$\cdot\exp\{\ j2\pi[f_0 + f_{c_k} + f_k(t-\tau_{kl}) + b_k(t-\tau_{kl})\Delta]t\}\ \} + n(t), \tag{11.4}$$

where $\psi_{kl}(t) = \mu_{kl} + \alpha_k(t-\tau_{kl}) - 2\pi[f_0 + f_{c_k} + f_k(t-\tau_{kl}) + b_k(t-\tau_{kl})\Delta]\tau_{kl}$ and C stands for the number of cells, each one containing K active users, and c_k denotes the c_kth cell (the integer portion of $1+(k-1)/K, c_k = 1,\cdots,C$). The first cell ($c_k$=1) is defined as the cell of interest, and P_k and $\xi(\gamma, c_k, k)$ are defined as

$$P_k = \hat{P}_k /(d_{c_k,k})^{\gamma} \tag{11.5}$$

and

$$\xi(\gamma,c_k,k) = (d_{c_k,k} / d_{1,k})^{\gamma} \tag{11.6}$$

respectively, where $d_{c_k,k}$ is the distance of the kth mobile user to its own base station (the c_kth cell) and $d_{c_k,k} \neq 0$.

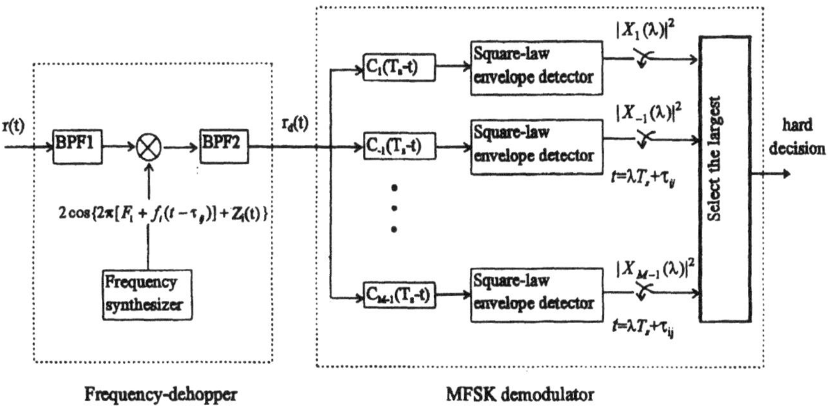

Fig. 11.3 Detailed model of frequency dehopper and MFSK demodulator.

The receiver consists of a frequency dehopper, a MFSK demodulator, a hard decision device, a de-interleaver, a Reed-Solomon decoder and a parallel-to-serial converter. A detailed model of the frequency dehopper and the MFSK demodulator is shown in Fig.11.3. The received signal r(t) enters a bandpass filter which removes out of band noise. The mixer of the dehopper performs the appropriate frequency translation, according to both the first-cell frequency F_1 and the hopping sequence $f_i(t)$ of user i. It is assumed that the hopping pattern of the receiver is synchronized with the hopping pattern of the signal associated with jth path of user i (which will be denoted as the reference path). The bandpass filter which follows the mixer removes both high frequency terms and terms corresponding to non-reference user hopping frequency. The dehopper output signal is given by

$$r_d(t)=\sum_{k=1}^{CK}\sqrt{2P_k\xi(\gamma,c_k,k)}\sum_{l=1}^{L}\beta_{kl}\delta[f_{c_k}+f_k(t-\tau_{kl}),F_1+f_i(t-\tau_{ij})]$$

$$\cdot\cos\{2\pi[f_0+b_k(t-\tau_{kl})\Delta]t+\Psi_{kl}(t)\}+\hat{n}(t), \qquad (11.7)$$

where $\delta(x,y)$ is a kronecker function, defined as $\delta(x,y)=1$ or 0 for $x=y$ or $x\neq y$, respectively. F_1 is the frequency of the first cell (or the cell of interest) and $\hat{n}(t)$ can be treated as band-limited Gaussian noise with spectral density $N_0/2$. The phase waveform $\Psi_{kl}(t)=\psi_{kl}(t)-z_i(t)$, where $z_i(t)$ is the phase introduced by the dehopper, which is constant over symbol interval; $z_i^{(\lambda)}$ stands for the constant phase during the λth symbol. Note that the dehopper suppresses, at any instant t, all path signals whose hopping frequency at instant t differs from $F_1+f_i(t-\tau_{ij})$. The reference path signal is not suppressed; the other path signals from the reference user are suppressed during a part of a symbol, depending on the relative delay of the considered path with respect to the reference path. Path signals from the users of the first cell contribute the dehopper output during those time intervals for which their hopping frequencies accidentally equal that of the reference path signal. When the frequency reuse factor N=7 (see Fig. 11.4(a)), there is no multi-access (or adjacent-cell) interference from all cells of the first and second layers, and 12 of 18 cells of the third layer, since all their frequencies are different from F_1. However, the frequencies of the remaining 6 (shadowed cells) out of 18 cells of the third layer are the same as F_1. Therefore, the six cells may cause multiple-access interference, when their hopping frequencies accidentally equal that of the reference path signal. On the other hand, when the frequency reuse factor N=3 (see Fig. 11.4(b)), there is no multi-access (or adjacent-cell) interference from all cells of the first layer, 6 of 12 cells of the second layer and 12 of 18 cells of the third layer, since their frequencies are different from F_1. However, the remaining 6 of 12 cells of the second layer and 6 of 18 cells of the third layer may cause multiple-access interference, since their cell frequencies are the same as F_1.

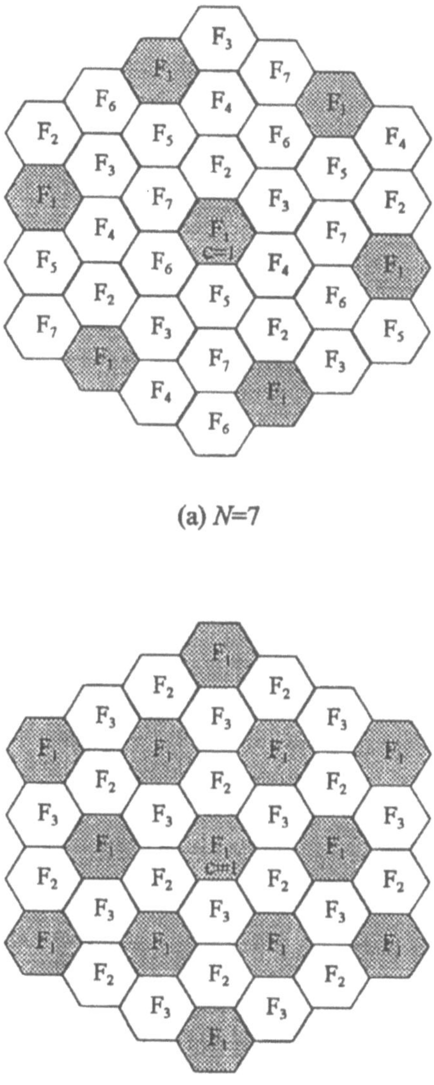

(a) N=7

(b) N=3

Fig. 11.4 The cellular model (N is frequency reuse factor) [7].

The dehopper output signal $r_d(t)$ enters a non-coherent MFSK demodulator. It consists of M bandpass matched filters with impulse responses $\{C_\varepsilon(T_s - t)\}$, where $C_\varepsilon(t) = 2\cos[2\pi(f_0 + \varepsilon\Delta)t]$, $0<t<T_s$ and ε

= ±1, ±3, ..., ±(M-1), followed by square-law envelope detectors. A square-law envelope detector consists of a squarer, followed by a lowpass filter; the output of the lowpass filter is the square of the envelope of the bandpass signal at the input of the squarer. The M square-law detector outputs are sampled at the instants $\lambda T_s + \tau_{ij}$; this yields the M random variables $\{|X_\varepsilon(\lambda)|^2\}$ for ε = ±1, ±3, ..., ±(M-1), where

$$X_\varepsilon(\lambda) = \int_{(\lambda-1)T_s+\tau_{ij}}^{\lambda T_s+\tau_{ij}} 2r_d(t)\exp[j2\pi(f_0 + \varepsilon\Delta)t]dt \tag{11.8}$$

The receiver bases its hard decision about the coded symbol $b_i^{(\lambda)}$ on the M random variables $\{|X_\varepsilon(\lambda)|^2\}$ for ε = ±1, ±3, ..., ±(M-1), by selecting the largest $|X_\varepsilon(\lambda)|^2$, and declaring that the symbol with the corresponding value of ε has been transmitted.

11.3 STATISTICAL DESCRIPTION

In the following, it will be assumed that $E[\beta_{kl}^2]$ equals 2ρ for all k and l. The random variable $X_\varepsilon(\lambda)$, given by (11.8), consists of the following terms:

- A complex-valued reference path term $D_{j,\varepsilon}(\lambda)$ which is due to the jth path signal from the reference user. As this reference path signal passes through the dehopper and only through the MFSK demodulator branch with $\varepsilon = b_i^{(\lambda)}$, this useful term is given by

$$D_{j,\varepsilon}(\lambda) = \sqrt{2P_i}\{\beta_{ij}\exp[jV(i,j,\lambda)]\}T_s\delta(\varepsilon, b_i^{(\lambda)}), \tag{11.9}$$

where $V(k,l,\zeta)$ is a phase angle and is defined as

$$V(k,l,\zeta) = \mu_{kl} + \alpha_k^{(\zeta)} - z_i^{(\lambda)} + 2\pi[f_0 + f_{c_k} + f_k^{(\zeta)} + b_k^{(\zeta)}\Delta]\tau_{kl}$$
$$- 2\pi[f_0 + F_1 + f_i^{(\lambda)} + b_i^{(\lambda)}\Delta]\tau_{ij} \tag{11.10}$$

Strictly speaking, the probability density function of $V(k,l,\zeta)$ should be triangular, since $V(k,l,\zeta)$ is a sum of uniform random variables. However, because for $\exp[jV(k,l,\zeta)]$ any value of $V(k,l,\zeta)$ outside $[0,2\pi]$ is equivalent to a value inside $[0,2\pi]$, $V(k,l,\zeta)$ is assumed uniform over $[0,2\pi]$. It is clear from (11.9) that $D_{j,\varepsilon}(\lambda) = 0$ for $\varepsilon \neq b_i^{(\lambda)}$, and for $\varepsilon = b_i^{(\lambda)}$ the real and imaginary parts of $D_{j,\varepsilon}(\lambda)$ are uncorrelated and have the same variance $2P_i\rho\, T_s^2$. This variance is obtained by averaging over the gain and the phase of the reference path signal. Therefore, the variance of the real and imaginary parts of the useful term is given by

$$S_1 = 2P_i\rho\, T_s^2\delta(\varepsilon, b_i^{(\lambda)}). \tag{11.11}$$

- A complex-valued multipath term, which is due to the L-1 other path signals of the reference user. The contributions from different paths are uncorrelated.
- A complex-valued multi-access interference term, which is due to the path signals from the K-1 non-reference users of the cell of interest.
- A complex-valued adjacent-cell interference term, which is due to the signals from the adjacent cells, whose frequencies are the same as the frequency (F_1) of the cell of interest.
- A complex-valued Gaussian noise term, which is due to the noise $\hat{n}(t)$ at the dehopper output. The real and imaginary parts of the noise term are uncorrelated and have the same variance N_0T_s. The noise contribution of different matched filter outputs are uncorrelated.

During the λth hop (symbol) of the reference path signal, part of the signal, corresponding to the same hop of another path of the reference user, passes through the dehopper, which is synchronized to the reference path signal. During the considered hop of the reference path signal, the signals, corresponding to the next hop of an earlier path signal of the reference user or the previous hop of a later path signal of the reference user, are suppressed by the dehopper. This is because two consecutive hopping frequencies are always different for the first-order Markov hopping sequence. The dehopper output signal passes through the same MFSK demodulator branch as the reference path signal (i.e. the branch with $\varepsilon = b_i^{(\lambda)}$), because they carry the same symbols. The contribution from the lth path is given by

$$D_{l,\varepsilon}(\lambda) = \sqrt{2P_i}\,\beta_{il}\exp[jV(i,l,\lambda)](T_s - |\tau_{il} - \tau_{ij}|)\delta(\varepsilon, b_i^{(\lambda)}), \tag{11.12}$$

where $V(i,l,\lambda)$ is given by (11.10). It is known from the channel model that the path delay τ_{kl} for any k and l is uniformly distributed in [0, T_s]. Therefore, the delay difference between any two paths from the reference user $\tau_{il} - \tau_{ij}$ has a symmetric, triangular distribution in [-T_s, T_s]. When $\varepsilon = b_i^{(\lambda)}$, the real and imaginary parts of $D_{l,\varepsilon}(\lambda)$ are uncorrelated and have the same variance, equal to $P_i\rho\, T_s^2$. This variance is obtained by averaging over the gain, the phase and the delay difference (with respect to the reference path signal) of the non-reference path signal from the reference user. When $\varepsilon \neq b_i^{(\lambda)}$, $D_{l,\varepsilon}(\lambda)$ is identically zero. The variance of the real and imaginary part of the total multipath term in $X_\varepsilon(\lambda)$ is given by

$$S_2 = P_i \rho T_s^2 (L-1)\delta(\varepsilon, b_i^{(\lambda)}). \tag{11.13}$$

The contribution of the multipath term to the nearest demodulator branch is shown in the Appendix with the variance equal to $\frac{P_i \rho \cdot T_s^2}{\pi^2 \cdot n_0^2}$, where n_0 is an integer. When $n_0 \geq 1$, this variance is much smaller than the variance of $D_{l,\varepsilon}(\lambda)$. Therefore, when the spacing between adjacent tones is sufficiently large ($n_0 \geq 1$), the contribution of the multipath term to other MFSK demodulator branches is negligible. Hence, the multipath term is a useful term.

The contribution from the *l*th path signal of non-reference user *k* of the cell of interest, for which the λth symbol starts earlier than the λth symbol of the reference path signal, is given by

$$I_{k,l}(\varepsilon,\lambda) = \sqrt{2P_k}\Big\{ \beta_{kl}\{\exp[jV(k,l,\lambda)]\delta[f_k^{(\lambda)}, f_i^{(\lambda)}]\delta[\varepsilon, b_k^{(\lambda)}]\cdot[T_s - (\tau_{ij} - \tau_{kl})]$$

$$+\exp[jV(k,l,\lambda+1)]\delta[f_k^{(\lambda+1)}, f_i^{(\lambda)}]\delta[\varepsilon, b_k^{(\lambda+1)}]\cdot(\tau_{ij} - \tau_{kl})\}\Big\} \tag{11.14}$$

Fig.11.5 schematically represents the λth hop of the reference path signal, the earliest and the latest path signals from non-reference user *k*. Let us consider the case where the λth hop of the reference user starts during the λth symbol of the earliest path signal from user *k*.

(a) when the frequencies of the (λ-1)th, λth and (λ+1)th hops of user *k* are all different from the frequency of the λth hop of the considered reference path, there is no interference from user *k*.

(b) when the frequency of the λth hop from user *k* equals the frequency of the λth hop of the considered reference path signal, the considered reference path hop is partially hit by all *L* path signals of user *k*. The interfering path signals pass through the branch of the MFSK demodulator with $\varepsilon = b_k^{(\lambda)}$. When the spacing 2Δ between adjacent tones is sufficiently large, the effect on other MFSK demodulator branches can be ignored. The variance of the real (or imaginary) part of such interference from user *k* is

$$\sigma_1^2 = P_k \rho T_s^2 L\delta(\varepsilon, b_k^{(\lambda)}). \tag{11.15}$$

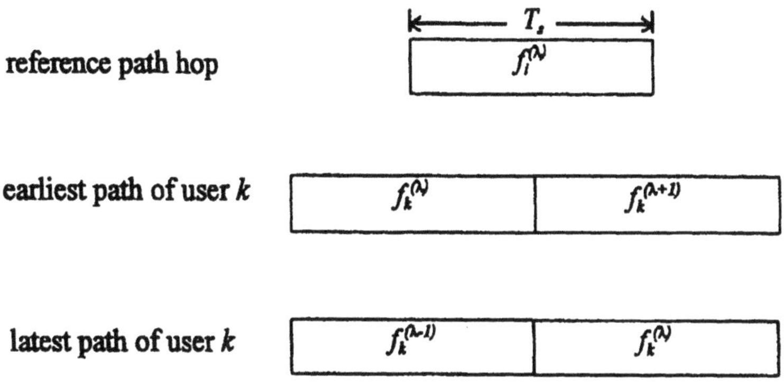

Figure 11.5 Illustration of multiaccess interference.

when the frequency of the λth hop from user k equals the frequency of the λth hop of the considered reference path signal, the considered reference path hop is partially hit by all L path signals of user k. The interfering path signals pass through the branch of the MFSK demodulator with $\varepsilon = b_k^{(\lambda)}$. When the spacing 2Δ between adjacent tones is sufficiently large, the effect on other MFSK demodulator branches can be ignored. The variance of the real (or imaginary) part of such interference from user k is

$$\sigma_1^2 = P_k \rho T_s^2 L \delta(\varepsilon, b_k^{(\lambda)}) . \tag{11.15}$$

Note that when $f_k^{(\lambda)} = f_i^{(\lambda)}$, neither $f_k^{(\lambda-1)}$ nor $f_k^{(\lambda+1)}$ is equal to $f_i^{(\lambda)}$, since two consecutive hopping frequencies are different. For a large number (q) of available hopping frequencies, the probability of a hit from a non-reference user is small. Therefore, in the following, we shall consider only two dominating events, and ignore the other events. The two dominating events are as follows:

(i) The considered reference path symbol is not hit by any non-reference user from the cell of interest. The total multi-access interference is similarly zero.

(ii) The considered reference path symbol is hit by only one non-reference user from the cell of interest; the joint occurrence of hits by more non-reference users has a much smaller probability than the occurrence of a single hit (roughly $1/q^2$ and $2/q$, respectively), and therefore, will be ignored.

There are M single-hit events (M possible values for the non-reference symbol involved in the hit), and the probability of each such event is approximated by

$$P_h = 2(K-1)/(qM). \tag{11.16}$$

The contribution from the lth path signal of user k of an adjacent (or the c_kth) cell, for which the λth symbol starts earlier than the λth symbol of the reference path signal of the first cell, is given by

$$W_{k,l}(\varepsilon,\lambda) = \sqrt{\xi(\gamma,c_k,k)}I_{k,l}(\varepsilon,\lambda)\delta[f_{c_k},F_1], \tag{11.17}$$

where $\xi(\gamma,c_k,k)$ and $I_{k,l}(\varepsilon,\lambda)$ are given by (11.6) and (11.14), respectively. When the frequency of the c_kth cell is equal to F_1 (the frequency of the first cell), there may be adjacent cell interference. Otherwise, there is no interference from the c_kth cell. $W_{k,l}(\varepsilon,\lambda)$ has the conditional variance, conditioned on $\xi(\gamma,c_k,k)$, given by

$$E[W_{k,l}^2(\varepsilon,\lambda)]\Big|_{\xi(\gamma,c_k,k)} = \xi(\gamma,c_k,k)E[I_{k,l}^2(\varepsilon,\lambda)]\delta[f_{c_k},F_1]. \tag{11.18}$$

In order to average the variance over the area of the c_kth cell, we approximate the hexagonal cell with a circular cell of equal area, as in [1]. As shown in Fig. 11.4(a), when the frequency reuse factor N=7, it is only possible for adjacent cell interference to be from 6 (shadowed cells) out of 18 third-layer cells, whose frequencies are the same as F_1. However, when the frequency reuse factor N=3, adjacent cell interference could be from 6 of 12 second-layer cells and 6 of 18 third-layer cells (see Fig. 11.4(b)). Referring to Table I of [4], the average of $\xi(\gamma,c_k,k)$ over the areas of the shadowed cells, indicated in Figs. 11.4(a) and 11.4(b), is shown in Table 11.1 for different values of propagation exponent (γ). Therefore, when the frequency of the λth hop from user k equals the frequency of the λth hop of the considered reference path signal, the considered reference path hop is partially hit by all L path signals of user k. The interfering path signals pass through the branch of the MFSK demodulator with $\varepsilon = b_k^{(\lambda)}$. The variance of the real (or imaginary) part of such interference from user k is

$$\sigma_2^2(\gamma,x) = P_k\rho T_s^2 L\xi(\gamma,x)\delta(\varepsilon,b_k^{(\lambda)}). \tag{11.19}$$

11.4 BIT ERROR RATE (BER)

It follows from Section III that the real and imaginary parts of the complex-valued random variables $\{X_\varepsilon(\lambda)|\varepsilon = \pm 1,\cdots,\pm(M-1)\}$ are uncorrelated and have the same variance. This variance can be written as $S(\varepsilon)+\sigma^2(\varepsilon)$, where S(ε) denotes the useful part, caused by the reference path signal term ad the multipath term, and $\sigma^2(\varepsilon)$ is the contribution from the additive noise, the multi-access and adjacent-cell interference. In the next subsection, it will be assumed that the random variables $\{X_\varepsilon(\lambda)|\varepsilon = \pm 1,\cdots,\pm(M-1)\}$ are Gaussian, this will allow us to obtain analytical results in closed form. The Gaussian approximation is shown accurate [1] when the number of paths L is large (L >> 1), because the channel parameters (gain, phase, delay) are assumed independent from path to path and because each MAI term (11.14) looks like Gaussian (Rayleigh gain and uniform phase).

Table 11.1 Adjacent cell interference for a shadowed cell.

γ	*N*=3		*N*=7
	ξ(γ, 1)	ξ(γ, 2)	ξ(γ, 3)
	second layer	third layer	third layer
2	0.049	0.0202	0.0156
3	0.013	0.0033	0.0023
4	0.0037	0.0006	0.0003

11.4.1 Symbol error probability before decoding

The receiver bases its hard decision about the symbol $b_i^{(\lambda)}$ on the *M* decision variables $|X_\varepsilon(\lambda)|^2$ $(\varepsilon = \pm 1,\cdots,\pm(M-1))$. The receiver declares that the symbol corresponding to the largest $|X_\varepsilon(\lambda)|^2$ has been transmitted. The symbol error rate (SER) P_s after hard decision is approximated by the *union bound.* Let us consider the case where $b_i^{(\lambda)} = 1$. Then this approximation yields

$$P_s = \sum_{\varepsilon \neq 1} \sum_{h \in H} P_s(\varepsilon \mid h) P_h(h), \tag{11.20}$$

where

$$P_s(\varepsilon \mid h) = \Pr[| X_1(\lambda) |^2 < X_\varepsilon(\lambda) |^2 \, |h], \tag{11.21}$$

$P_h(h)$ is the probability of even h, and H is the set of the joint occurrence of the following considered events:

$$H = \{h_0, h_{\varepsilon_0}, h_{\varepsilon_1}, \cdots, h_{\varepsilon_I} \big| h_{\varepsilon_i} = \pm 1, \cdots, \pm(M-1)\}. \tag{11.22}$$

In the above, h_0 is the no-hit event, whereas $\{h_{\varepsilon_i}, i = 0, \cdots, I\}$ are single-hit events both with ε_i as the interfering symbol and with I as the total interfering cells, whereas h_{ε_0} is a single-hit event from the cell of interest. When the frequency reuse factor of a cellular model N=7, I=6 (shadowed cells of Fig. 11.4(a)), whereas when N=3 (Fig. 11.4(b)), I =12.

For the joint occurrence of the events, detail consideration at the outputs of the M branches of the MFSK demodulator is very complicated, since there are too many combinations of the joint occurrence and the M branches. In order to obtain simple analytical results, we use the following approximation: all joint occurrence of two or more single-hit events is assumed with the identical interfering symbols (i.e., $\hat{\varepsilon} = \varepsilon_1 = \varepsilon_2 = \cdots = \varepsilon_I$). This assumption can be justified mathematically for interfering symbols from adjacent cells, based on the fact that interference from any adjacent cell with respect to useful signal power is much less than 1 (see Table 11.1).

Suppose two events h_{ε_1} and h_{ε_2} jointly occur with symbol $\varepsilon_1 = \hat{\varepsilon}_1 \neq 1$, symbol $\varepsilon_2 = \hat{\varepsilon}_2 \neq 1$ and $\hat{\varepsilon}_1 \neq \hat{\varepsilon}_2$. Since the error probability in a Rayleigh fading channel is $1/(2+SNR)$ [5], the probability of error caused by the two different interfering symbols is given by

$$A_1 = \frac{1}{2 + \frac{S}{N_0T + x}} + \frac{1}{2 + \frac{S}{N_0T + y}} \tag{11.23a}$$

$$\approx \frac{2N_0T + x + y}{S}, \qquad \tfrac{S}{N_0T} >> 2, \tag{11.23b}$$

where S is the power of a useful signal, N_0T is noise variance, x and y are the variances of two interfering symbols (or users), respectively, from adjacent cells. Note that either S/x or S/y is always much greater than 1.

Now suppose two events h_{ε_1} and h_{ε_2} jointly occur with $\varepsilon_1 = \hat{\varepsilon}_1 \neq 1$, $\varepsilon_2 = \hat{\varepsilon}_2 \neq 1$ and $\hat{\varepsilon}_1 = \hat{\varepsilon}_2$. Then, the probability of error caused by the two-identical symbols and a non-interfering symbol is given by

$$A_2 = \frac{1}{2+\frac{S}{N_0T+x+y}} + \frac{1}{2+\frac{S}{N_0T}} \tag{11.24a}$$

$$\approx \frac{2N_0T + x + y}{S}, \qquad \frac{S}{N_0T} >> 2, \tag{11.24b}$$

It is clear from (11.23b) and (11.24b) that the two error probabilities are approximately the same. However, when one of the interfering symbols is from the intracell, S and x should be comparable. Both (11.23a) and (11.24a) are plotted in Fig. 11.6. It can been seen that when S/x is small (i.e., the interfering symbol from intracell), A_2 is only 10% smaller than A_1. However, when S/x is large (i.e., the interfering symbol possibly from an adjacent cell), both A_2 and A_1 are very close. Therefore, A_2 is a good approximation of A_1. Note that it is impossible for the two interfering symbols to be from intracell, since only a single hit is considered in any one cell. The above approximation can be applied to the joint occurrence of more than two single-hit events.

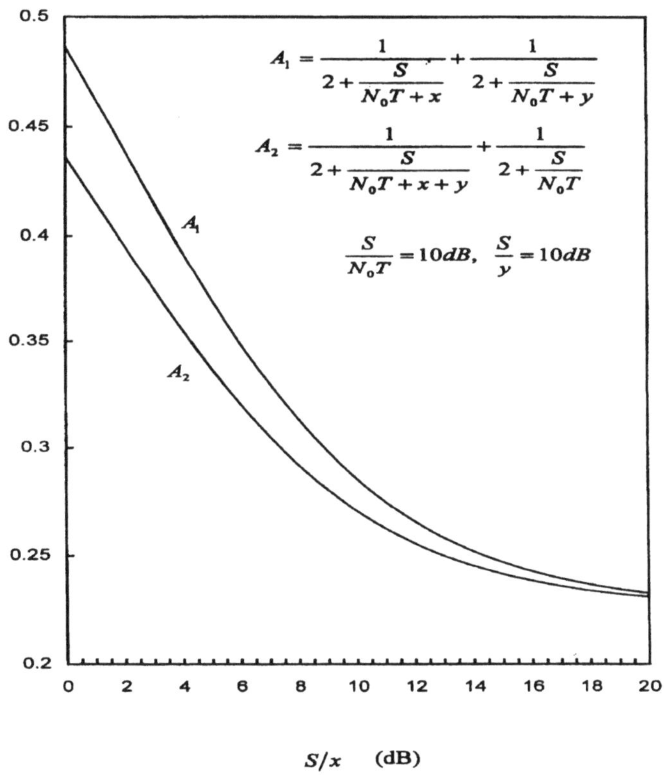

Fig. 11.6 (11.23a) and (11.24a) as a function of S/x.

It is assumed that random variables, y_1, y_2, and y_3 represent the numbers of hits from the cell of interest, the second and the third layers of cells, respectively, for N=3, whereas for N=7, the random variables y_1 and y_4 are assumed to be the numbers of hits from the cell of interest and the third layer of cells, respectively. Note that $y_1 \le 1$, $y_2 \le 6$, $y_3 \le 6$, and $y_4 \le 6$. Also, it is assumed that random variables, x_1, x_2, x_3 and x_4 , corresponding to y_1, y_2, y_3 and y_4, respectively, represent the numbers of hits, which has the same symbols as the reference symbol $(\varepsilon = 1)$ and $x_1 \le y_1$, $x_2 \le y_2$, $x_3 \le y_3$ and $x_4 \le y_4$.

Therefore, there are two cases for error probability.

---- For those branches (M-1 branches), which have no interfering symbol passing, the error probability is given by

$$P(\varepsilon \ne \varepsilon_i) = \begin{cases} \sum_{y_1=0}^{1}\sum_{y_2=0}^{6}\sum_{y_3=0}^{6} P_e(y_1,y_2,y_3)P_h(y_1,y_2,y_3), & N=3 \\ \sum_{y_1=0}^{1}\sum_{y_4=0}^{6} P_e(y_1,y_4)P_h(y_1,y_4), & N=7 \end{cases} \tag{11.25}$$

where $P_e(y_1,y_2,y_3)$ and $P_h(y_1,y_2,y_3)$ are the error probability and probability of the joint occurrence, respectively, of (y_1, y_2, y_3) for N=3 (see Fig.11.4(b)). $P_e(y_1,y_4)$ and $P_h(y_1,y_4)$ are the error probability and probability of the joint occurrence of (y_1, y_4), respectively for N=7 (see Fig. 11.4(a)). When N=3, those probabilities are given by

$$P_h(y_1,y_2,y_3) = \binom{6}{y_2}\binom{6}{y_3} P_h^{y_1+y_2+y_3}(1-P_h)^{13-y_1-y_2-y_3}, \tag{11.26}$$

where P_h is given by (11.16) and

$$P_e(y_1,y_2,y_3) = \sum_{x_1=0}^{y_1}\sum_{x_2=0}^{y_2}\sum_{x_3=0}^{y_3} \hat{P}_e(x_1,x_2,x_3)\hat{P}_h(x_1,x_2,x_3), \tag{11.27}$$

where

$$\hat{P}_h(x_1,x_2,x_3) = \binom{y_2}{x_2}\binom{y_3}{x_3}\left(\tfrac{1}{M}\right)^{x_1+x_2+x_3}\left(1-\tfrac{1}{M}\right)^{y_1+y_2+y_3-x_1-x_2-x_3}, \tag{11.28}$$

and

$$\hat{P}_e(x_1,x_2,x_3) = \frac{1}{2+r_1(x_1,x_2,x_3)}, \tag{11.29}$$

where

$$r_1(x_1,x_2,x_3)=\frac{S_1+S_2+x_1\sigma_1^2+x_2\sigma_2^2(\gamma,1)+x_3\sigma_2^2(\gamma,2)}{N_0T_s}$$
$$=\frac{\overline{E}_b}{N_0}(\log_2 M)\left\{\,[1+\alpha x_1+\alpha x_2\xi(\gamma,1)+\alpha x_3\xi(\gamma,2)]L+1\,\right\}/2, \tag{11.30}$$

where S_1, S_2, σ_1^2, $\sigma_2^2(\gamma,\cdot)$ are given by (11.11), (11.13), (11.15), (11.19), respectively. $\overline{E}_b=2P_iT_b\rho_0 \;=2P_iT_s\rho_0\log_2 M$ is the average energy per coded bit. $\alpha=P_1\,/\,P_0$ is ratio of the received non-reference user power to the received reference user power. $\xi(\gamma,x)$ is given by Table 11.1.

When N=7, in (11.25), we have

$$P_h(y_1,y_4)=\binom{6}{y_4}P_h^{y_1+y_4}(1-P_h)^{7-y_1-y_4}, \tag{11.31}$$

$$P_e(y_1,y_4)=\sum_{x_1=0}^{y_1}\sum_{x_4=0}^{y_4}\hat{P}_e(x_1,x_4)\hat{P}_h(x_1,x_4), \tag{11.32}$$

where

$$\hat{P}_h(x_1,x_4)=\binom{y_4}{x_4}\left(\tfrac{1}{M}\right)^{x_1+x_4}\left(1-\tfrac{1}{M}\right)^{y_1+y_4-x_1-x_4}, \tag{11.33}$$

and

$$\hat{P}_e(x_1,x_4)=\frac{1}{2+r_2(x_1,x_4)}, \tag{11.34}$$

where

$$r_2(x_1,x_4)=\frac{S_1+S_2+x_1\sigma_1^2+x_4\sigma_2^2(\gamma,3)}{N_0T_s}$$
$$=\frac{\overline{E}_b}{N_0}(\log_2 M)\left\{\,[1+\alpha x_1+\alpha x_4\xi(\gamma,3)]L+1\,\right\}/2 \tag{11.35}$$

---- For the branch (only one branch), which has interfering symbols passing, the discussion of the error probability is also the same as the above, only by substituting $r_1(x_1,x_2,x_3)$ with $r_3(x_1,x_2,x_3)$ and $r_2(x_1,x_4)$ with $r_4(x_1,x_4)$, given by

$$r_3(x_1,x_2,x_3)=\frac{S_1+S_2+x_1\sigma_1^2+x_2\sigma_2^2(\gamma,1)+x_3\sigma_2^2(\gamma,2)}{N_0T_s+(y_1-x_1)\sigma_1^2+(y_2-x_2)\sigma_2^2(\gamma,1)+(y_3-x_3)\sigma_2^2(\gamma,2)}$$

$$=\frac{1+\alpha(1-\frac{1}{L+1})[x_1+x_2\xi(\gamma,1)+x_3\xi(\gamma,2)]}{\frac{2}{\frac{\bar{E}_b}{N_0}(\log_2 M)(L+1)}+\alpha(1-\frac{1}{L+1})[(y_1-x_1)+(y_2-x_2)\xi(\gamma,1)+(y_3-x_3)\xi(\gamma,2)]}$$

(11.36)

and

$$r_4(x_1,x_4)=\frac{S_1+S_2+x_1\sigma_1^2+x_4\sigma_2^2(\gamma,3)}{N_0T_s+(y_1-x_1)\sigma_1^2+(y_4-x_4)\sigma_2^2(\gamma,3)}$$

$$=\frac{1+\alpha(1-\frac{1}{L+1})[x_1+x_4\xi(\gamma,3)]}{\frac{2}{\frac{\bar{E}_b}{N_0}(\log_2 M)(L+1)}+\alpha(1-\frac{1}{L+1})[(y_1-x_1)+(y_4-x_4)\xi(\gamma,3)]}$$

(11.37)

Once the SER P_s after hard decision has been obtained, the corresponding bit error rate (BER) P_b is given by

$$P_b=\frac{M}{2(M-1)}P_s. \tag{11.38}$$

The above expression assumes that, given a symbol error, all M-1 erroneous symbols are equally likely to be chosen; this assumption is valid because of the orthogonality of the MFSK signals and the statistical properties of the disturbance (noise + multiaccess interference + adjacent cell interference).

11.4.2 Error Probability after decoding

Reed-Solomon (RS) codes are non-binary, linear, cyclic symbol-error-correcting block codes. The length of a RS code is $n=M-1$ M-ary symbols, m of which are information symbols and the remaining n-m of which are redundancy. This code is referred to as an RS(n,m) code; it can correct up to [(n-m)/2] symbol errors, where [x] denotes the largest integer contained in x. We have selected $m=(M-2)/2$ so that the coding rate m/n is nearly 1/2, and the code can correct up to $t=M/4$ symbol errors. When decoding of RS codes with hard decision is employed, the symbol error probability P_{SC} after RS decoding is well approximated by [5]

$$P_{SC}=\frac{1}{n}\sum_{i=t+1}^{n}i[{}^{n}_{i}]P_s^i[1-P_s]^{n-i} \tag{11.39}$$

where $t=M/4$, $n=M-1$ and P_s is the symbol error probability before decoding. The above expression assumes that the symbol errors before decoding occur independently; this assumption is valid when an interleaving/de-interleaving technique is used even though a hit or more hits may be possible. The P_{bc} after decoder is approximated by substituting in P_b and P_s by P_{bc} and P_{sc}.

11.5 NUMERICAL RESULTS

The BERs of one-hop/symbol cellular FHMA systems with MFSK modulation, noncoherent square-law envelope detection and Reed-Solomon coding for a multipath Rayleigh channel are now evaluated. Representative numerical results are presented. It is assumed that the value of propagation exponent equals 3 (or $\gamma = 3$), so that $\xi(\gamma, x) = 0.013, 0.0033$ and 0.0023, for x=1,2,3, respectively. Unless indicated otherwise, we assume $\alpha = 1$. For a fair comparison of the BERs of uncoded and coded systems, it is assumed that these systems have the same information bit rate R_i, the same system bandwidth B, given by (11.2), and the same energy per uncoded bit $\overline{E}_i$ ($\overline{E}_i = \overline{E}_b n / m$).

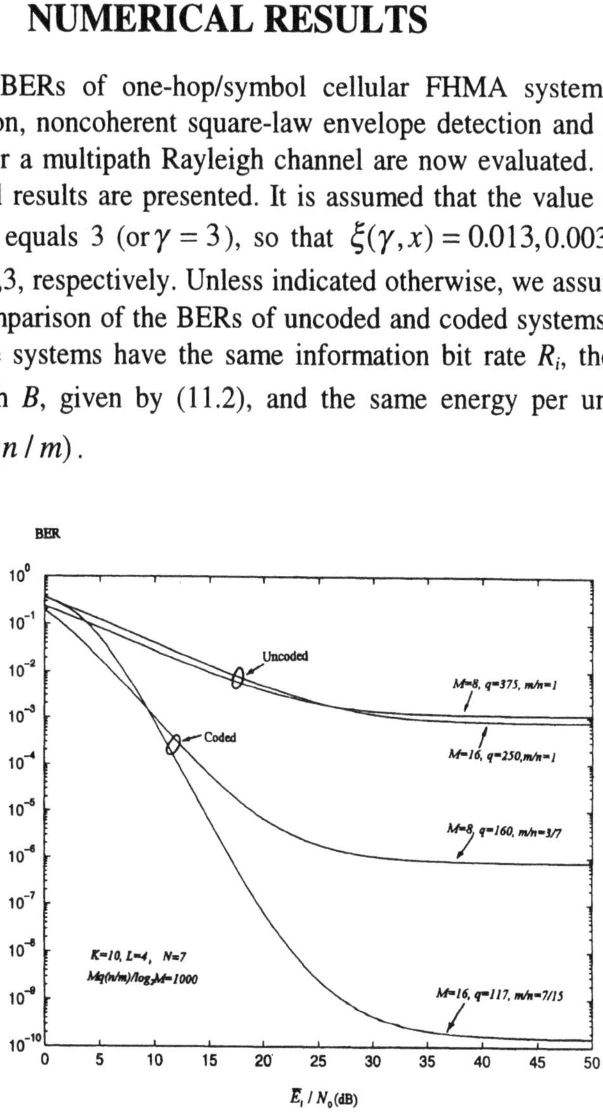

Fig. 11.7 BER with and without RS coding for fixed system bandwidth.

First, in Fig. 11.7, the RS-coded BERs are illustrated for different M, given a fixed system bandwidth B, given by (11.2). Also, uncoded (n/m=1) BERs are shown for comparison. As denoted in equs. (11.16) and (11.26), the asymptotic uncoded BER increases as q decreases (or as M increases for a fixed values of $(Mq / \log_2 M)$, whereas the performance advantage obtained by using RS coding becomes larger when M increases. This is because the RS code error-correcting capability increases when M is increased.

The effect of the number of paths, L, on the uncoded BER performance is considered in Fig. 11.8. It is seen that for $\overline{E}_i / N_0 < 30dB$, additive Gaussian noise is the dominating disturbance; the BER performance improves with increasing L: the useful variance, due to the L paths of the reference user, increases with L, whereas the noise variance does not depend on L. For $\overline{E}_i / N_0 \rightarrow \infty$, the multiaccess and adjacent cell interference is the dominating disturbance; in this case the BER is only weakly dependent on the number of paths (L), because both the useful variance and the interference variance are essentially proportional with L. As described in Section IV for asymptotic performance, it is clear from equ. (11.36-11.37) that the BER is almost independent of L for L>1. This is contrast with DS systems; the performance of a DS system is degraded as L increases. Note that this discussion is not involved with any diversity (or Rake receiver).

The asymptotic BER for FHMA systems with RS coding is shown in Fig. 11.9 for the reuse factor of 3 (N=3) as a function of the nonreference user to reference user power ratio, α, for a given system bandwidth. For comparison, the BER of a DS-CDMA system with the same system bandwidth (processing gain $\hat{N}$=1000) is also shown in Fig. 11.9. Note that the reuse factor of the DS-CDMA system is unitary. A powerful coding Golay (23, 12) code with half code rate and equal gain combining of the second order for the DS-CDMA system [4] are used. Its uncoded error probability is given by (19) of [4] with signal to noise ratio $r_c = 3\hat{N} / [2KL(1+\xi(\gamma))]$, where $\xi(\gamma)$ stands for the effect of adjacent cell interference and $\xi(\gamma)$=0.9 (see [4]) when the propagation exponent is 3. Then, the coded BER is given by (18) of [4] with n=23 and t=3 for Golay (23,12) code, which can correct 3 errors in one block code. It can been seen that when α is very small, the DS-CDMA system outperforms the FHMA system. However, when α is very large, the performance of DS-CDMA degrades dramatically, whereas the FHMA system can keep acceptable performance. Therefore, the frequency hopping system is more robust than the DS-CDMA system in overcoming the near/far problem.

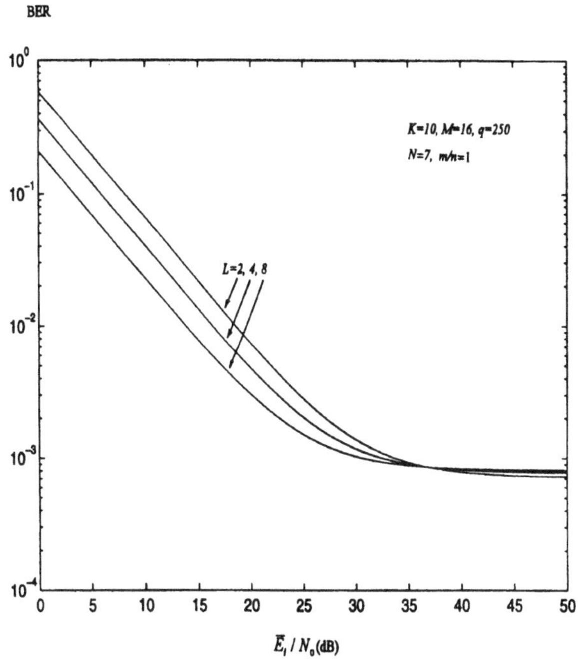

Fig. 11.8 Effect of number of paths (L) on uncoded BER.

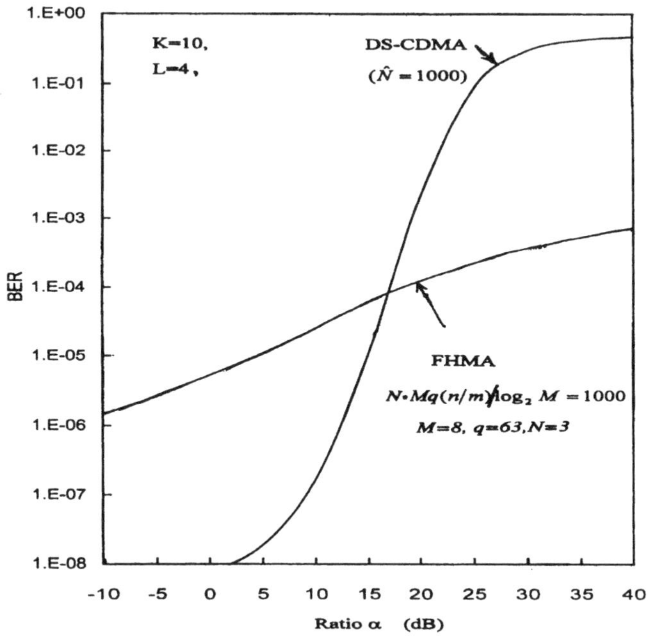

Fig. 11.9 Asymptotic BER with RS coding for fixed system bandwidth.

Finally, Fig. 11.10 illustrates the BER of a cellular FHMA system as a function of the number of active users per cell for different values of frequency reuse factor, N, when the system bandwidth is fixed (qN=1000). It is seen from this figure that for a given BER, a smaller value of N provides larger overall capacity of the cellular system. Although a smaller value of N causes more adjacent cell interference, it increases the total number, q, of hopping frequencies per cell, which decreases the hit probability (11.16) more and therefore reduces BER tremendously. Basically, this conclusion is consistent with the general conclusion of cellular FDMA systems, where system capacity increases with decreasing the number of cells in a cellular cluster.

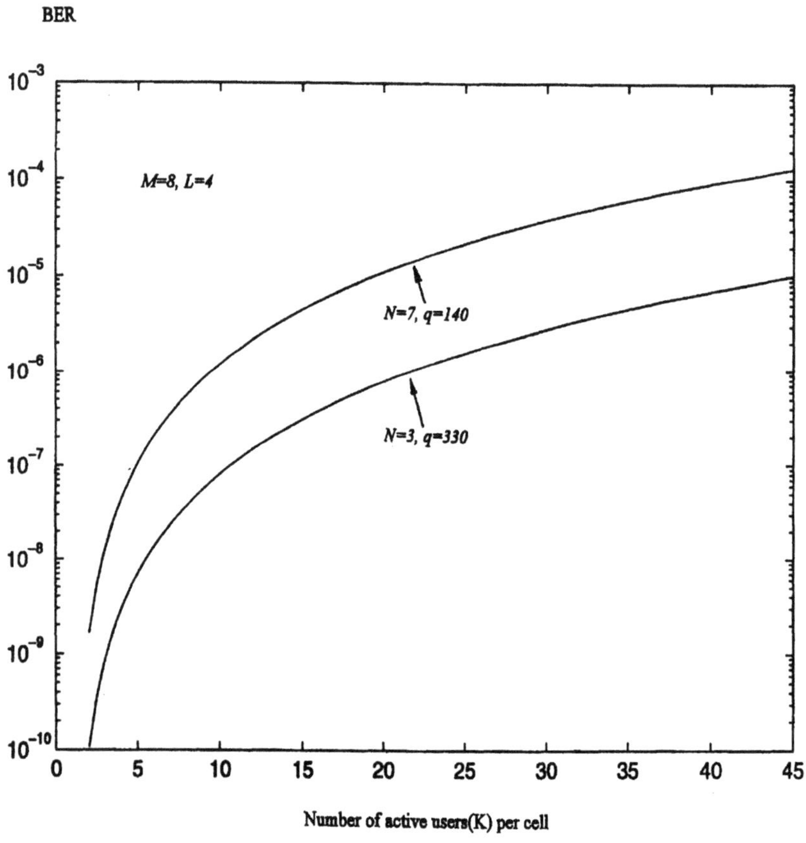

Fig. 11.10 BER comparison of N=3 and N=7 for fixed system bandwidth.

11.6 CONCLUSIONS

In this study, the bit error rate (BER) performance of one hop per symbol FHMA with MFSK modulation, noncoherent envelope detection and Reed-Solomon coding has been investigated for cellular applications. The following results have been obtained:

(a) The FHMA system is superior to a DS-CDMA system in overcoming the near/far problem.

(b) In order to have high capacity of the cellular FHMA system, a small value of the frequency reuse factor should be used.

REFERENCES

[1] J. Wang and M. Moeneclaey, "Multihops/symbol FFH-SSMA with MFSK modulation and Reed-Solomon coding for indoor radio," IEEE Trans. Commun., vol. COM-41, pp. 793-801, May 1993.

[2] W. E. Stark, "Coding for frequency hopped spread spectrum communications with partial-band interference, Part II: Coded performance," IEEE Trans. Commun., vol. COM-33, pp. 1045-1057, Oct. 1985.

[3] J. Wang and M. Moeneclaey, "Hybrid DS/SFH-SSMA with predetection diversity and coding over indoor radio multipath Rician fading channels," *IEEE Transactions on Communications*, vol. COM-40, pp. 1654-1662, October 1992.

[4] J. Wang and L. B. Milstein, "CDMA overlay situations for microcellular mobile communications," *IEEE Trans. Commun.*, vol. COM-43, pp. 603-614, February 1995.

[5] J. G. Proakis, Digital Communications, McGraw Hill, 1995.

[6] J. Chen, "Applications of frequency hopping systems" Ph.D. Thesis, University of Hong Kong, December 1999.

[7] J. Wang, "Performance of one-hop per symbol FHMA for cellular mobile communications," *IEEE Trans. Vehicular Technology.*, vol. VT-49, March 2001.

Chapter 12

A Novel FHMA Receiver

Abstract A novel FH/MFSK receiver with additional partial correlators is proposed by means of side information for asynchronous frequency hopping multiple-acess (FHMA) systems in the presence of Rayleigh fading. Performance analysis is carried out and compared for both the novel and conventional receivers. One hop per symbol and Reed-Solomon (RS) coding are used in the FHMA systems. It is shown that the novel receiver performs much better than the conventional receiver for a wide range of signal to noise ratio. The capacity of the FHMA system with novel receivers is almost as twice as that of the system with conventional receivers.

Keywords Frequency Hopping, Multiple Access, Fading Channels

12.1 INTRODUCTION

The increasing demand for bandwidth in mobile communication services has fueled the search for higher capacity wireless systems. This search has created a commercial interest in spread spectrum (SS) systems, which offer significant capacity improvement in multiple access (MA) systems. Although a great deal of emphasis has been placed on direct-sequence spread spectrum (DS/SS), recent developments in frequency hopping multiple access (FHMA) technology have resulted in commercial applications such as mobile cellular communications, personal communications and wireless local area networks (LANs). On the other hand, FHMA systems have been considered for a variety of military applications due to their frequency diversity and resistance to the near-far problem.

In FHMA systems, each active user transmits MFSK modulated signals with independent hopping patterns over a given common spread bandwidth. Therefore, detection of symbol from the reference transmitter is affected by the signals from other non-reference users transmitting simultaneously at the same frequency. This phenomenon is known as a "hit". The performance measure that one may be interested in is the number of users for which communication is possible with the error probability below a certain desired value. This is called the achievable region. This performance measure allows one to determine the maximum number of users that can simultaneously transmit.

In this chapter, a novel receiver with additional correlators and an noncoherent M-ary frequency shift keying (MFSK) demodulator is presented, which offers the performance improvement without incurring much of the implementation complexity. In FHMA systems, synchronous hopping of all users is useful to mitigating the effect of multiple-access interference under some special conditions, but it is technically very difficult to achieve this kind of synchronization in practice, specially in decentralized systems. Therefore, our system is assumed to operate in an asynchronous manner. That is, the transmission times between various users are not coordinated. When the observed signal is hit, it is expected to utilize the delay between the desired signal and interference signals of non-reference users to reduce the effect of interference. In order for the error probability of the receiver to be below a certain desired value over Rayleigh channel and further mitigating the effect of multiple-access interference, the Reed-Solomon (RS) coding with fixed rate is considered because of its good burst error-correcting capability.

In Section 12.2, we establish the system model and present the construction of the novel FH/MFSK receiver. Section 12.3 concentrates on the analysis of bit error rate of the novel MFSK receiver with Rayleigh fading. Numerical results and comparison between the novel and conventional receivers are given in Section 12.4. Finally, this thesis is concluded in Section 12.5.

12.2 SYSTEM MODELS

Let K_u denote the number of active transmitter-receiver pairs of the FHMA system under consideration. The transmitter for the *k*th user $(1 \le k \le K_u)$ consists of the following parts:

-- A Reed-Solomon encoder, which converts block of N_l source symbols into a code-word of N_n coded symbols. The sequence of coded symbols is represented as $b_k(t)$; the λ th coded symbol of $b_k(t)$ has amplitude $b_k^{(\lambda)}$, taking values from the set $\{0, 1, \cdots, M-1\}$. The coded symbol rate R_S is given by $R_S = R_b / \log_2 M$, where R_b denotes the rate of the coded bits. The information bit rate R_i (i.e., the rate of source bits) is given by $R_i = R_b(N_l / N_n)$, where N_l / N_n is the rate of the RS code.

-- A MFSK modulator, which translates each coded symbol $b_k^{(\lambda)}$ into a sinusoidal tone with frequency $f_m + b_k^{(\lambda)}\Delta f$, where Δf stands for the frequency spacing among adjacent MFSK tones. In order that these tones are orthogonal, $\Delta f T_S$ must be an integer, where $T_S = 1/R_S$ is the symbol duration.

-- A frequency hopper, operating at only one hop per coded symbol. *N* is defined as the number of total hopping frequencies. The frequency hopper translates the signal symbol into a sinusoidal tone with frequency $f_k^{(\lambda)} + b_k^{(\lambda)}\Delta f$, where $f_k^{(\lambda)}$ is first-order Markov sequence so that two consecutive hopping frequencies are always different. This assumption guarantees that two adjacent hops of the reference user are not hit together by an non-reference user. It is assumed that the hits due to multiple-access interference are independent from hop to hop. In fact, Hegde and Stark [1] has proven that though the hits exhibit an underlying Markov structure, the independence assumption is quit accurate. When the RS coding is employed, the minimum system bandwidth is given by

$$W_{Total} = R_i N(M / \log_2(M))(N_n / N_l). \tag{12.1}$$

It is assumed that the channel between any transmitter and the reference receiver is a slow Rayleigh fading channel, where "slow" means that the signal amplitude can be assumed to be fixed in one symbol.

Corresponding to the transmitter, the receiver consists of a frequency dehopper, an noncoherent MFSK demodulator, a hard decision device and a

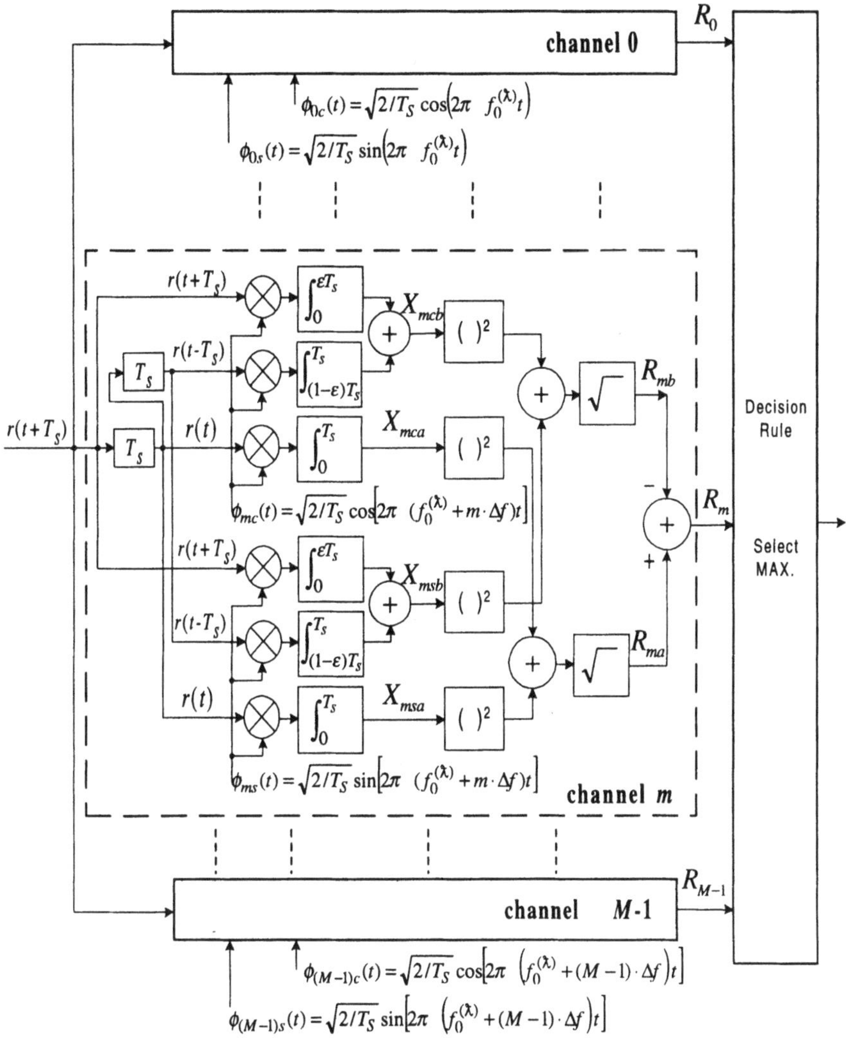

Fig.12.1 A detail block diagram of MFSK receiver with partial correlation, where $\boxed{T_S}$ is the delay element [4].

Reed-Solomon (RS) decoder. To reduce the effect of multi-access interference, a number of partial correlators in the demodulator are used (Fig.12-1), where parameter ε can be selected in the region of [0,1]. In the novel receiver, the earlier and later side information of interfering signals is utilized to reduce the effect of multi-access interference for asynchronous transmission. When $\varepsilon = 0$, the proposed receiver reduces to the conventional FH/MFSK receiver (Fig.12-2). Assuming that the user "0" is the reference user, the received signal is the sum of the reference signal, multiple-access interference and channel noise, given by

$$r(t) = \sqrt{2}\beta_0 \cos\left[2\pi\left(f_0^{(\lambda)} + b_0^{(\lambda)} \cdot \Delta f\right)t + \theta_0\right] + \sum_{k=1}^{K_u-1} \sqrt{2}\beta_k \cos\left[2\pi\left(f_k^{(\lambda)} + b_k^{(\lambda)} \cdot \Delta f\right)(t - \tau_k) + \theta_k\right] + n(t) \tag{12.2}$$

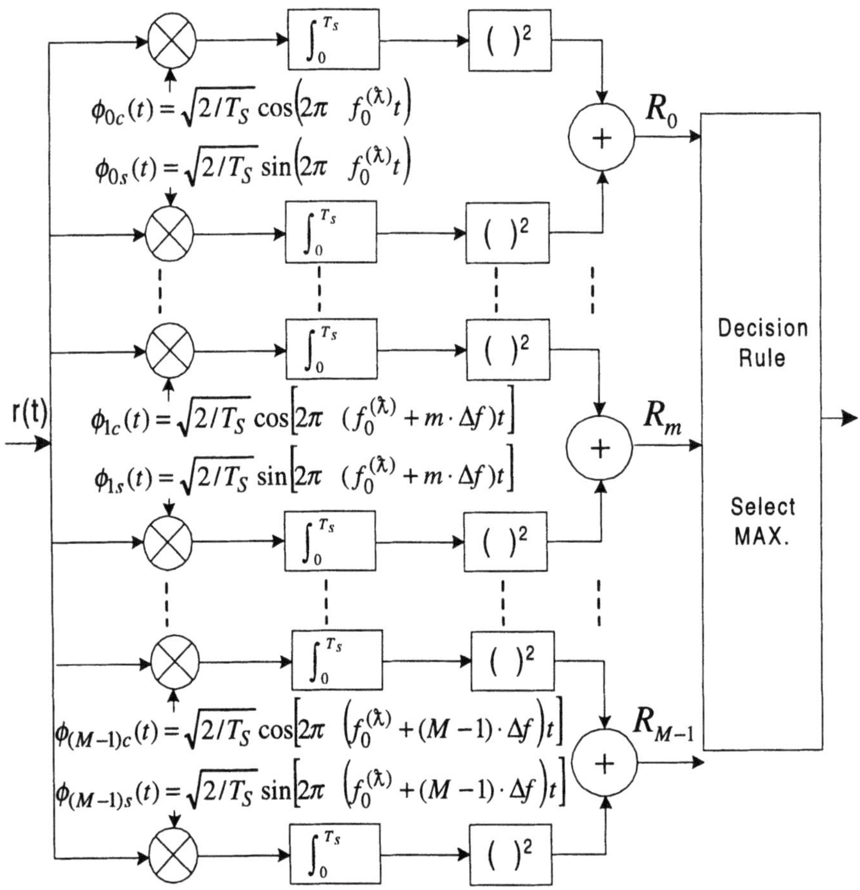

Fig.12.2 Conventional MFSK noncoherent receiver.

where the first term represents the reference signal (or desired signal). The remaining terms consist of the multiple-access interference from non-reference users and white noise with two-sided power spectral density of height $N_0/2$. The phases θ_0 and θ_k are independent random variables uniformly distributed in $[0, 2\pi)$. β_0 and β_k are the signal amplitudes of reference and non-reference users, respectively, and are independent Rayleigh random variables with variances $P_S = E[\beta_0^2]/2$ and $P_k = E[\beta_k^2]/2$, respectively. For simple analysis, we assume that all signals are received with the same average power $P_k = P_S$. Furthermore, we assume that the delays (τ_k) of non-reference signals are random variables. During the λth signal symbol, the signal frequency is $f_0^{(\lambda)} + b_0^{(\lambda)} \cdot \Delta f$ and the corresponding pair of orthogonal basis functions $\phi_{mc}(t)$ and $\phi_{ms}(t)$ for channel m of the demodulator (Fig.12.1 and Fig.12.2) are given by

$$\begin{cases} \phi_{mc}(t) = \sqrt{\dfrac{2}{T_S}} \cos\left[2\pi\left(f_0^{(\lambda)} + m \cdot \Delta f\right)t\right] \\ \phi_{ms}(t) = \sqrt{\dfrac{2}{T_S}} \sin\left[2\pi\left(f_0^{(\lambda)} + m \cdot \Delta f\right)t\right] \end{cases} \tag{12.3}$$

where $\lambda = \lfloor t/T_S \rfloor$.

12.3 ANALYSIS

As shown in Fig.12.1, the random variable R_m, $m = 0, 1, \ldots, M-1$, represents the output of the mth channel of the demodulator and is the subtraction of the square-roots of two sums, given by

$$\begin{aligned} R_m &= R_{ma} - R_{mb} \\ &= \sqrt{X_{mca}^2 + X_{msa}^2} - \sqrt{X_{mcb}^2 + X_{msb}^2} \end{aligned} \tag{12.4}$$

where

$$X_{mca} = \int_0^{T_S} r(t)\phi_{mc}(t)dt, \tag{12.5}$$

$$X_{msa} = \int_0^{T_S} r(t)\phi_{ms}(t)dt, \tag{12.6}$$

$$X_{mcb} = \int_0^{\varepsilon T_S} r(t+T_S)\phi_{mc}(t)dt + \int_{(1-\varepsilon)T_S}^{T_S} r(t-T_S)\phi_{mc}(t)dt \tag{12.7}$$

and

$$X_{msb} = \int_0^{\varepsilon T_S} r(t+T_S)\phi_{ms}(t)dt + \int_{(1-\varepsilon)T_S}^{T_S} r(t-T_S)\phi_{ms}(t)dt. \tag{12.8}$$

The bit error probability of receiver can be obtained by assuming that the desired signal is present in channel 0 of the demodulator (i.e., $b_0^{(\lambda)} = 0$). Since the cases of hit occurred in all M channels of the receiver due to multi-access interference are assumed independent of each other when the number of active users $K_u >> 1$, the probability density functions $\{ f_{R_m}(r_m) \}$ of random variables $\{ R_m \}$ are also independent. Thus, the probability of symbol error is given by

$$P_{SE} = 1 - \int_0^{\infty} f_{R_0}(r_0) \left[\int_0^{r_0} f_{R_m}(r_m) dr_m \right]^{M-1} dr_0, \qquad m = 1,2,\ldots,M-1. \tag{12.9}$$

In any channel of the receiver, the occurrence of the multi-access interference can be described in the following three cases (see Fig.12.3). In case I, channel m is not hit by any non-reference user during the observed symbol period. In case II, channel m of the demodulator is hit by an non-reference user (or user k) during the observed symbol period. The delay difference between the desired signal and interference signal is uniformly distributed in $[-T_S, T_S]$. In case III, channel m is not hit by an non-reference user during the observed symbol period, but is hit during the adjacent symbol of the observed signal symbol. The delay difference between the desired signal and interference signal is uniformly distributed in $[-(1+\varepsilon)T_S, -T_S]$ or $[T_S, (1+\varepsilon)T_S]$. Note that an interference symbol whose delay is less than $-(1+\varepsilon)T_S$ or greater than $(1+\varepsilon)T_S$ does not cause any interference at the output of all correlators due to partial integration (see Fig.12.1). For a large number (N) of available hopping frequencies, the probability of a hit occurred in one of M channels is small. Therefore, we only consider the case that channel m of the reference demodulator is hit by one non-reference user. Two or more hits in one channel at same time are neglected. The probabilities of cases I, II and III occurred in one channel of the receiver are given by

$$P_{CaseI} = 1 - \frac{2(1+\varepsilon)(K_u - 1)}{NM}, \tag{12.10a}$$

$$P_{CaseII} = \frac{2(K_u - 1)}{NM} \tag{4.10b}$$

and

$$P_{CaseIII} = \frac{2\varepsilon(K_u - 1)}{NM}. \tag{4.10c}$$

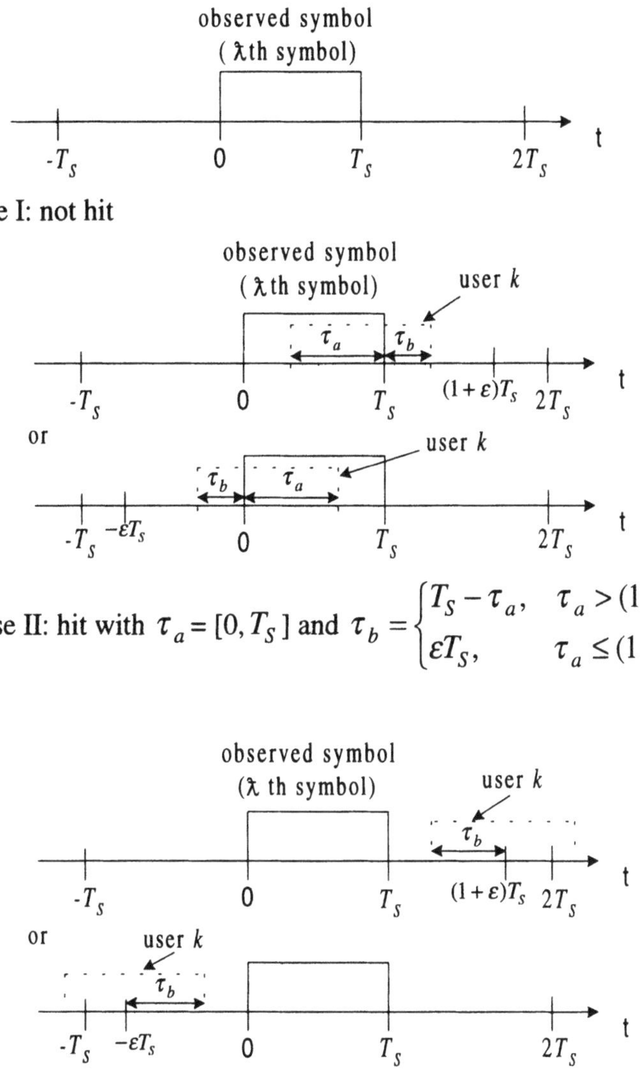

(a) Case I: not hit

(b) Case II: hit with $\tau_a = [0, T_S]$ and $\tau_b = \begin{cases} T_S - \tau_a, & \tau_a > (1-\varepsilon)T_S \\ \varepsilon T_S, & \tau_a \le (1-\varepsilon)T_S \end{cases}$

(c) Case III: hit with $\tau_b = [0, \varepsilon T_S]$

Fig.12.3 The cases of hit by user k [4].

Since the probability density functions $\{ f_{R_m}(r_m) \}$ are independent of each other, $f_{R_m}(r_m)$ can be obtained by averaging over all cases of hit occurred in channel m

$$f_{R_m}(r_m) = P_{Case\ I} \cdot f_{R_m|Case\ I}(r_m) + P_{Case\ II} \cdot \frac{1}{T_S}\int_0^{T_S} f_{R_m|Case\ II}(r_m|\tau_a)d\tau_a$$
$$+ P_{Case\ III} \cdot \frac{1}{\varepsilon T_S}\int_0^{\varepsilon T_S} f_{R_m|Case\ III}(r_m|\tau_b)d\tau_b \tag{12.11}$$

where $f_{R_m|Case\ I}(r_m)$, $f_{R_m|Case\ II}(r_m|\tau_a)$ and $f_{R_m|Case\ III}(r_m|\tau_b)$ are the conditional probability density functions of R_m, conditioned on the hit case I, II and III, respectively, and are discussed and derived in the following:

- Case I:

In case I, channel m is not hit by any non-reference user during the observed symbol period. The random variables X_{mca} and X_{msa} in (12.5) and (12.6) are given by

$$X_{mca} = \int_0^{T_S} \left[\beta_0 \sin\left(2\pi f_0^{(\lambda)}t + \theta_0\right)\delta(m,0) + n(t)\right] \cdot \sqrt{\frac{2}{T_S}} \cos\left[2\pi\left(f_0^{(\lambda)} + m\cdot\Delta f\right)t\right]dt$$
$$= \sqrt{\frac{T_S}{2}}\beta_0 \sin(\theta_0)\delta(m,0) + \int_0^{T_S} n(t) \cdot \sqrt{\frac{2}{T_S}} \cos\left[2\pi\left(f_0^{(\lambda)} + m\cdot\Delta f\right)t\right]dt \tag{12.12}$$

and

$$X_{msa} = \int_0^{T_S} \left[\beta_0 \sin\left(2\pi f_0^{(\lambda)}t + \theta_0\right)\delta(m,0) + n(t)\right] \cdot \sqrt{\frac{2}{T_S}} \sin\left[2\pi\left(f_0^{(\lambda)} + m\cdot\Delta f\right)t\right]dt$$
$$= \sqrt{\frac{T_S}{2}}\beta_0 \cos(\theta_0)\delta(m,0) + \int_0^{T_S} n(t) \cdot \sqrt{\frac{2}{T_S}} \sin\left[2\pi\left(f_0^{(\lambda)} + m\cdot\Delta f\right)t\right]dt \tag{12.13}$$

where $\delta(m,0) = 1$ and 0 for m=0 and $m \neq 0$, respectively. Since the two terms on the right hand side of (12.12) or (12.13) are Gaussian, both X_{mca} and X_{msa} are Gaussian random variables with the same variance

$$\sigma_{ma}^2 = \sigma_0^2\delta(m,0) + \sigma_N^2 \tag{12.14}$$

where $\sigma_0^2 = \frac{1}{2}E\left[\beta_0^2\right]T_S = P_S T_S$ and $\sigma_N^2 = N_0/2$. Therefore, the output of the square root $R_{ma} = \sqrt{X_{mca}^2 + X_{msa}^2}$ is a Rayleigh random variable with the following probability density function

$$f_{R_{ma}|Case\ I}(r_a) = \begin{cases} \frac{r_a}{\sigma_{ma}^2} \exp\left(-\frac{r_a^2}{2\sigma_{ma}^2}\right), & r_a \geq 0 \\ 0, & r_a < 0. \end{cases} \tag{12.15}$$

From (12.7) and (12.8), X_{mcb} and X_{mcb} are given by

$$\begin{aligned} X_{mcb} &= \int_0^{\varepsilon T_S} n(t+T_S) \cdot \sqrt{\frac{2}{T_S}} \cos\left[2\pi\left(f_0^{(\lambda)} + m \cdot \Delta f\right)t\right] dt \\ &+ \int_{(1-\varepsilon)T_S}^{T_S} n(t-T_S) \cdot \sqrt{\frac{2}{T_S}} \cos\left[2\pi\left(f_0^{(\lambda)} + m \cdot \Delta f\right)t\right] dt \end{aligned} \tag{12.16}$$

and

$$\begin{aligned} X_{msb} &= \int_0^{\varepsilon T_S} n(t+T_S) \cdot \sqrt{\frac{2}{T_S}} \sin\left[2\pi\left(f_0^{(\lambda)} + m \cdot \Delta f\right)t\right] dt \\ &+ \int_{(1-\varepsilon)T_S}^{T_S} n(t-T_S) \cdot \sqrt{\frac{2}{T_S}} \sin\left[2\pi\left(f_0^{(\lambda)} + m \cdot \Delta f\right)t\right] dt \end{aligned} \tag{12.17}$$

where both X_{mcb} and X_{msb} are zero mean Gaussian random variables with variance

$$\sigma_b^2 = 2\varepsilon\sigma_N^2. \tag{12.18}$$

Therefore, $R_{mb} = \sqrt{X_{mcb}^2 + X_{msb}^2}$ is a Rayleigh random variable with the following probability density function

$$f_{R_{mb}|Case\ I}(r_b) = \begin{cases} \frac{r_b}{\sigma_b^2} \exp\left(-\frac{r_b^2}{2\sigma_b^2}\right), & r_b \geq 0 \\ 0, & r_b < 0. \end{cases} \tag{12.19}$$

The probability density function of the mth channel output ($R_m = R_{ma} - R_{mb}$) is convolution between (12.15) and (12.19),

$$f_{R_m|Case\ I}(r) = \begin{cases} \int_0^\infty f_{R_{ma}|Case\ I}(x+r) f_{R_{mb}|\ Case\ I}(x)dx, & r \geq 0 \\ \int_0^\infty f_{R_{ma}|Case\ I}(x) f_{R_{mb}|Case\ I}(x-r)dx, & r < 0. \end{cases} \tag{12.20}$$

Substituting (12.15) and (12.19) into (12.20), one obtains

$$f_{R_m|Case\ I}(r)=\begin{cases}\dfrac{\sigma_{ma}^2 r}{\left(\sigma_{ma}^2+\sigma_b^2\right)^2}\exp\left(-\dfrac{r^2}{2\sigma_{ma}^2}\right)+\dfrac{\sqrt{\pi}\sigma_{ma}\sigma_b}{\sqrt{2\left(\sigma_{ma}^2+\sigma_b^2\right)^3}}\left(1-\dfrac{r^2}{\sigma_{ma}^2+\sigma_b^2}\right) \\ \cdot\exp\left(-\dfrac{r^2}{2(\sigma_{ma}^2+\sigma_b^2)}\right)\cdot erfc\left(\sqrt{\dfrac{\sigma_b}{2(\sigma_{ma}^2+\sigma_b^2)\sigma_{ma}^3}}\cdot r\right), & r\geq 0 \\ \dfrac{\sigma_b^2(-r)}{\left(\sigma_{ma}^2+\sigma_b^2\right)^2}\exp\left(-\dfrac{r^2}{2\sigma_{ma}^2}\right)+\dfrac{\sqrt{\pi}\sigma_{ma}\sigma_b}{\sqrt{2\left(\sigma_{ma}^2+\sigma_b^2\right)^3}}\left(1-\dfrac{r^2}{\sigma_{ma}^2+\sigma_b^2}\right) \\ \cdot\exp\left(-\dfrac{r^2}{2(\sigma_{ma}^2+\sigma_b^2)}\right)\cdot erfc\left(\sqrt{\dfrac{\sigma_{ma}}{2(\sigma_{ma}^2+\sigma_b^2)\sigma_b^3}}\cdot(-r)\right), & r<0\end{cases} \tag{12.21}$$

where $erfc(x)=\frac{2}{\sqrt{\pi}}\int_x^{\infty}e^{-t^2}dt$.

- Case II:

In case II, channel m is hit by one non-reference user (user k) and the delay difference between the desired signal and the interference signal is uniformly distributed in $\left[-T_S,T_S\right]$. In this case, the same interference user (β_k) may cause interference for both R_{ma} and R_{mb}. From (12.7) and (12.8), the random variables X_{mca} and X_{msa} are given by

$$\begin{aligned}X_{mca}&=\sqrt{\frac{T_S}{2}}\beta_0\sin(\theta_0)\delta(m,0)+\frac{\tau_a}{\sqrt{2T_S}}\beta_k\sin(\theta_k)\\&+\int_0^{T_S}n(t)\cdot\sqrt{\frac{2}{T_S}}\cos\left[2\pi\left(f_0^{(\lambda)}+m\cdot\Delta f\right)t\right]dt\end{aligned} \tag{12.22}$$

and

$$\begin{aligned}X_{msa}&=\sqrt{\frac{T_S}{2}}\beta_0\cos(\theta_0)\delta(m,0)+\frac{\tau_a}{\sqrt{2T_S}}\beta_k\cos(\theta_k)\\&+\int_0^{T_S}n(t)\cdot\sqrt{\frac{2}{T_S}}\sin\left[2\pi\left(f_0^{(\lambda)}+m\cdot\Delta f\right)t\right]dt\end{aligned} \tag{12.23}$$

where β_k represents the amplitude of the interference signal of the kth non-reference user and is the Rayleigh random variable with mean $\sqrt{\frac{\pi}{T_S}}\sigma_I$, second moment $\frac{4}{T_S}\sigma_I^2$, where σ_I^2 is the energy of one interference symbol, (i.e., $\sigma_I^2 = P_k T_S$). θ_k is a random variable uniformly distributed in $[0, 2\pi)$. We define the random variable $Y = \sqrt{\frac{T_S}{2}}\beta_k$. The probability density function of Y is given by

$$f_Y(y) = \begin{cases} \frac{y}{\sigma_I^2}\exp\left(-\frac{y^2}{2\sigma_I^2}\right), & y \geq 0 \\ 0, & y < 0. \end{cases} \tag{12.24}$$

When $Y = y$, both X_{mca} and X_{msa} are Gaussian random variables with mean $\frac{\tau_a}{T_S} y\sin(\theta_k)$ and $\frac{\tau_a}{T_S} y\cos(\theta_k)$, respectively. Therefore, R_{ma} is a Ricean random variable with the pdf

$$f_{R_{ma}|Case\ II}\left(r_a|\{\tau_a, y\}\right) = \begin{cases} \frac{r_a}{\sigma_{ma}^2}\exp\left(-\frac{r_a^2 + (\tau_a y/T_S)^2}{2\sigma_{ma}^2}\right) \cdot I_0\left(\frac{\tau_a y r_a}{T_S \sigma_{ma}^2}\right), & r_a \geq 0 \\ 0, & r_a < 0 \end{cases} \tag{12.25}$$

where variance σ_{ma}^2 is defined in (12.14). On the other hand, the random variables X_{mcb} and X_{msb} are given by

$$X_{mcb} = \frac{\tau_b}{\sqrt{2T_S}}\beta_k \sin\left(\hat{\theta}_k\right) + \int_0^{\varepsilon T_S} n(t+T_S) \cdot \sqrt{\frac{2}{T_S}} \cos\left[2\pi\left(f_0^{(\lambda)} + m\cdot\Delta f\right)t\right]dt$$
$$+ \int_{(1-\varepsilon)T_S}^{T_S} n(t-T_S) \cdot \sqrt{\frac{2}{T_S}} \cos\left[2\pi\left(f_0^{(\lambda)} + m\cdot\Delta f\right)t\right]dt \tag{12.26}$$

and

$$X_{msb} = \frac{\tau_b}{\sqrt{2T_S}} \beta_k \cos(\hat{\theta}_k) + \int_0^{\varepsilon T_S} n(t+T_S) \cdot \sqrt{\frac{2}{T_S}} \sin\left[2\pi\left(f_0^{(\lambda)} + m \cdot \Delta f\right)t\right] dt$$
$$+ \int_{(1-\varepsilon)T_S}^{T_S} n(t-T_S) \cdot \sqrt{\frac{2}{T_S}} \sin\left[2\pi\left(f_0^{(\lambda)} + m \cdot \Delta f\right)t\right] dt$$

(12.27)

where $\tau_b = \begin{cases} T_S - \tau_a, & \tau_a > (1-\varepsilon)T_S \\ \varepsilon T_S, & \tau_a \le (1-\varepsilon)T_S \end{cases}$. $\hat{\theta}_k = \theta_k \pm 2\pi T_S \left(f_0^{(\lambda)} + m \cdot \Delta f\right)$ is

a random variable uniformly distributed in $[0, 2\pi)$. When $Y = y$, the random variables X_{mcb} and X_{msb} are Gaussian random variables with mean $\frac{\tau_b}{T_S} y \sin(\hat{\theta}_k)$ and $\frac{\tau_b}{T_S} y \cos(\hat{\theta}_k)$, respectively. Similarly, R_{mb} is also a Ricean random variable with pdf

$$f_{R_{mb}|Case\ II}\left(r_b|\{\tau_a, y\}\right) = \begin{cases} \frac{r_b}{\sigma_b^2} \exp\left(-\frac{r_b^2 + (\tau_b y/T_S)^2}{2\sigma_b^2}\right) \cdot I_0\left(\frac{\tau_b y r_b}{T_S \sigma_b^2}\right), & r_b \ge 0 \\ 0, & r_b < 0 \end{cases}$$

(12.28)

where variance σ_b^2 is defined as (12.18). Furthermore, we average $f_{R_m|Case\ II}(r|\{\tau_a, y\})$ over Y to obtain

$$f_{R_m|Case\ II}(r|\tau_a) = \int_0^\infty f_Y(y) f_{R_m|Case\ II}\left(r|\{\tau_a, y\}\right) dy. \qquad (12.29)$$

For $R_m = R_{ma} - R_{mb}$,

$$f_{R_m|Case\ II}(r|\tau_a) = \begin{cases} \int_0^\infty f_Y(y) \int_0^\infty f_{R_{ma}|Case\ II}\left(x + r|\{\tau_a, y\}\right) f_{R_{mb}|Case\ II}\left(x|\{\tau_a, y\}\right) dx dy, & r \ge 0 \\ \int_0^\infty f_Y(y) \int_0^\infty f_{R_{ma}|Case\ II}\left(x|\{\tau_a, y\}\right) f_{R_{mb}|Case\ II}\left(x - r|\{\tau_a, y\}\right) dx dy, & r < 0. \end{cases}$$

(12.30)

Substituting (12.24), (12.25) and (12.28) into (12.30) and using (6.633.4) of [25], (12.30) can be simplified as

$$f_{R_m|Case II}(r|\tau_a)=\begin{cases}\dfrac{T_S^2}{\Omega}\int_0^\infty x(x+r)\exp\left(-\dfrac{x^2\left(\tau_a^2\sigma_I^2+T_S^2\sigma_{ma}^2\right)+(x+r)^2\left(\tau_b^2\sigma_I^2+T_S^2\sigma_b^2\right)}{2\Omega}\right)\\ \cdot I_0\left(\dfrac{x(x+r)\tau_a\tau_b\sigma_I^2}{\Omega}\right)dx, r\geq 0\\ \dfrac{T_S^2}{\Omega}\int_0^\infty x(x-r)\exp\left(-\dfrac{(x-r)^2\left(\tau_a^2\sigma_I^2+T_S^2\sigma_{ma}^2\right)+x^2\left(\tau_b^2\sigma_I^2+T_S^2\sigma_b^2\right)}{2\Omega}\right)\\ \cdot I_0\left(\dfrac{x(x-r)\tau_a\tau_b\sigma_I^2}{\Omega}\right)dx, r< 0\end{cases} \quad (12.31)$$

where $\Omega=\tau_a^2\sigma_b^2\sigma_I^2+\tau_b^2\sigma_{ma}^2\sigma_I^2+T_S^2\sigma_{ma}^2\sigma_b^2$.

- Case III:

In case III, channel m is not hit by an non-reference user during the observed symbol period, but is hit during the adjacent symbol of the observed signal symbol. The delay difference between the desired signal and interference signal is uniformly distributed in $[-(1+\varepsilon)T_S, -T_S]$ or $[T_S, (1+\varepsilon)T_S]$. In this case, the random variables X_{mca} and X_{msa} are the same as that in case I, and the random variables X_{mcb} and X_{msb} are same as (12.26) and (12.27) in case II but with τ_b uniformly distributed on $[0, \varepsilon T_S]$ (see Fig.12.3). Since both X_{mca} and X_{msa} do not include common multiple access interference unlike (12.22) and (12.23) in case II, $\{X_{mca}, X_{msa}\}$ are independent of $\{X_{mcb}, X_{msb}\}$. The pdf of R_{ma} is given by (12.15). However, given τ_b, since both X_{mcb} and X_{msb} are Gaussian random variables with variance

$$\hat{\sigma}_b^2(\tau_b)=\left(\frac{\tau_b}{T_S}\sigma_I\right)^2+2\varepsilon\sigma_N^2, \quad (12.32)$$

whereas the pdf of R_{mb} is a Rayleigh distribution, given by (12.19) with the substitution of σ_b^2 by $\hat{\sigma}_b^2(\tau_b)$. Furthermore, given τ_b, the pdf $f_{R_m|Case\ III}(r|\tau_b)$ of R_m in case III is given by (12.21) with the substitution of σ_b^2 by $\hat{\sigma}_b^2(\tau_b)$.

$$f_{R_m|Case\ III}(r|\tau_b)=\begin{cases}\frac{\sigma_{ma}^2 r}{\left(\sigma_{ma}^2+\hat{\sigma}_b^2\right)^2}\exp\left(-\frac{r^2}{2\sigma_{ma}^2}\right)+\frac{\sqrt{\pi}\sigma_{ma}\hat{\sigma}_b}{\sqrt{2\left(\sigma_{ma}^2+\hat{\sigma}_b^2\right)^3}}\left(1-\frac{r^2}{\sigma_{ma}^2+\hat{\sigma}_b^2}\right) \\ \cdot\exp\left(-\frac{r^2}{2(\sigma_{ma}^2+\hat{\sigma}_b^2)}\right)\cdot erfc\left(\sqrt{\frac{\hat{\sigma}_b}{2(\sigma_{ma}^2+\hat{\sigma}_b^2)\sigma_{ma}^3}}\cdot r\right), & r\geq 0 \\ \frac{-\hat{\sigma}_b^2 r}{\left(\sigma_{ma}^2+\hat{\sigma}_b^2\right)^2}\exp\left(-\frac{r^2}{2\sigma_{ma}^2}\right)+\frac{\sqrt{\pi}\sigma_{ma}\hat{\sigma}_b}{\sqrt{2\left(\sigma_{ma}^2+\hat{\sigma}_b^2\right)^3}}\left(1-\frac{r^2}{\sigma_{ma}^2+\hat{\sigma}_b^2}\right) \\ \cdot\exp\left(-\frac{r^2}{2(\sigma_{ma}^2+\hat{\sigma}_b^2)}\right)\cdot erfc\left(\sqrt{\frac{\sigma_{ma}}{2(\sigma_{ma}^2+\hat{\sigma}_b^2)\hat{\sigma}_b^3}}\cdot(-r)\right), & r<0.\end{cases} \tag{12.33}$$

Finally, substituting (12.10), (12.21), (12.31) and (12.33) into (12.11), we can obtain the average pdf $f_{R_m}(r)$ of random variable R_m, and obtain the uncoded symbol error probability P_{SE} by (12.9). Once P_{SE} after hard decision has been obtained, assuming symbol errors are random when interleaving technique is used, the corresponding bit error rate (BER) P_{BE} after hard decision is given by

$$P_{BE}=\frac{M/2}{M-1}P_{SE}\,. \tag{12.34}$$

Reed-Solomon (RS) codes are non-binary, linear, cyclic symbol-error-correcting block codes. The length of an RS code is $N_n=M-1$ M-ary symbols, N_l of which are redundancy symbols; N_l can be chosen between 1 and $M-2$. In this paper, we select $N_l=(M-2)/2$ so that the coding rate N_l/N_n is nearly $1/2$, and the code can correct up to $t=M/4$ symbol errors. When decoding of RS codes with hard decision is employed, the symbol error probability P_{SEC} after RS decoding is well approximated by [3]

$$P_{SEC}=\frac{1}{N_n}\sum_{j=t+1}^{N_n} j\begin{bmatrix}N_n\\ j\end{bmatrix}[P_{SE}]^j[1-P_{SE}]^{N_n-j} \tag{12.35}$$

where $t=M/4$ and $n=M-1$. The BER P_{BEC} after decoding is approximated by substituting in (12.34) P_{BE} and P_{SE} by P_{BEC} and P_{SEC}.

12.4 NUMERICAL RESULTS

The probability of bit error of the novel MFSK receiver in a FHMA system with the Rayleigh fading channel is now numerically computed. For fair comparison, all of the following results are obtained under the condition that $\frac{W_{Total}}{R_i} = N\left(\frac{M}{\log_2 M}\right)\left(\frac{N_n}{N_l}\right) = 2000$ is fixed where $(N_n / N_l) \approx 0.5$ for coding.

Firstly, the effects of the parameter ε of the additional partial correlators on the performance of the MFSK receiver are shown in Figs.12.4 and 12.5. The probability of bit error is shown in Fig.12.4 for various values of E_b / N_0 when the modulation order M=32, whereas it is shown in Fig.12.5 for various values of M when $E_b / N_0 = 16$ dB. Note that the performance of a conventional receiver is shown in the both figures when $\varepsilon = 0$. It can be seen from Fig.12.4 that when E_b / N_0=8dB, the performance of a conventional receiver is better. However, when E_b / N_0 is large, i.e., $E_b / N_0 \geq 12$ dB, the performance of a novel receiver is better. The probability of bit error decreases when ε increases for a given value of E_b / N_0. When ε takes an optimum value ($\varepsilon \approx 0.4$), the error probability reaches the minimum. When ε keeps increasing after the optimum value, the probability of error steadily increases. The optimum value of ε is a function of E_b / N_0, but it falls in the region from 0.3 to 0.5, approximately. The optimum region of ε from 0.3 to 0.5 can also be observed in Fig.12.5 when $M \geq 8$. It can also be seen from the both figures that given E_b / N_0 or M, the performance is not very sensitive to ε when ε takes any value from the region of 0.3 to 0.5. Therefore, the medium value of $\varepsilon = 0.4$ is used for following figures.

The BER performance is plotted in Fig.12.6 as a function of E_b / N_0 when the modulation order M=64. The performance of a conventional receiver (or $\varepsilon = 0$) is also shown for comparison. It can be seen that when E_b / N_0 is small, the performance of a novel receiver is slightly poorer than that of a conventional receiver since the white noise dominates the multiple access interference so that partial correlators of the novel receiver introduces more noise. However, when E_b / N_0 is large, the multiple access interference dominates so that the novel receiver is better because it uses the side information of the interfering signal to reduce the interference. When E_b / N_0 is very large, the significant improvement in performance has been achieved with the novel receiver.

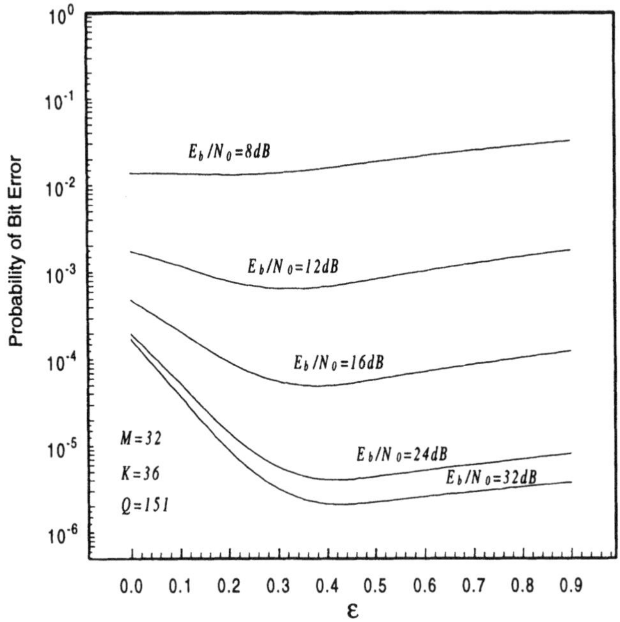

Fig. 12.4 The probability of bit error with RS coding.

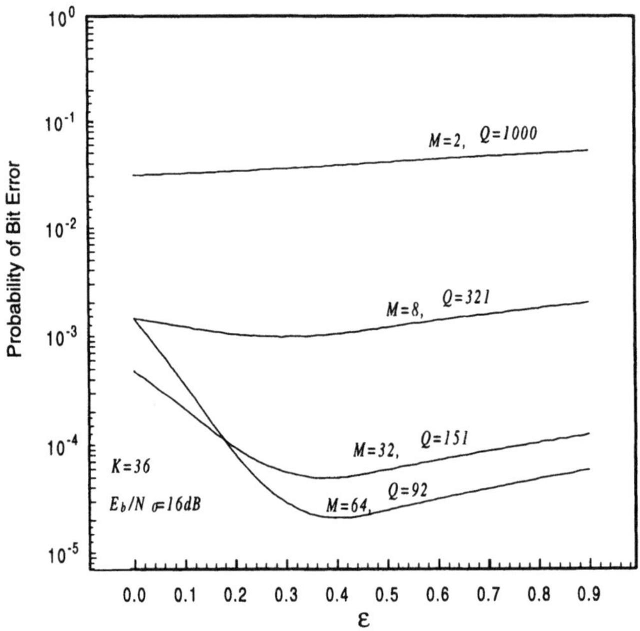

Fig. 12.5 The probability of bit error with RS coding.

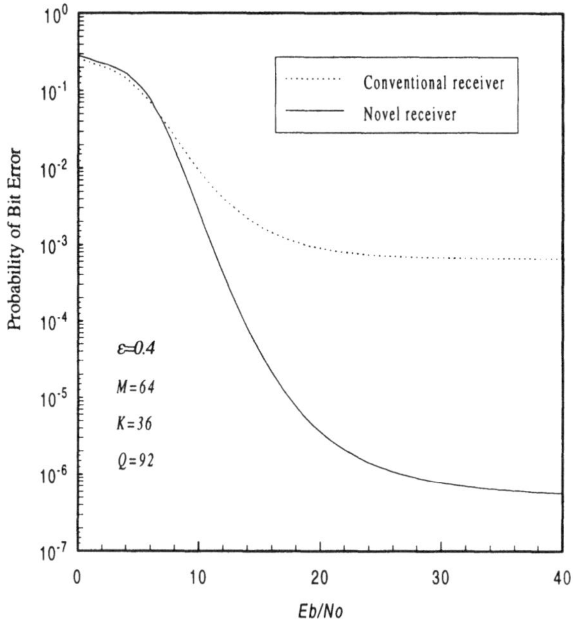

Fig. 12.6 The probability of bit error with RS coding.

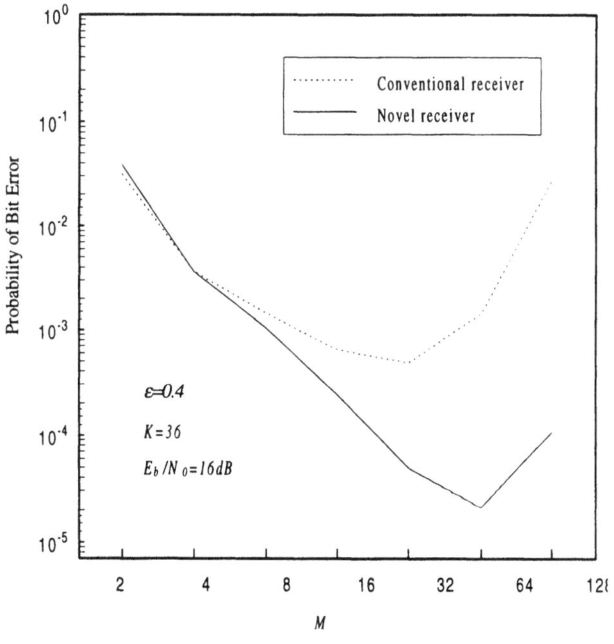

Fig. 12.7 The probability of bit error versus the modulation order with RS coding.

Fig.12.7 illustrates the BER for both receivers as a function of modulation order M. It can be seen that when $M \leq 4$, the conventional receiver performances better than novel receiver. However, when $M > 4$, the novel receiver is better. For a fixed system bandwidth, the best performance is achieved when M=32 for a conventional receiver, while the optimum value of M is 64 for a novel receiver. Actually, this is consistent with Fig.12.5. When M is larger than the optimum value, the performance degrades.

Finally, the BER is shown in Fig.12.8 as a function of number of active users (K_u). It can be seen that for a BER of 10^{-4}, the system capacity with novel receivers is almost as twice as that with conventional receivers.

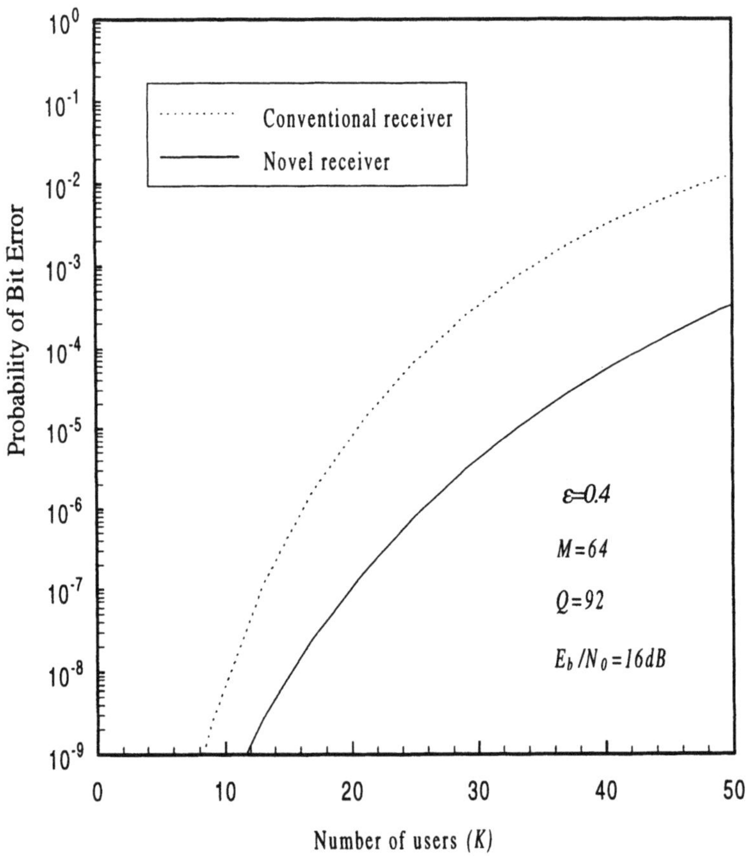

Fig. 12.8 The probability of bit error with RS coding.

12.5 CONCLUSIONS

In this chapter, the performance of the novel FHMA/MFSK receiver has been investigated. The following conclusions have been drawn:

(1) The novel receiver performs much better than a conventional receiver for a wide range of signal to noise ratio.
(2) The optimum value of parameter ε of the partial correlators in the novel receiver is in the region of 0.3 to 0.5.
(3) For a fixed system bandwidth, the optimum modulation order M is 64 for novel receivers, while the optimum M is 32 for conventional receivers.
(4) The capacity of the system with novel receivers is almost as twice as that of the system with conventional receivers.

REFERENCES

[1] M. Hegde and W. E. Stark, "Capacity of frequency-hop spread-spectrum multiple-access communication systems," *IEEE Trans. Commun.*, vol.38, pp.1050-1059, July 1990.

[2] I. S. Gradshteyn and I. M. Ryzhik, "Table of Integration, Series and Productions," 5nd ed. CD-Rom version 1.0, Academic Press, 1996.

[3] J. Wang and M. Moeneclaey, "Multiple hopes/symbol FFH-SSMA with MFSK modulation and Reed-Solomon coding for indoor radio," *IEEE Trans. Commun.,* vol. 41, No.5, pp.793-801, May 1993.

[4] J. Chen and J. Wang, "A novel receiver for FHMA systems," Submitted to *IEEE Trans. Vehicular Technology.*

Chapter 13

4G MOBILE COMMUNICATIONS

Abstract 3G mobile systems either are currently being or will be soon deployed (from 2001 to 2004) in different regions of the world. Certainly, there will be a huge potential market in 3G in the world, simply since it can provide wideband multimedia services with data rate up to 2Mbps. However, with both economic growth and technology development of the world, multimedia services with even higher data rate (i.e. 20Mbps) will be necessary. Discussions on 4G mobile communications have already been started today in the world. Therefore, extensive international R&D and standardization activities are required to solve many technical challenges. In this chapter, the forecast of mobile communications and key techniques of 4G will be presented.

Keywords: OFDM, Multiple Access, Software Radio, Smart Antennas, CDMA

13.1 MARKET TREND

Mobile radio communications systems have been successfully deployed since early 1980's in different regions of the world to extend telephone services to mobile users. In 1G, analog systems were deployed to support only voice telephony. These narrowband systems are based on frequency modulation (FM) with frequency division duplex (FDD). The success of the mobile communications is mainly due to the introduction of cellular concept, which was proposed in 1970's. High capacity and low power consumption can be achieved by means of cellular model. The cellular concept has made a great contribution to mobile communications. The main representative of 1G mobile systems is US advanced mobile phone service (AMPS) at 800 MHz, which was the first cellular system in the world and which was deployed in 1983 in USA.

However, in Europe [1], there were several 1G analog systems, developed by different countries. Those 1G European analog systems are generally incompatible with one another, because of the different frequencies and communications protocols used. Therefore, international roaming was impossible. These analog systems were later being replaced by the 2G global systems for mobile (GSM), which was the first 2G system in the world and which was deployed in 1991 in Europe. The main step was the introduction of digital technology (TDMA) for similar services as in 1G analog systems. The GSM standard has gained worldwide acceptance (more than 100 countries). In addition to GSM, other 2G digital systems were deployed in some countries, such as US TDMA (IS-54), US CDMA (IS-95), and Japanese PDC (personal digital cellular). However, the deployment of all the 2G systems started later than GSM.

Based on the big worldwide popularity of GSM, this system is being further developed to offer more advanced data services [2]. General packet radio service (GPRS) and enhanced data rates for GSM evolution (EDGE) or (E-GPRS) as well as WAP will allow the transmission of higher data rates as well as speech. The enhancement of PDC is the i-mode, which is becoming very popular in Japan. Starting out from speech-dominated services we are now experiencing massive growth in applications involving short message service together with the start of Internet applications prior to the introduction of 3G.

Although the evolved 2G systems can provide data services, the data rates are very limited (up to around 100kbps). In order to provide high speed wireless data service, 3G systems are currently being standardized and developed worldwide. The main goals are to support broadband data services up to 2Mbps with a wideband radio interface and international roaming. The radio air interface standards are based on WCDMA, multicarrier CDMA2000 and EDGE. 3G systems aim to integrate different

2G services. In other words, 3G is about a diversity of terminal types, including many non-voice terminals.

3G is intended to operate in wider bandwidth allocations at 2GHz. These new frequency bands will provide wider bandwidths for some multimedia services and the first spectrum allocations have been made in some countries via spectrum auctions in Europe or beauty contests in Japan. The opportunity has been taken to increase competition by allowing new operators into the bands as well as extending existing operator licenses. 3G mobile systems will be deployed between 2001 and 2004 in different regions of the world. 2G and 3G will continue to coexist for some time with optimization of service provision between them.

While 3G systems will significantly improve the spectral efficiency and possibly the cost of the system, a more profound feature is the significant improvement of its data and multimedia capabilities. While this feature seems to be a mere evolution from a technical viewpoint, its potential lies in the promotion of communications not only from person to person, but also from person to machine and from machine to machine.

In addition, wireless local area network (LAN) type systems like HIPERLAN 2, IEEE 802.11, digital audio broadcast (DAB), and digital video broadcasting (DVB) are becoming available. For short-range connectivity, systems like Bluetooth are being developed. Internationally, different frequency bands, such as ISM bands are already allocated to wireless LAN-type systems. The wireless LAN systems are designed in particular for high data rate access and low range and in general for low mobility. They are applicable to corporate networks and public access as a complement to 2G or 3G systems for hot spot applications such as company campuses, conference centers and airport.

Even though 3G can provide multimedia services with data rate as high as 2Mbps, demands for even higher speed (i.e. 20Mbps) data services are necessary in the future. Also, it is difficult to provide a full range of multi-rate services in 3G all with different QoS and performance requirements due to the constraints imposed on the core network by the air interface standard [3]. For example, it is not a fully integrated system (different 3G systems and different wireless LANs). These technologies represent a very flexible and powerful platform to support future requirements of services and applications. However, most of these systems have been designed in isolation without taking into account possible interworking with other access technologies. Their system design is mainly based on the traditional vertical approach to support a certain set of services with a particular technology. This situation is the starting point for the discussion of mobile radio systems in 4G. Depending on future user requirements and economic demands, new radio access schemes such as wideband radio access for high mobility might be developed. In addition, 4G systems have to be transparent with fixed network. The network of 4G may be new one with IP capability. Seamless service area should be supported. Flexible mobile terminal with multi-mode function will be needed.

Although 4G is necessary in the future for very high speed multimedia services and for integrating different sets of services, It is predicted that the deployment of 4G will happen only after 2015. There are three main reasons for this prediction

- 3G WCDMA might provide multimedia services with data rates much higher than 2Mbps in downlink [4]. Using adaptive modulation and coding scheme with interference cancellation, high data rates up to 8Mbps can be supported. The MS sends channel SNR information to the BS. After BS detects that information, it decides what modulation scheme is used. When the downlink SNR is large, multi-level modulation (i.e. 64QAM) plus low rate coding is employed. However, when the downlink SNR is small, the simple modulation such as QPSK plus high rate coding is used. High level modulation should be used in conjunction with other advanced techniques, like smart antennas and interference cancellation in order to achieve high data rates.

- Possibility that wireless multimedia services of 20Mbps are needed within 10 years is low. How many people require wireless data services of 20Mbps in 10 years? We think that the R&D and deployment of 4G systems will take much longer than expected.

- It is anticipated that the service charge for 3G and future 4G will be very high, simply since the license fee in auction is too high. If service charge is too high, how many people can afford the services? That is, whether any technology or service is popular is mainly up to economic capability of customers. After the bubble of "telecommunications" stocks in stock exchange markets of the world has burst, the development and prediction of the whole wireless are becoming more realistic.

Figure 13.1 shows the data rate of mobile communications as a function of year for different generations. It can be seen that the data rates are increased one order when the generation is increased by one. Table 13.1 summarizes the features of different generations of mobile communications.

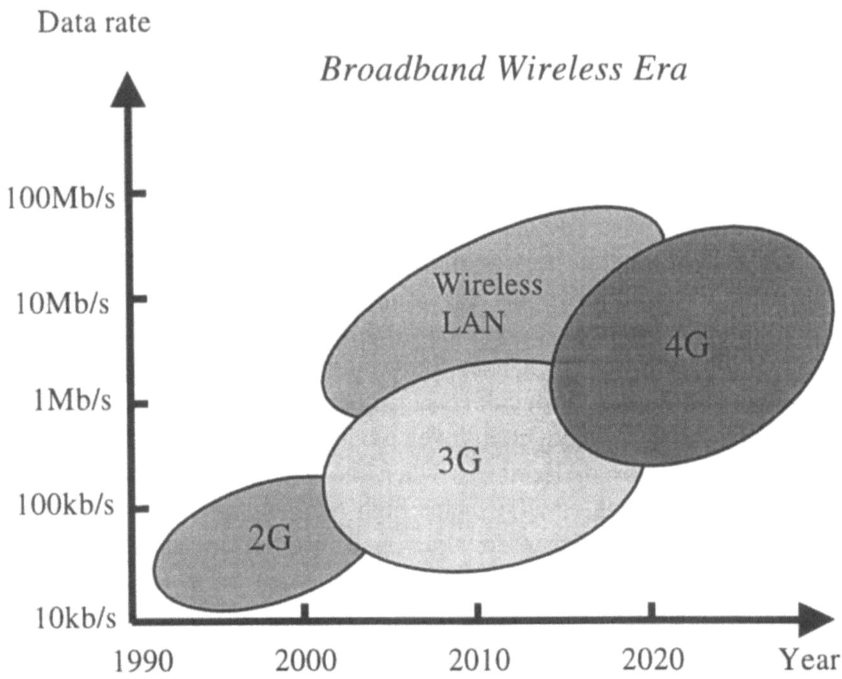

Figure 13.1 Data rate for different generations as a function of year

Table 13.1 Major mobile standards

Generation	Year of service	Major systems	Peak data rate
1G	1983	AMPS	
2G	1991	GSM, PDC, IS-54 IS-95	13kbps 8kbps 9.6kbps
3G	2001-2004	WCDMA CDMA2000 EDGE	2Mbps
4G	2015 or after?	OFDM-based?	> 20Mbps?

13.2 KEY TECHNOLOGIES

Frequency spectrum is a very scarce resource, which requires more efficient use of spectrum and the exploration of new frequency bands for wireless applications. Increasing computing power at lower cost enables new possibilities for more sophisticated signal processing algorithms for coding, decoding, detection, advanced antenna concepts and software-defined radio implementations.

13.2.1 Multiple Access

4G must be able to offer the broadband services to mobile users, presumably with the maximum information bit rates of more than 2 Mbps in a vehicular environment and 20 Mbps in indoor-to-pedestrian environments. However, this is quite adverse due to a large amount of resolved multipaths. In such an environment, the signal power of each resolved path becomes weaker. In order to achieve high data rates with reasonable frequency bandwidth, the required received signal power should be increased. The cell radius of 4G cellular systems must be very small [3] (i.e. 20 to 30-meter radius). Large coverage will be severely degraded without adding a large number of BSs per unit area, especially for indoor deployments. Therefore, advanced multiple access techniques are needed to provide high data rate services and large capacity.

In 4G, wide-range variable-speed transmission is needed with asymmetric traffic for up- and down- links. The existing well-developed CDMA technologies might be helpful for the 4G transmissions [6]. One of choices for multiple access of 4G is OFDM+CDMA. [6-8]. One distinctive advantage of OFDM is its characteristic of high data rate transmission with low system complexity by FFT in a stable channel where orthogonality between subcarriers is guaranteed. In particular, costly equalizers required in signal-carrier systems are dramatically simplified or even omitted in the case that different modulation schemes are used. However, OFDM is vulnerable to synchronization errors and is very sensitive to frequency offset and nonlinear amplifications. The combination of OFDM signaling with CDMA scheme has tackled this problem and it can lower the symbol rate in each subcarrier so that a longer symbol duration makes it easier to approximately synchronize the transmissions among different users. However, it has been shown in the Appendix that the capacity of OFDM+CDMA is the same as that of single carrier CDMA for a given system bandwidth.

13.2.2 Software Radio

The software radio is emerging from military applications. Because of the recent advances in high speed digital signal processors (DSP) and analog-to-digital converters (A/D), the commercial implementation of software radios has become viable [16]. In software radio systems, the IF signal is digitized using wideband A/D and all of the subsequence processing is implemented in software by means of digital signal processing hardware such as field programmable gate arrays (FPGAs) and DSPs. The main advantage of software radio is its great flexibility such that it can be programmed for emerging wireless standards. It also can be dynamically updated with new software without any changes in hardware and infrastructure. Rapid deployment is another important feature of the software radio. In wireless applications where different standards might be deployed, users' roaming can be a big issue in existing platforms. In the software radio system, users just need to download the new air interface on entering the new territory.

The advances in A/D and D/A conversions have made it possible to directly convert signals closer to the antenna at high speed with adequate dynamic range. This reduces the radio components needed and greatly facilitates digital implementation [13]. With the signal converted to digital in the very early stages of the system at high speed, a wideband radio approach can be adopted that provides inherent flexibility to support different standards that operate with different frequency bandwidth. As an example [14], in the wireless communications systems, advances in software radio developments have provided impetus for A/D performance improvements, especially for sampling rates of approximately 100 million samples per second. More generally, the large number of signal types to be digitized has led to a diverse selection of data converters in terms of architectures, resolution and sampling rates.

13.2.3 Smart Antennas

Due to continued rapid growth of wireless mobile applications, more system capacity and high data rate are needed. Smart antennas are a promising technique that can significantly increase systems capacity. Smart antennas have intelligent functions such as suppressions of interference signals, auto tracking of desired signals, and digital beam forming with adaptive space-time processing algorithms [3]. One type of smart antennas, adaptive array antennas [15] are expected to reduce much interference, with respect to conventional antenna diversity. Using beamforming techniques at the receiver two or more transmitters can share the same traffic channel to communicate with the base station at the same time. An adaptive antenna array is used at the base station to form several antenna beams simultaneously. Each beam captures one transmitter by automatically pointing its pattern toward the transmitter while mulling other cochannel transmitters and multipath signals. In this way, the cochannel interference is minimized and therefore the carrier-to-interference ratio for the signal of interest is maximized. In urban wireless environments, the signal transmitted by a user is reflected by its surrounding buildings and terrain. Therefore, several copies of the transmitted signal are received at the base station with different delays and different attenuation. It has been shown in [15] that multipath interference can be rejected effectively by the smart antennas.

A smart antenna is a multi-beam adaptive array with its gain pattern adjusted dynamically [16]. In a smart antenna system, each beam tracks its user within the coverage of the cell. Moreover, smart antenna systems provide diversity gain against multipath signals and strong interfering users by placing antenna nulls toward those sources. Since the beamforming in adaptive arrays is done in IF and baseband [15], these systems are well suited for software radio implementation. In general, the objective is to form the main beam toward the desired user and nulls toward the interference sources as shown in Figure 13.3.

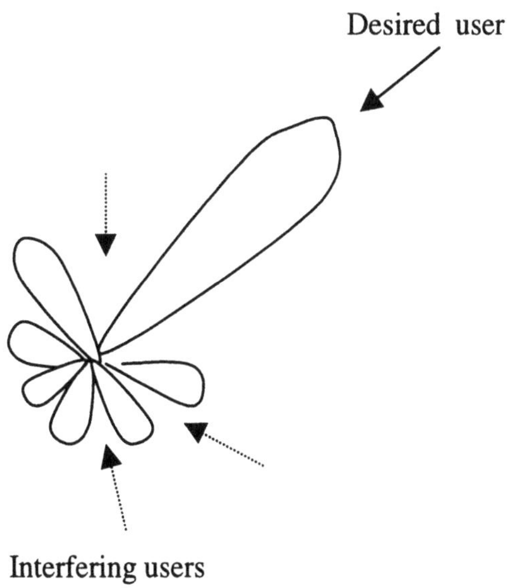

Figure 13.3 Beam pattern of adaptive antennas

APPENDIX 13: OFDM+CDMA

OFDM+CDMA technique [6-9] combines both orthogonal frequency division multiplexing (OFDM) and direct-sequence spread-spectrum (DSSS) technique in such a way that, by increasing the number of subcarriers, the processing gain of a OFDM+CDMA system can be made substantially greater than that of a conventional (single-carrier) DSSS system under the same-bandwidth condition. It has been expected that this greater processing gain could lead to a greater reduction in the multiple-access interference (MAI) so that a OFDM+CDMA system could support a larger number of users than a conventional DSSS system [10]. However, based on the bit error probability to be derived, we show that, under equal-bandwidth and equal-bit-rate conditions, the multiple-access (MA) capacity of an asynchronous OFDM+CDMA system is identical to that of a conventional DSSS system and does not increase by an increase of the number of subcarriers used. Despite the MA capacity is not increased, the greater processing gain of a OFDM+CDMA signal does have other advantages such as more effective suppression of the narrowband interference. The result of this letter thus indicates that, for an *effective* utilization of the greater processing gain, system designers should focus on these advantages rather than on increasing the MA capacity.

Consider a OFDM+CDMA system with K users transmitting their signals at equal power and equal bit rate. At the transmitter (Fig. 13.4), the incoming bit stream having a rate of M/T_b bits per second is first serial-to-parallel converted into M lower-rate bit substreams each of which has a bit duration of T_b seconds. Assume that rectangular symbol waveform and binary phase shift keying are used. The M substreams are respectively modulated onto M orthogonal subcarriers that are equally spaced by f_Δ Hz. Orthogonality among subcarriers is maintained by setting $f_\Delta = 1/T_b$. The data-modulated subcarriers are multiplexed and the resultant multiplexed signal is spread by a spreading sequence of length N. The chip duration, T_c, is given by $T_c = T_b/N$. The spread signal is then transmitted over the channel. Note that in the special case of $M = 1$, the OFDM+CDMA system is essentially a conventional DSSS system. The spectrum densities of OFDM+CDMA and pure CDMA are shown in Figure 13.5.

The received signal is the sum of K transmitted signals corrupted by additive white Gaussian noise (AWGN) having a two-sided power spectral density $N_0/2$. It is assumed that the receiver is intended to detect the Lth-user signal. The complex envelope of the received signal is given by

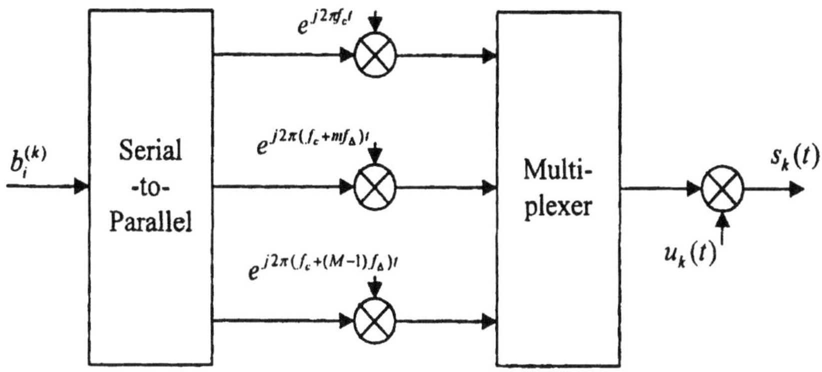

Figure 13.4 Transmitter of OFDM+CDMA

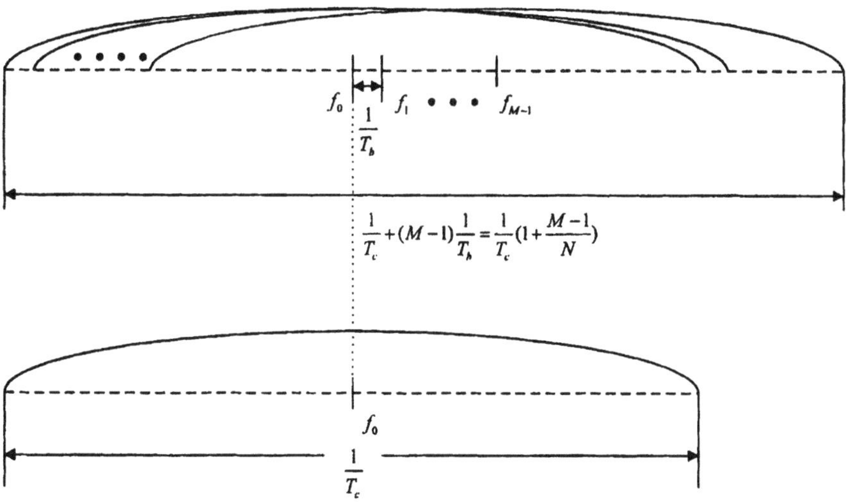

Figure 13.5 Spectrum densities of OFDM+CDMA and single carrier CDMA

$$r(t)=n(t)+\sum_{k=1}^{K}\sqrt{2P_b}e^{j\theta_k}\sum_{i=-\infty}^{\infty}\sum_{m=0}^{M-1}b_{i,m}^{(k)}e^{j2\pi mf_\Delta(t-\tau_k)}u_k(t-iT_b-\tau_k) \quad (13.1)$$

where (i) $n(t)$ is the baseband-equivalent AWGN, (ii) P_b is the signal power per bit (so that MP_b is the transmitted signal power), (iii) θ_k is the random phase introduced by the kth-user transmitter, (iv) $b_{i,m}^{(k)} \in \{+1,-1\}$ is the ith bit transmitted on the mth subcarrier of the kth-user signal, (v) τ_k is the delay of the kth-user signal relative to the Lth-user signal, and (vi)

$$u_k(t)=\sum_{n=0}^{N-1}a_n^{(k)}\psi(t-nT_c) \quad (13.2)$$

is the kth-user spectral spreading waveform. In (13.2), $(a_0^{(k)}, a_1^{(k)}, \dots, a_{N-1}^{(k)})$ is the kth-user spreading sequence satisfying $\left|a_n^{(k)}\right| = 1$, and $\psi(t)$ is the chip waveform given by $\psi(t) = 1$ for $t \in [0,T_c)$ and $\psi(t) = 0$ otherwise. Asynchronous signal transmission is considered so that τ_L is given by $\tau_L = 0$ and τ_k's, $k \neq L$, are modeled as mutually independent random variables uniformly distributed over $[0,T_b)$. The received signal is processed by a bank of M matched filters each of which detects the bits transmitted on a particular subcarrier. Assume that coherent detection is used and $b_{\tilde{i},\tilde{m}}^{(L)}$ is to be detected. The desired matched-filter output, $Z_{\tilde{i},\tilde{m}}$, is given by

$$Z_{\tilde{i},\tilde{m}}=\frac{1}{\sqrt{2P_bT_b}}\int_{\tilde{i}T_b}^{(\tilde{i}+1)T_b}r(t)u_L^*(t-\tilde{i}\,T_b)e^{-j2\pi\tilde{m}f_\Delta t}e^{-j\theta_L}dt. \quad (13.3)$$

The decision variable for $b_{\tilde{i},\tilde{m}}^{(L)}$ is the sign of the real part of $Z_{\tilde{i},\tilde{m}}$. Although we use M matched filters to process $r(t)$ for theoretical modeling, in practical implementations efficient DFT algorithms are often employed. Let $E_b = P_bT_b$ denote the bit energy. Substituting (13.1) into (13.3) gives

$$Z_{\tilde{i},\tilde{m}}=\eta+b_{\tilde{i},\tilde{m}}^{(L)}+\sum_{k=1,k\neq L}^{K}J_k \quad (13.4)$$

where η is the zero-mean complex Gaussian noise with a variance $E\{\eta\eta^*\} = (E_b/N_0)^{-1}$, and

$$J_k = \frac{1}{N} e^{j(\theta_k - \theta_L)} \sum_{m=0}^{M-1} e^{-j2\pi m f_\Delta \tau_k} e^{j2\pi(m-\tilde{m}) f_\Delta \tilde{i} T_b} \cdot \left[b_{\tilde{i}-1,m}^{(k)} S_{k,L}\left((m-\tilde{m}) f_\Delta\right) + b_{\tilde{i},m}^{(k)} \hat{S}_{k,L}\left((m-\tilde{m}) f_\Delta\right) \right] \tag{13.5}$$

is the MAI due to the kth user. In (13.5),

$$S_{k,L}(\mu) = T_c^{-1} \int_0^{\tau_k} e^{j2\pi\mu t} u_k(t + T_b - \tau_k) u_L^*(t) dt \tag{13.6}$$

and

$$\hat{S}_{k,L}(\mu) = T_c^{-1} \int_{\tau_k}^{T_b} e^{j2\pi\mu t} u_k(t - \tau_k) u_L^*(t) dt. \tag{13.7}$$

Let

$$R(\tau;\mu) = T_c^{-1} \int_0^{\tau} e^{j2\pi\mu t} dt, \tag{13.8}$$

$$\hat{R}(\tau;\mu) = T_c^{-1} \int_{\tau}^{T_c} e^{j2\pi\mu t} dt \tag{13.9}$$

and

$$C_{k,L}(q;\xi) = \begin{cases} \sum_{i=0}^{N-1-q} a_i^{(k)} a_{i+q}^{(L)*} e^{j2\pi(i+q)\xi} & 0 \le q \le N-1 \\ \sum_{i=-q}^{N-1} a_i^{(k)} a_{i+q}^{(L)*} e^{j2\pi(i+q)\xi} & -(N-1) \le q < 0 \\ 0 & |q| \ge N \end{cases} . \tag{13.10}$$

Let $w_k \in \{0, 1, \cdots, N-1\}$ and $\hat{\tau}_k \in [0, T_c)$ be the quotient and remainder, respectively, when τ_k is divided by T_c. Note that $\tau_k = w_k T_c + \hat{\tau}_k$. Substituting (13.2) into (13.6) and (13.7) followed by some algebraic manipulations, we get

$$S_{k,L}(\mu) = R(\hat{\tau}_k;\mu) \cdot C_{k,L}\left(-(N - w_k - 1); \mu T_c\right) + \hat{R}(\hat{\tau}_k;\mu) \cdot C_{k,L}\left(-(N - w_k); \mu T_c\right) \tag{13.11}$$

and

$$\hat{S}_{k,L}(\mu) = R(\hat{\tau}_k;\mu)\cdot C_{k,L}(w_k+1;\mu T_c) + \hat{R}(\hat{\tau}_k;\mu)\cdot C_{k,L}(w_k;\mu T_c). \quad (13.12)$$

We apply the Gaussian approximation on the MAI in the computation of the bit error probability, P_e. This approximation is justified for a moderate value of K. It follows from (13.4) that

$$P_e = Q\left(\left\{(\frac{2E_b}{N_0})^{-1} + \frac{1}{2}\sum_{k=1,k\neq L}^{K} \mathrm{var}\{J_k\}\right\}^{-1/2}\right) \quad (13.13)$$

where $Q(x) = (2\pi)^{-1/2}\int_x^\infty e^{-t^2/2}dt$ is the standard Q function, and

$$\mathrm{var}\{J_k\} = \frac{1}{N^2}\sum_{m=0}^{M-1}\left[V_1(V_4+V_6)+V_2(V_5+V_7)+2\,\mathrm{Re}(V_3V_8+V_3V_9)\right]. \quad (13.14)$$

In (13.14),

$$V_1 = E\{|R(\hat{\tau}_k;\mu)|^2\},$$

$$V_2 = E\{|\hat{R}(\hat{\tau}_k;\mu)|^2\},$$

$$V_3 = E\{R(\hat{\tau}_k;\mu)\cdot\hat{R}^*(\hat{\tau}_k;\mu)\},$$

$$V_4 = E\{|C_{k,L}(-(N-w_k-1);\mu T_c)|^2\},$$

$$V_5 = E\{|C_{k,L}(-(N-w_k);\mu T_c)|^2\},$$

$$V_6 = E\{|C_{k,L}(w_k+1;\mu T_c)|^2\},$$

$$V_7 = E\{|C_{k,L}(w_k;\mu T_c)|^2\},$$

$$V_8 = E\{C_{k,L}(-(N-w_k-1);\mu T_c)\cdot C^*_{k,L}(-(N-w_k);\mu T_c)\},$$

$$V_9 = E\{C_{k,L}(w_k+1;\mu T_c)\cdot C^*_{k,L}(w_k;\mu T_c)\}$$

where $\mu = (m-\tilde{m})f_\Delta$ and $E\{\cdot\}$ is the expectation taken over random variables $\hat{\tau}_k$ and w_k. Notice that P_e is dependent on the choice of $\tilde{m}$.

The MA capacity may be defined as the number of users that can be supported at a given error performance. We are interested in comparing the MA capacity of a OFDM+CDMA system and that of a conventional DSSS system under conditions of equal bit transmission rate and equal bandwidth. This comparison is made easy since a OFDM+CDMA system reduces to a conventional DSSS system in the special case of $M = 1$. We are also interested in comparing MA capacities for OFDM+CDMA systems using different values of M. We want to find out whether the MA capacity could be increased by an increase in the processing gain through the use of more subcarriers.

We assume that the OFDM+CDMA system and the corresponding conventional DSSS system have the same chip duration T_c. As a result, the bandwidth of a OFDM+CDMA signal is slightly greater than the signal bandwidth required for a conventional DSSS system. The difference in their bandwidths is small (see [9]) so that the equal-bandwidth condition can be considered satisfied. As M subcarriers are used in the OFDM+CDMA system, under equal-bit-rate condition the bit duration T_b of the OFDM+CDMA system is required to be M times of that of the conventional DSSS system. Hence, the spreading-code length N of the OFDM+CDMA system should also be M times of that of the corresponding conventional DSSS system. The increase in the spreading-code length constitutes an increase in the processing gain of a OFDM+CDMA system. Define the *basic processing gain*, N_{ba}, by

$$N_{ba} = \frac{N}{M}, \tag{13.15}$$

which is a constant invariant to the choice of M. Note that N_{ba} is also the spreading-code length of the conventional DSSS system. Since M bits are transmitted concurrently, the power per bit P_b for a OFDM+CDMA system is required to be $1/M$ times of that of the conventional DSSS system. The net result of an M-fold increase in the bit duration and an M-fold reduction in the power per bit is that the bit energy values are the same for both OFDM+

CDMA and conventional DSSS systems. Thus, we can compare the two systems at the same E_b/N_0 value.

Some insight can be obtained by considering random binary sequences as spreading sequences. It can be shown that, for random binary sequences and $k \neq L$,

$$E\left\{\left|C_{k,L}(q;\mu T_c)\right|^2\right\} = \begin{cases} N-|q| & |q| < N \\ 0 & |q| \geq N \end{cases} \tag{13.16}$$

$$E\left\{C_{k,L}(q_1;\mu T_c) \cdot C^*_{k,L}(q_2;\mu T_c)\right\} = 0 \quad \text{for } q_1 \neq q_2. \tag{13.17}$$

It follows that $\mathrm{var}\{J_k\} = I/N_{ba}$ where

$$I = \frac{1}{M}\sum_{m=0}^{M-1}\left[E\left\{\left|R(\hat{\tau}_k;(m-\tilde{m})f_\Delta)\right|^2\right\} + E\left\{\left|\hat{R}(\hat{\tau}_k;(m-\tilde{m})f_\Delta)\right|^2\right\}\right]. \tag{13.18}$$

Table 13.1 lists the maximum and minimum values of I obtained over $\tilde{m} \in \{0, 1, \ldots, M-1\}$ for various combinations of M and N under the condition of same N_{ba}. It is apparent that $I \approx 2/3$ for all combinations. Therefore, P_e in (13.13) can be closely approximated by

$$P_e \approx Q\left(\left\{\left(\frac{2E_b}{N_0}\right)^{-1} + \frac{K-1}{3N_{ba}}\right\}^{-1/2}\right). \tag{13.19}$$

Eqn. (13.19) is identical to the one obtained by Pursley [12] for computing the error probability of a conventional DSSS system. It follows that the MA capacity of a OFDM+CDMA system is the same as that of a conventional DSSS system when asynchronous signal transmission and random sequences are used. Also notice that P_e for a OFDM+CDMA system depends only on the basic processing gain. Thus, using a larger number of subcarriers to increase the processing gain of the system does not reduce P_e and hence does not increase the MA capacity.

Table 13.1. Maximum and minimum values of I for various combinations of M and N under the condition of same N/M.

N/M	I min.	max.
16/1	0.6667	0.6667
32/2	0.6660	0.6660
64/4	0.6655	0.6662
128/8	0.6653	0.6662
256/16	0.6651	0.6662
512/32	0.6650	0.6662
1024/64	0.6650	0.6662
128/1	0.6667	0.6667
256/2	0.6667	0.6667
512/4	0.6667	0.6667
1024/8	0.6667	0.6667

REFERENCES

[1] W. Mohr and W. Konhauser, "Access network evolution beyond third generation mobile communications," IEEE Communications Magazine, December 2000.

[2] B. G. Evans and K. Baughan, "Vision of 4G", Electronics and Communications Engineering Journal, December 2000.

[3] S. Ohmori, Y. Yamao and N. Nakajima, "The future generations of mobile communications based on broadband access technologies," IEEE Communications Magazine, December 2000.

[4] K. Higuchi, A. Fujiwara and M. Sawahashi, "Throughput performance of high speed packet transmission with adaptive modulation and coding scheme using multipath interference canceller in WCDMA forward link," VTC2001-Spring.

[5] Q. Bi, G. I. Zysman, and H. Menkes, "Wireless mobile communications at the start of the 21st century," IEEE Communications Magazine, January 2001.

[6] S. Abeta, H. Atarashi, M. Sawahashi and F. Adachi, "Coherent multicarrier/DS-CDMA and MC-CDMA for broadband packet wireless access," VTC2000-Spring, Japan, May 2000.

[7] S. Abeta, H. Atarashi and M. Sawahashi, "Forward link capacity of coherent DS-CDMA and MC-CDMA broadband packet wireless access in a multi-cell environment," VTC2000-Fall, USA, Sept. 2000

[8] K. W. Yip, X. Zhang, T. S. Ng and J. Wang, "On the multiple access capacity of multitone-CDMA communications," IEEE Communications Letters, February 2000.

[9] L. Vandendorpe, "Multitone spread spectrum multiple access communications system in a multipath Rician fading channel," *IEEE Trans. Veh. Technol.*, vol. 44, pp. 327-337, May 1995.

[10] S. Hara and R. Prasad, "Overview of multicarrier CDMA," *IEEE Commun. Mag.*, vol. 35, no. 12, pp. 126-133, Dec. 1997.

[11] B. Hirosaki, "An orthogonally multiplexed QAM system using the discrete Fourier transform," *IEEE Trans. Commun.*, vol. 29, pp. 982-989, Jul. 1981.

[12] M. B. Pursley, "Performance evaluation for phase-coded spread-spectrum multiple-access communication — part I: system analysis," *IEEE Trans. Commun.*, vol. 25, pp. 795-799, Aug. 1977.

[13] Q. Bi, G. I. Zysman, and H. Menkes, "Wireless mobile communications at the start of the 21st century," *IEEE Communications Magazine, January 2001.*

[14] R. H. Walden, "Analog-to-digital converter survey and analysis," *IEEE Journal on Selected Areas in Communications, April 1999.*

[15] S. Tanaka, H. Harada, M. Sawahashi and F. Adachi, "Experiments on coherent antenna array diversity for wideband DS-CDMA mobile radio", *IEEE J. Select. Areas Comm.*, August 2000.

[16] J. Razavilar, F. R. Farrokhi and K. Liu, "Software radio architecture with smart antennas: a tutorial on algorithms and complexity", *IEEE J. Select. Areas Comm.*, April 1999.

ABBREVIATIONS

1G	First generation
2G	second generation
3G	third generation
3GPP	third generation partner project
4G	fourth generation
A/D	analog to digital
ALE	adaptive line enhancer
AMPS	advanced mobile phone systems
ATF	adaptive time frequency
AWGN	additive white gaussian noise
BCH	broadcast channel
BER	bit error rate
BFSK	binary frequency shift keyed
BPF	band pass filter
BPSK	binary Phase Shift Keying
BS	base station
BW	bandwidth
BWC	broadband wireless communications
CCPCH	common control physical channel
CCTrCh	coded composite transport channel
CDF	cumulative density function
CDMA	code Division Multiple Access
CIR	carrier to interference ratio

CLPC	closed loop power control
CPICH	Common downlink pilot channels
CRC	cyclic redundancy check
D/A	digital to analog
DAB	digital audio broadcast
DCH	dedicated transport channel
DPCCH	dedicated physical control channel
DPCH	Dedicated Physical Channel
DPDCH	dedicated physical data channel
DLL	delay-locked loop
DM	delta modulation
DS	direct sequence
DS-CDMA	direct sequence code division multiple access
DSP	digital signal processing
DS-SS	direct sequence spread spectrum
DVB	digital video broadcast
EDGE	enhanced data rates for GSM evolution
FBI	feedback information
FDD	frequency division duplex
FDMA	frequency division multiple access
FFH	fast frequency hopping
FFT	fast Fourier transform
FIR	finite impulse response
FH	frequency hopping
FHMA	frequency hopping multiple access
FM	frequency modulation
FPGA	field programmable gate arrays
GI	group identification
GPRS	general packet radio service

GPS	Global position systems
GSM	global systems for mobile
I	in phase
IC	Interference cancellation
ID	identification
IID	independent identically distributed
i-mode	internet mode
IMT-2000	international mobile telecommunications 2000
ISDN	integrated service digital network
ISI	inter-symbol- interference
ISM	industrial scientific and medical
I/Q	in-phase/quadrature
J/S	jammer to signal power ratio
Kbps	kilo bits per second
LAN	local area network
LMS	least mean square
LPF	lowpass filter
MA	multiple Access
MAI	multiple Access Interference
Mbps	mega bits per second
MC	multicarrier CDMA
Mcps	mega chips per second
MCI	multi-code interference
MF	matched filter
MFSK	*M*-ary frequency shift keyed
MIP	multipath intensity profile
MRC	maxim ratio combining/combiner
MS	mobile station
MTJ	multitone jamming

MTLL	mean time to lose lock
MUX	multiplexing
NBI	narrowband Interference
ND	normalized deviation
OFDM	orthogonal frequency division multiplexing
OLPC	open loop power control
OVSF	orthogonal variable spreading factor
PAR	peak-to-average ratio
PARCOR	partial correlation
PBJ	partial-band jamming
PC	power control
PCCC	Parallel Concatenated Convolutional Code
PCM	pulse code modulation
PDC	personal digital cellular
PSC	primary synchronization code
PSH	primary synchronization channel
P-CPICH	primary Common downlink pilot channels
PG	processing gain
PLL	phase-locked lock
PN	pseudonoise
PSD	power spectral density
Q	quadrature
QAM	quadrature amplitude modulation
QoS	quality of service
QPSK	quadrature phase shift keying
RF	radio frequency
RLS	recursive least square
RS	Reed Solomon

S-CPICH	Common downlink pilot channels
SF	spreading factor
SFH	slow frequency hopping
SGL	stochastic gradient lattice
SINR	signal-to-interference plus noise ratio
SIR	signal-to-interference ratio
SNR	signal to noise ratio
SS	spread-spectrum
SSCH	second synchronization channel
SSDT	Site selection diversity transmit
TDMA	Time division multiple access
TFCI	transport-format combination indicator
TPC	transmit power control
TrCH	transport channel
VCO	voltage controlled oscillator
WAP	wireless access protocol
WCDMA	wideband code division multiple access
WMSA	weighted multi-slot averaging
WSSUS	wide-sense-stationary uncorrelated-scattering

INDEX

D

E

F

G

H

I

L

M

N

THE AUTHOR

Jiangzhou Wang received the B.S. and M.S. degrees from Xidian University, Xian, China, in 1983 and 1985, respectively, and the Ph.D. degree (with Greatest Distinction) from the University of Ghent, Belgium, in 1990, all in electrical engineering.

From 1990 to 1992, he was a Postdoctoral Fellow at the University of California at San Diego, USA, where he worked on the research and development of cellular CDMA systems. From 1992 to 1995, he was a Senior System Engineer at Rockwell International Corporation, Newport Beach, California, where he worked on the development and system design of wireless communications. Since 1995, he has been with the University of Hong Kong, where he is currently an Associate Professor. Dr. Wang has held a Visiting Professor position at NTT DoCoMo, Japan. He was a Research Fellow for two years in Southeast University, China, in 1980's. He was a Technical Chairman of the IEEE Workshop in 3G Mobile Communications, 2000. Dr. Wang has published over 100 papers, including more than 20 IEEE Transactions/Journal papers in the areas of wireless mobile and spread spectrum communications. He has written/edited two books, including this book.

Dr. Wang is an Editor for IEEE Transactions on Communications and a Guest Editor on Wideband CDMA for IEEE Journal on Selected Areas in Communications. He holds one US patent in the GSM system. He is a Senior Member of IEEE and listed in Marquis Who's Who in the World.

MIX
Papier aus verantwortungsvollen Quellen
Paper from responsible sources
FSC® C105338

If you have any concerns about our products,
you can contact us on
ProductSafety@springernature.com

In case Publisher is established outside the EU,
the EU authorized representative is:
Springer Nature Customer Service Center GmbH
Europaplatz 3, 69115 Heidelberg, Germany

Printed by Libri Plureos GmbH
in Hamburg, Germany